ALLE · ZEIT · WACH · 1842

Y. Hirasawa · C.B. Sledge · S.L.-Y. Woo (Eds.)

Clinical Biomechanics
and Related Research

With 270 Figures

Springer-Verlag
Tokyo Berlin Heidelberg
New York London Paris
Hong Kong Barcelona Budapest

Yasusuke Hirasawa, M.D., Ph.D.
Professor and Chairman, Department of Orthopaedic Surgery, Kyoto Prefectural University of Medicine, Kawaramachi-hirokoji, Kamigyo-ku, Kyoto, 602 Japan

Clement Blount Sledge, M.D.
Professor, Department of Orthopedic Surgery, Brigham and Women's Hospital, Harvard Medical School, 75 Francis Street, Boston, MA 02115, USA

Savio Lau-Yuen Woo, Ph.D.
Ferguson Professor of Orthopaedic Surgery/Vice Chairman for Research.
Professor, Department of Mechanical Engineering, University of Pittsburgh, Kaufman Building, Suite 1010, 3471 Fifth Avenue, Pittsburgh, PA 15213, USA

ISBN 978-4-431-66861-9 ISBN 978-4-431-66859-6 (eBook)
DOI 10.1007/978-4-431-66859-6

Library of Congress Cataloging-in-Publication Data
Clinical biomechanics and related research/Y. Hirasawa, C.B. Sledge, S.L.-Y. Woo (eds.). p. cm. Based on the 20th Annual Meeting of the Japanese Society for Clinical Biomechanics and Related Research, held Nov. 5–6, 1993, in Kyoto, Japan. Includes bibliographical references and index. ISBN 4-431-70144-3 (alk. paper). ISBN 978-4-431-66861-9
1. Biomechanics—Congresses. 2. Human mechanics—Congresses. 3. Arthroplasty—Congresses. I. Hirasawa, Yasusuke, 1937- . II. Sledge, Clement B., 1930- . III. Woo, Savio L.-Y., 1942- . IV. Japanese Society for Clinical Biomechanics and Related Research. Meeting (20th:1993:Kyoto, Japan) [DNLM: 1. Biomechanics—congresses. 2. Orthopedics—congresses. WE 103 C6406 1994] QH513.C55 1994 612.7′6—dc20 DNLM/DLC for Library of Congress. 94-36332

Printed on acid-free paper

Typesetting: Best-set Typesetter Ltd., Hong Kong

Preface

The Japanese Society for Clinical Biomechanics and Related Research has developed significantly since its first meeting in 1974. The late professor Kengo Yamada of Tokushima University, who sadly is no longer with us, chaired the inaugural meeting. Through the successful leadership and great efforts of past chairmen and members of the organizing committees over the last 20 years, a dynamic atmosphere has been fostered in which to present our ideas on the field of biomechanics, both in Japan and overseas. We are very proud of these esteemed individuals and humbly salute them. We also wish to dedicate this book to the past successes which have lead up to the Twentieth Annual Meeting held on November 5–6, 1993, in Kyoto, Japan.

The Twentieth Annual Meeting also marked the historic inclusion of engineers in our society. We now wish to build on this significant progress and broaden the cooperation between physicians and bioengineers. We hope that researchers from other fundamental science disciplines will join us in the future to further the development of medical sciences in Japan.

YASUSUKE HIRASAWA, M.D.
CLEMENT B. SLEDGE, M.D.
SAVIO L.-Y. WOO, PH.D.

Contents

Part 1. Bone

Mechanical Behavior of Cortical Bone Under Static and Substatic Compressive Loading

Hidetake Yamamoto[1], Tsuneo Hirai[1], Tsutao Katayama[1], Yasusuke Hirasawa[2], Nozomu Inoue[2], and Yoshinobu Watanabe[2]

Summary. Elucidation of the deformation mechanism of cortical bone is essential to estimating the biomechanical strength of long bones. Cortical bone consists of osteon and interstitial lamellae; the osteon has a composite multi-layer structure consisting of collagen fibers such as helical structures and hydroxyapatite crystals. In this article, the deformation mechanism of cortical bone as a composite is clarified. A typical model of cortical bone is as a unidirectional, fiber-reinforced composite, assuming that the osteon is the reinforcing material and the interstitial lamella is the matrix. Cylindrical cortical bone specimens at angles of both 0° and 90° to the axis were obtained from bovine tibias. Static and substatic compression tests were carried out using an acoustic emission technique, and the elastic index of work done was measured in each strain step under the static test condition to estimate microfracture of the specimens. It appeared that microfracture of the specimens occurred as delamination at the interface between the osteons and interstitial lamellae or as debonding in the lamellae by deflective deformation of osteons. For static tests, the modulus of 0° direction specimens was higher than that of 90° direction samples. This was explained by the large shear force occurring in osteons of 0° direction. For substatic tests, the relationship was the inverse.

Key words: Cortical bone—Composite material—Compressive behavior —Microfracture

Introduction

For the purpose of estimating the biomechanical strength in curing fractures such as the diaphyseal fracture of long bones, a great interest has been taken in

---

[1] Department of Mechanical Engineering, Doshisha University, Kyoto, 610-03 Japan
[2] Department of Orthopedic Surgery, Kyoto Prefectural University of Medicine, Kyoto, 602 Japan

the response behavior of cortical bone to dynamic loading. It is certain that the growth and remodeling of living bone are closely related to the stress conditions of their biomechanical surroundings, and there are many reports that loading by suitable exercise accelerates fracture healing [1–3]. However, as a starting point, it is essential to understand the nonlinear deformation mechanisms of cortical bone, which has a composite structure. When analyzing stress wave propagation through cortical bone, the deformation behavior of heterogeneous structures within the bone should first be taken into account.

In this study, static and substatic compression tests were carried out using acoustic emission technique, and the elastic index of the work done was measured to define the deformation behavior of cortical bone as a composite material. The histological structure of cortical bone is composed of the osteon, which has a composite multilayer structure consisting of helical collagen fiber bundles and hydroxyapatite crystals, and the interstitial lamella, which is an irregular section of the osteon. The simplified modeling of cortical bone is typically as unidirectional, continuous-fiber-reinforced composite materials, assuming that the osteon is the reinforcing fiber and that the interstitial lamella is the matrix [4–7].

Methods

Cylindrical Specimens

From the diaphysis of 20 fresh bovine tibias, 50 cylindrical cortical bone specimens (not including the lamellar bone) 8.33 mm in diameter and 8.33 mm high were prepared for angles of 0° and 90° relative to the axis. They were termed 0° direction and 90° direction specimens, respectively, named by their angles to the axis. Both surfaces of the specimens were finished parallel to each other by using accurate jigs, and the specimens were kept in physiological saline until the experiments.

Static and Substatic Compression Tests

Compression tests were carried out using an Instron testing machine (model 4206) at four strain rates of 10^{-4}, 10^{-3}, 10^{-2}, and 10^{-1} per second (Fig. 1) to obtain the (nominal) stress–strain curve of the specimens. The compression tests at the strain rates of both 10^{-4} and 10^{-3}/s were termed static tests and the compression tests at the strain rates of 10^{-2} and 10^{-1}/s were termed substatic tests. Acoustic emission (AE) of the microfracture in the specimens was detected using two piezoelectric transducers (NF AE-9006-WB), one on both ends of the specimens. The signals were preamplified by NF AE-912 with a gain of 40 dB, and the event counts were sent to a personal computer system (PC9801Vm, NEC, Tokyo, Japan) through GP-IB cable after memorization by local processor (NF AE-9600) transiently; the threshold level, the nonmeasur-

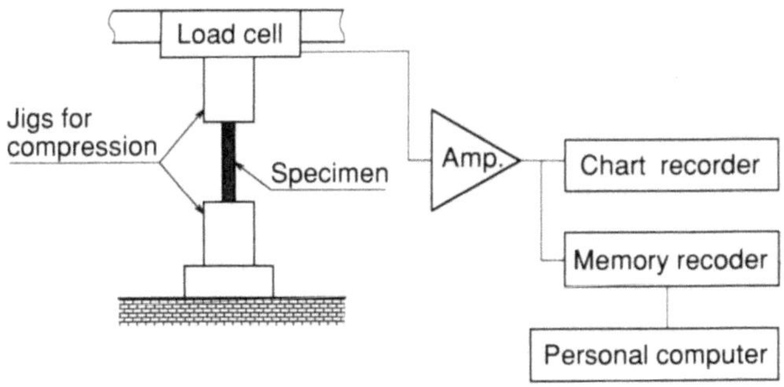

Fig. 1. Block diagram of compression testing system. The interfaces between the cylindrical specimen and the compression jigs were greased to reduce friction

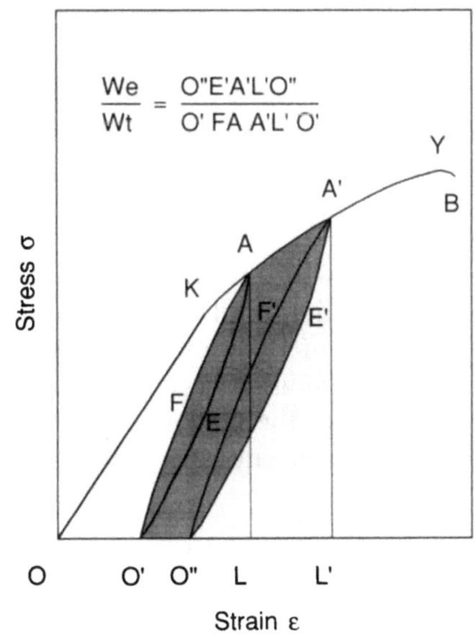

Fig. 2. Definition of elastic index of the work done (We/Wt). By unloading at point A and loading again at point 0′ on the stress–strain curve KY, the curve drawn is the hysteresis loop, AEO′FA. The elastic index of the work done at the strain step L′ (We/Wt) is then defined by the foregoing equation

ing time, and the minimum duration were set at 120 mV, 1 μs and 2 μs at the output level, respectively.

Measurement of the Elastic Index of the Work Done

To investigate the plastic behavior of the specimens after the microfractures originated, 10 specimens of each direction were repetitively loaded and unloaded in strain increments of 0.30% by the static test (at the strain rate of 10^{-4}/s); the elastic index of the work done (EIW) [8] was measured at each strain stage. Here, We/Wt expresses the ratio of the elastic work to the total work done at an arbitrary strain stage (Fig. 2). The influence of elastic aftereffect on the loaded–unloaded curves was disregarded in this study because it was only a negligible amount.

Results

The following results are either those judged to be typical or the mean value and standard deviation (SD) of groups judged to be significant by statistical analysis.

Compressive Behavior of the 0° Direction Specimen

Typical stress–strain curves and values of EIW of both direction specimens in the statical tests are shown in Fig. 3 and Fig. 4, respectively. The stress–strain curve of the 0° direction specimens (Fig. 3A) was in the elastic region until the strain reached 1.6% because it was linear; no AE events could be observed, and the value of EIW remained high. After a temporary increase, the tangent modulus decreased until the strain reached 2.4% and then began to increase again. It was clear that the tangent modulus decreased as a result of the microfractures in the specimens because some AE events were detected and the value of EIW remarkably decreased.

Compressive Behavior of the 90° Direction Specimen

The stress–strain curve of the 90° direction specimens (Fig. 3B) also was in the elastic region until the strain reached 1.7% because it was slightly convex and the value of EIW remained high. After a temporary increase and decrease, the tangent modulus began to increase again after the strain reached 2.4%. The decrease of the tangent modulus was probably caused by microfractures in the specimens because the value of EIW remarkably decreased at that time.

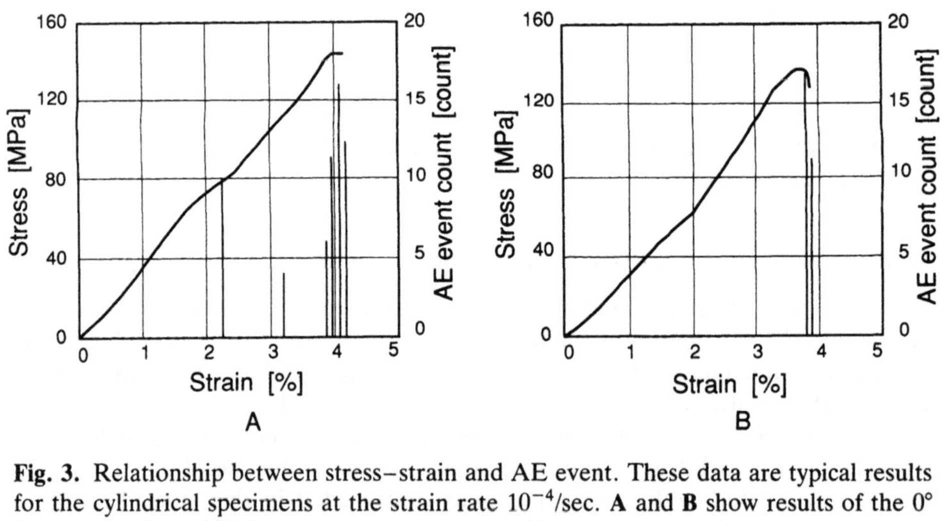

Fig. 3. Relationship between stress–strain and AE event. These data are typical results for the cylindrical specimens at the strain rate 10^{-4}/sec. **A** and **B** show results of the 0° direction and the 90° direction specimens, respectively

4

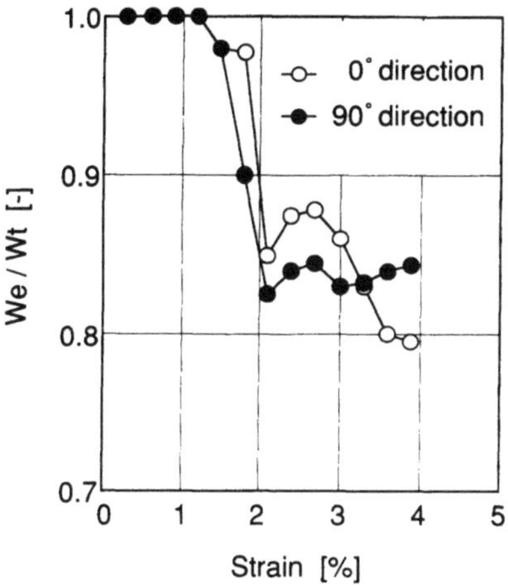

5

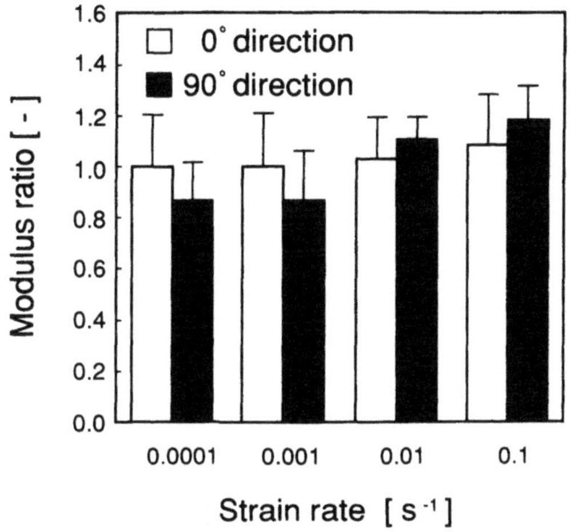

Fig. 4. Relationship between We/Wt and strain. These data are typical results for the strain rate of 10^{-4}/s

Fig. 5. Relationship between modulus ratio and strain ratio. The modulus ratio is defined as a proportion of each modulus to that of the 0° direction specimen at the strain rate 10^{-4}/s. The modulus is the tangent modulus at the initial strain stage of the stress–strain curve

Anisotropy of the Elastic Modulus

Figure 5 displays the relationship between the modulus ratio derived from the initial tangent modulus of the stress–strain curves and the strain rate. From the results of the static tests, it was observed that the initial tangent modulus of the 0° direction specimens was greater than that of the 90° direction specimens. For the substatic tests, however, this relationship was the inverse.

Discussion

Before discussing these results, to readily understand the compressive deformation condition of the cylindrical specimens, it was necessary to analyze the stress distribution at the end face of the specimen. It was observed from the estimated stress distribution (Fig. 6) that the greater stresses existed in the center part of the end surface. By taking the analytical result into account together with the experimental results, the compressive deformation behavior of cortical bone under static and substatic loads could be discussed.

First, the real stress–strain curve of the 0° direction specimens would be initially concave because the nominal stress–strain curve was then linear. It is believed that the differently oriented collagen fibers, which are concentrically wound around a Haversian canal, are loosened when the osteon, which has a

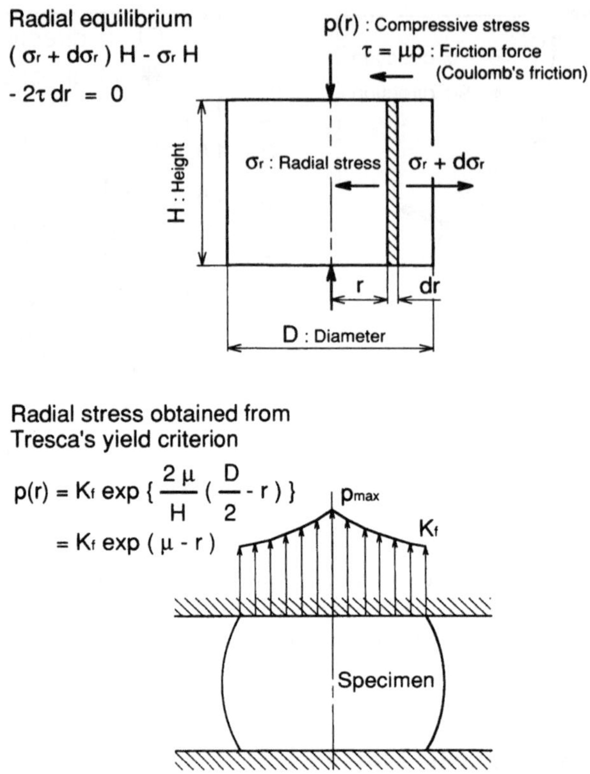

A

B

Fig. 6. Radial stress distribution at end face of the cylindrical specimen. Assuming that the friction force is constant along the symmetry axis of the specimen, the equation of radial equilibrium is as given in **A**. The friction between the specimen and the jig satisfies the condition of Coulomb's friction. Applying the maximum shear stress criterion to the fracture of the specimen, the radial stress distribution is obtained **B**

composite multilayer structure, is compressed on the axis. An increase of the tangent modulus after the strain reached 1.6% would result because a large resistance to the deformation arises from the deflexion or snaking of osteons in the center part of the specimens where the stresses are concentrated. Correspondingly, a decrease of the tangent modulus until the strain reached 2.4% would result from the deflection of the interstitial lamellae by containment of the deforming osteons; the delamination occurs at the interfaces between the osteon and the interstitial lamella. It is also probable that the tensile fracture of interstitial lamellae arose from the discontinuity of collagen fibers. However, an increase of the tangent modulus after the strain reached 2.4% was caused by evaluating the stress and strain using nominal values.

Second, the real stress–strain curve of the 90° direction specimens would initially be linear because the nominal stress–strain curve was then convex.

The circular cross section of the osteons would then be greatly deformed in the center part of the specimens. A temporary increase of the tangent modulus at a strain of 1.7% resulted from the large resistance that occurred there. Decrease of the tangent modulus after that was caused by delamination at the interfaces between the osteon and the interstitial lamella that resulted from the deformation. However, an increase of the tangent modulus after the strain reached 2.4% was caused by evaluating the stress and the strain using nominal values.

Further, considering anisotropy of the initial tangent modulus in the static tests by using the helical spring as a simplified model of the osteon, the collagen fibers in the osteon are twisted by the axial compression load on the osteon. As a result, resistance of osteons to the shearing deformation will occur in the 0° direction specimens. Further, resistance of osteons to the circumferential deformation will occur in the case of the 90° direction specimens. However, the results of the statical tests showed that the influence of the one was greater than that of the other. In the substatic tests, a satisfactory explanation might be obtained by using the models for viscoelastic phenomenon (i.e., by applying the Maxwell and Voigt model to the 0° direction and the 90° direction specimens, respectively).

In an earlier study, the authors elucidated the nonlinear tensile deformation behavior of cortical bone by applying the same model [6]. In the actual structure of the cortical bone specimens, not all osteons have continuity and a unidirectional orientation. However, the model devised in the present study gave an effective approximation of the deformation behavior of cortical bone.

Conclusions

The compressive deformation behavior of cortical bone as a composite material has been investigated with the following conclusions. It is very interesting that the biomechanical behavior of the logical model of cortical bone is compatible with that of composite materials.

1. For the 0° direction specimens, the deflexion or snaking of osteons and interstitial lamellae begins in the central part of the specimens at the initial stage of compressive deformation. Consequently, either delamination at the interfaces between osteons and lamellae or debonding in the lamellae occurs.
2. For the 90° direction specimens, the circular cross section of osteons and interstitial lamellae is greatly deformed in the central part of the specimens at the initial stage of compressive deformation. Consequently, delamination at the osteon–lamella interfaces or debonding in the lamellae occurs.
3. Cortical bone specimens show anisotropy of the tangent modulus. In the static strain mode, the modulus of the 0° direction is greater than that of the 90° direction because of the resistance to shearing deformation in the osteons. In the substatical strain mode, however, the modulus of the 90° direction is inversely greater than that of the 0° direction.

References

1. Albright JA, Brand RA (1979) The scientific basis of orthopaedics. pp 299–300. Appleton-Century-Crofts, New York
2. Samiento A, Schaeffer JF, Becherman L, Latta LL, Enis JE (1977) Fracture healing in rat femura as affected by functional weight bearing. J Bone Joint Surg 59A: 369–375
3. White AA (1975) Fracture treatment; the still unsolved problem. Clin Orthop 106:279–284
4. Curreey JD (1964) Biorheology 2:1
5. Tateishi T, Shirasaki Y, Kimura T, Ogawa K (1980) Visco-elastisity and strength of compact bone. Trans Jpn Soc Mech Eng [Series A] 46(401):97–105 (in Japanese)
6. Hirai T, Katayama T, Inoue N, Yamamoto H (1992) Nonlinear deformation mechanism of cortical bone (in Japanese). Trans Jpn Soc Mech Eng [Series A] 58(551): 34–39
7. Katz JL (1971) Hard tissue as a composite material: I. Bounds on the elastic behavior. J Biomech 4:455–473
8. Hirai T, Fujiwara S, Otuki S (1972) Elastic index of the work done of FRP. Bull Jpn Soc Mech Eng

Dynamic Viscoelastic Properties of Cortical Bone and Callus

Yoshiyuki Arai, Wookcheol Kim, Yasusuke Hirasawa, Nozomu Inoue, Munehiko Yoshida, and Yoshinobu Watanabe

Summary. The dynamic viscoelastic properties of rabbit tibial cortical bone and callus were investigated. The dynamic storage modulus (G′), dynamic loss modulus (G″), and dynamic loss tangent (tan δ) were measured under tensile and compressive strain at frequencies ranging from 0.1 to 70 Hz at body temperature. The initial strain was a 2000 microstrain, and the dynamic strain was set up at a 500 microstrain in all specimens. In wet cortical bone, G′, G″, and tan δ were dependent on frequency. There were clear differences between the immature cortical bone and the mature cortical bone in the shapes of the G″-frequency curves. However, these differences of frequency dispersion were eliminated by decalcification of specimens. The frequency dispersion also changed in dry cortical bone and in thin cortical bone in which the osteon structure was destroyed. Callus specimens were excised at 2, 4, 6, 9, and 12 weeks after making cortical defects in rabbit tibiae. G′ increased as the frequency increased in all specimens. In the case of G″, the shapes of the curves after 6 weeks differed significantly from those before 4 weeks. Tan δ of the 2-week specimens had a clear peak at 7–10 Hz, and the shape of the curve differed significantly from those of the other periods. The changes of frequency dispersion and the changes of the histological structure or the changes of ash content over the period of time coincided. However, other factors, such as changes of water content, collagen content, binding conditions between collagen and hydroxyapatite, and so on, should not be neglected.

Key words: Dynamic viscoelastic properties—Cortical bone—Callus—Dependence on frequency—Histological structure—Ash content

---

Department of Orthopaedic Surgery, Kyoto Prefectural University of Medicine, Kamigyo-ku, Kyoto, 602 Japan

Introduction

The biomechanical properties of bone have been studied since the nineteenth century. Wertheim reported the ultimate tensile strength of human fibula and femur in 1847 [1]. In Japan, Yamada summarized his studies into the mechanical properties of bone, extending over 20 years, in 1970 [2]. Investigation into the biomechanical properties of callus, however, is relatively recent. For the most part, these studies covered the static mechanical properties such as elastic modulus, ultimate stress, and ultimate strain measured by bending, tensile, and torsional tests.

Both bone and callus are intelligent materials, because they change their structures, constituent elements, and mechanical properties in response to external mechanical stimulation. Dynamic loading seems to be an important factor that regulates bone remodeling and fracture healing. The dynamic properties of bone were studied by Yasuda in 1956 [3] and by Narabayashi in 1969 [4]. Narabayashi analyzed bone vibration and reached the conclusion that bone was a viscoelastic material. Although there are a number of reports on the dynamic viscoelastic properties of bone [5–10], the authors are aware of only one paper on the dynamic viscoelastic properties of callus [11]. The essential biomechanical responses of cortical bone and callus under dynamic loading are still not well understood.

Materials and Methods

Dynamic Viscoelastic Properties of Cortical Bone (Fig. 1)

Twenty rabbits were divided into three groups according to their ages: group A (immature, $n = 5$), 1 month of age, 425 g average weight; group B (mature, $n = 10$), 12–18 months of age, 2984 g average weight; group C (aged, $n = 5$), over 46 months of age, 2870 g average weight. Rectangular cortical bone specimens with a thickness of 0.35 mm were excised from the bilateral medial side of the tibiae.

The dynamic storage modulus (G′), dynamic loss modulus (G″), and dynamic loss tangent (tan δ) of wet bone in 0.9% sodium chloride solution under tensile and compressive strain were measured at frequencies ranging from 0.1 to 70 Hz at body temperature. The initial strain was a 2000 microstrain, and the dynamic strain was set up at a 500 microstrain in all specimens using a Viscoelastic spectrometer (VES F-III, Iwamoto Seisakusho Co., Ltd., Kyoto, Japan).

After the dynamic viscoelastic tests, small fragments were taken from bilateral ends of each specimen. The dry weight of fragments heated to 100°C for 24 h and the ash weight of fragments heated again to 650°C for 12 h were measured. The ash content was defined as the ratio of ash weight to dry weight of the fragments. Ten specimens of each groups were decalcified in 10% formic acid solution with added sodium citrate. Five specimens of group B were dried

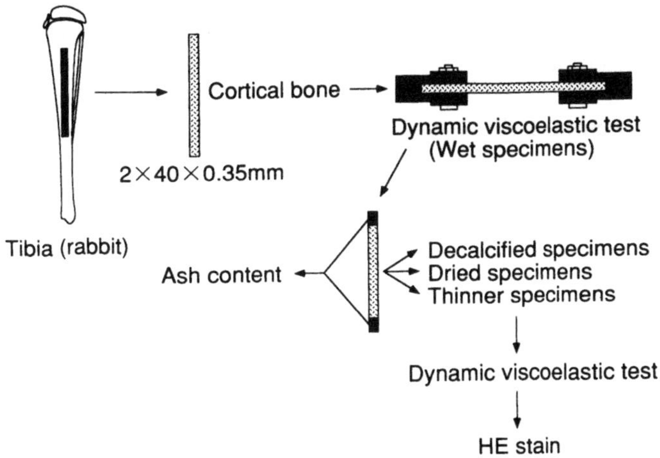

Fig. 1. Schematic diagram of the cortical bone specimens. The dynamic viscoelastic properties were measured in various conditions of specimens. *HE*, hematoxylin and eosin stain

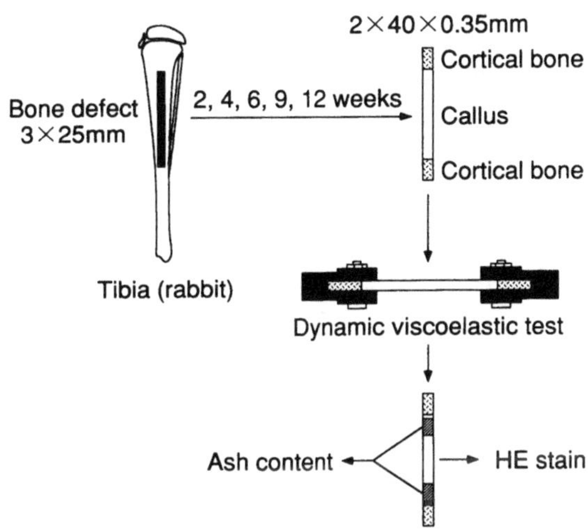

Fig. 2. Schematic diagram of the callus specimens. The dynamic viscoelastic properties were measured at 2, 4, 6, 9, and 12 weeks after making bone defects

at 100°C for 24 h. Thinner specimens, 0.30 mm and 0.10 mm thick, were formed from the remaining five specimens of group B. Dynamic viscoelastic tests on decalcified, dried, thinner specimens were performed as stated previously. The specimens of each group were stained with hematoxylin-eosin dye.

Dynamic Viscoelastic Properties of Callus (Fig. 2)

Cortical defects were made in the right tibiae of 34 rabbits that were 12 months of age (average weight, 2850 g). The periosteum was not removed and was sutured with fascia. The rabbits were free weight bearing in their cages. Callus specimens were excised at 2, 4, 6, 9, and 12 weeks after the initial operation. The dynamic viscoelastic tests on callus specimens were carried out as stated earlier in wet condition. After the dynamic viscoelastic tests, ash content was calculated and hematoxylin-eosin staining was performed.

Statistical Analysis

The age-related changes in dynamic viscoelastic properties were analyzed using a two-way analysis of variance; the change among the groups was one way and the frequency-related change (repeated measurement) was the other way. The corrected profile using the Geisser–Greenhouse method, at a confidence level of 95% ($P < 0.05$), was applied to the analysis of the difference in the shapes of the curves. The dependence on frequency of each group was analyzed using a one-way analysis of variance (repeated measurement). The changes between the decalcified specimens and the cortical bone specimens, the change among the three groups of different thickness, and the time-related changes of the callus specimens were analyzed as stated. The changes in ash content of cortical bone among the three groups and the changes in ash content of callus during the healing period were analyzed using a one-way analysis of variance. If a significant change was observed, the Sheffe F test was used to analyze the group and the time when the significant change occurred. The correlation analysis of ash content and G' or ash content and $\tan \delta$ was also carried out.

Results

Dynamic Viscoelastic Properties of Wet Cortical Bone

G' was dependent on frequency. G' increased as the frequency increased in all groups. G' of group A had a lower value than for the other groups at the frequency range 0.1–70 Hz. However, there were no significant differences among the three groups in the shapes of the G'-frequency curves (Fig. 3a). G'' and $\tan \delta$ were also dependent on frequency. G'' and $\tan \delta$ decreased as the frequency increased in all groups. There were clear differences between group A and groups B or C in the shapes of the G''-frequency curves (Fig. 3b). Tan δ of group A had a higher value than that of the other groups at the frequency range 0.1–70 Hz (Fig. 3c).

Dynamic Viscoelastic Properties of Cortical Bone in Various Conditions

In decalcified cortical bone, G', G'', and $\tan \delta$ were dependent on frequency. There were no significant differences among the three groups in the shapes of the curves. G'' had a peak at about 2 Hz, and $\tan \delta$ had a clear peak at 2 Hz (Fig. 4).

In dry cortical bone, G′ increased rapidly at a frequency greater than 40 Hz. G″ and tan δ had low values at a frequency range of 7–40 Hz. There was a significant difference between dry specimens and wet specimens in the shapes of the curves (Fig. 5).

In thinner cortical bone, G′ of the 0.10-mm-thick specimens had a lower value than the other thick specimens. However, there were no significant differences in the shapes of the curves (Fig. 6a). G″ and tan δ of the 0.10-mm-thick specimens increased rapidly at a frequency greater than 20 Hz. There were significant differences between the 0.10-mm-thick specimens and the other thick specimens in the shapes of the G″-frequency curves and the tan δ-frequency curves (Fig. 6b,c).

Dynamic Viscoelastic Properties of Callus

G′ was dependent on frequency. G′ increased as the frequency increased for all the periods. G′ increased rapidly before 6 weeks and increased gradually after 6 weeks at the frequency range 0.1–70 Hz. The average G′ of callus of 12 weeks was about 75% of the G′ of cortical bone (Fig. 7a). G″ and tan δ were also dependent on frequency. In the case of G″, the shapes of the curves after 6 weeks significantly differed from those before 4 weeks (Fig. 7b). Tan δ of 2 weeks had a clear peak at the frequency range 7–10 Hz, and the shapes of the curve significantly differed from those of the other periods (Fig. 7c).

Histological Structures

There were well-oriented osteons in the cortical bone of groups B and C. However, there were also well-oriented but immature osteons in group A. In group C, evidence of osteoporosis was not observed (Fig. 8). In 2-week and 4-week callus specimens, low-density and immature osteons were observed. After 6 weeks, mature osteons were observed and the density increased. However, even after 12 weeks the osteons were not oriented (Fig. 9).

Changes in Ash Content

Figure 10 shows the changes in ash content of cortical bone and callus. The average values of ash content in groups A, B, and C were 48.0%, 68.1%, and 71.8%, respectively. Group A had a significantly lower value than groups B and C (Fig. 10a). The average value of ash content for callus of 2, 4, 6, 9, and 12 weeks and for cortical bone were 22.1%, 42.1%, 59.0%, 60.3%, 62.0%, and 64.0%, respectively. The values after 2 weeks and 4 weeks were significantly lower than after 6 weeks (Fig. 10b).

The Relationship Between Ash Content and G′ or tan δ

In the three groups, the ash content of cortical bone significantly correlated with G′ and with tan δ at 0.1 Hz. There was a significant correlation between ash content, and G′ and tan δ, at 70 Hz. However, the distribution of

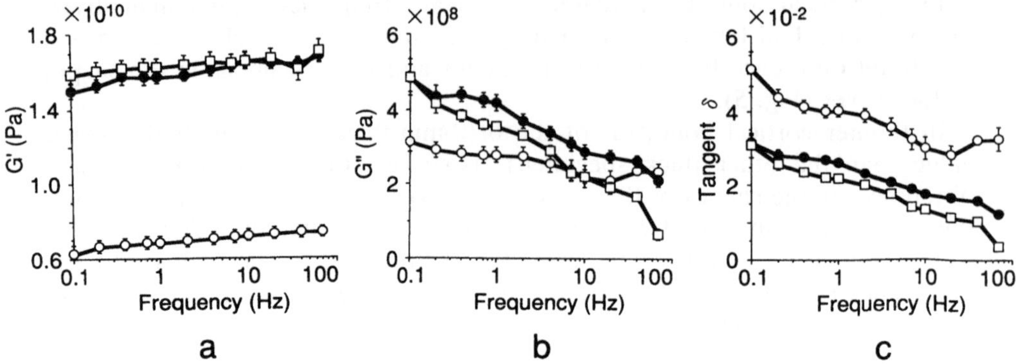

Fig. 3a–c. Dynamic viscoelastic properties of wet cortical bone (mean ± SEM). **a** Dynamic storage modulus. **b** Dynamic loss modulus. **c** Dynamic loss tangent. Group A, *open circles* ($n = 10$); group B, *solid circles* ($n = 20$); group C, *open squares* ($n = 10$). G′, G″, and tan δ were dependent on frequency (one-way ANOVA repeated measures; each group, $P < 0.001$). In G″, the shapes of the curves in group A significantly differed from the others (two-way ANOVA repeated measures; profile, $P < 0.05$). In tan δ, the shapes of the curves in group A differed at the frequency range 40–70 Hz from the other groups but not significantly ($P > 0.05$). *SEM*, standard error of mean; *ANOVA*, analysis of variance

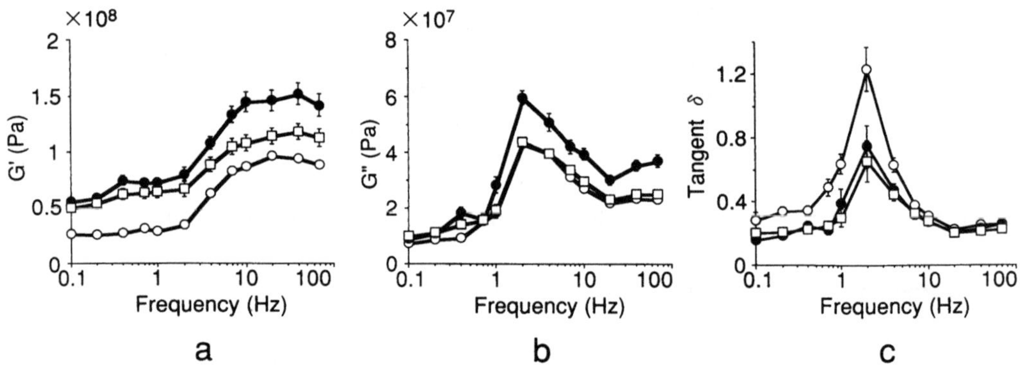

Fig. 4a–c. Dynamic viscoelastic properties of decalcified cortical bone (mean ± SEM). **a** Dynamic storage modulus. **b** Dynamic loss modulus. **c** Dynamic loss tangent. Group A, *open circles* ($n = 10$); group B, *solid circles* ($n = 10$); group C, *open squares* ($n = 10$). G′, G″, and tan δ were dependent on frequency (one-way ANOVA repeated measures; each group, $P < 0.001$). There were no significant differences among the three groups in the shapes of the curves (two-way ANOVA repeated measures; profile, $P > 0.05$). G″ had a peak at about 2 Hz, and tan δ had a clear peak at 2 Hz

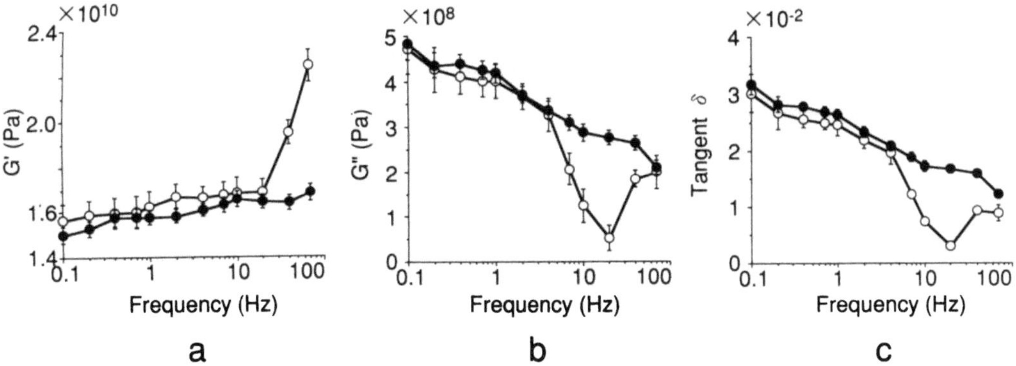

Fig. 5a–c. Dynamic viscoelastic properties of wet and dry cortical bone (mean ± SEM). **a** Dynamic storage modulus. **b** Dynamic loss modulus. **c** Dynamic loss tangent. Dry bone, *open circles* ($n = 5$); wet bone, *solid circles* ($n = 20$). G′, G″, and tan δ were dependent on frequency (one-way ANOVA repeated measures; each group, $P < 0.001$). There was a significant difference between dry specimens and wet specimens in the shapes of the curves for G′, G″, and tan δ (two-way ANOVA repeated measures; profile, $P < 0.05$)

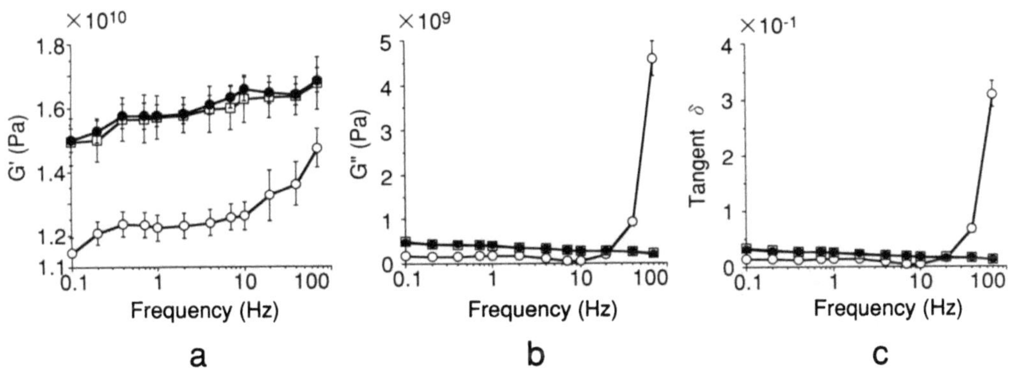

Fig. 6a–c. Dynamic viscoelastic properties of thin cortical bone (mean ± SEM). **a** Dynamic storage modulus. **b** Dynamic loss modulus. **c** Dynamic loss tangent. Thicknesses: 0.10 mm, *open circles* ($n = 5$); 0.30 mm, *open squares* ($n = 5$); 0.35 mm, *solid circles* ($n = 20$). G′, G″, and tan δ were dependent on frequency (one-way ANOVA repeated measures; each group, $P < 0.001$). There were significant differences between the 0.10-mm-thick specimens and the other thick specimens in the shapes of the G″-frequency curves and tan δ-frequency curves (two-way ANOVA repeated measures; profile, $P < 0.05$)

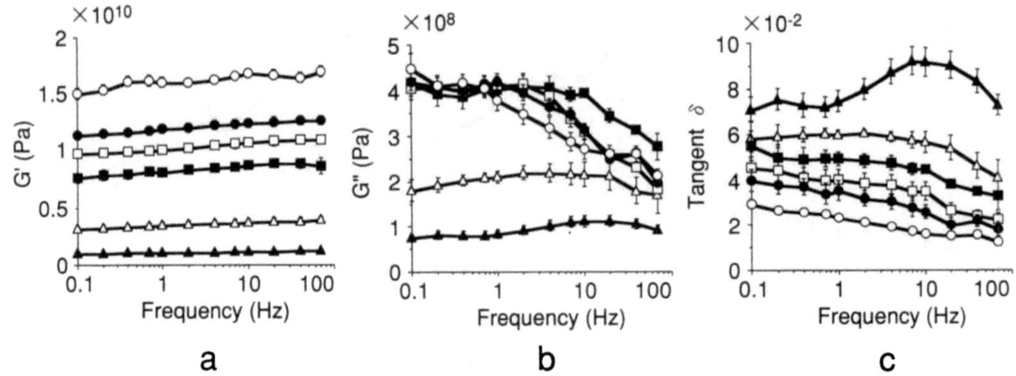

Fig. 7a–c. Dynamic viscoelastic properties of callus (mean ± SEM). **a** Dynamic storage modulus. **b** Dynamic loss modulus. **c** Dynamic loss tangent. Two weeks, *solid triangles* ($n = 8$); 4 weeks, *open triangles* ($n = 8$); 6 weeks, *solid squares* ($n = 6$); 9 weeks, *open squares* ($n = 6$); 12 weeks, *solid circles* ($n = 6$); and cortical bone, *open circles* ($n = 34$). G′, G″, and tan δ were dependent on frequency (one-way ANOVA repeated measures; each period of time, $P < 0.001$). In the case of G″, the shapes of the curves after 6 weeks significantly differed from those before 4 weeks (two-way ANOVA repeated measures; profile, $P < 0.05$). Tan δ of 2 weeks had a clear peak at the frequency range 7–10 Hz, and the shapes of the curve significantly differed from those of the other periods (two-way ANOVA repeated measures; profile, $P < 0.05$)

a, b
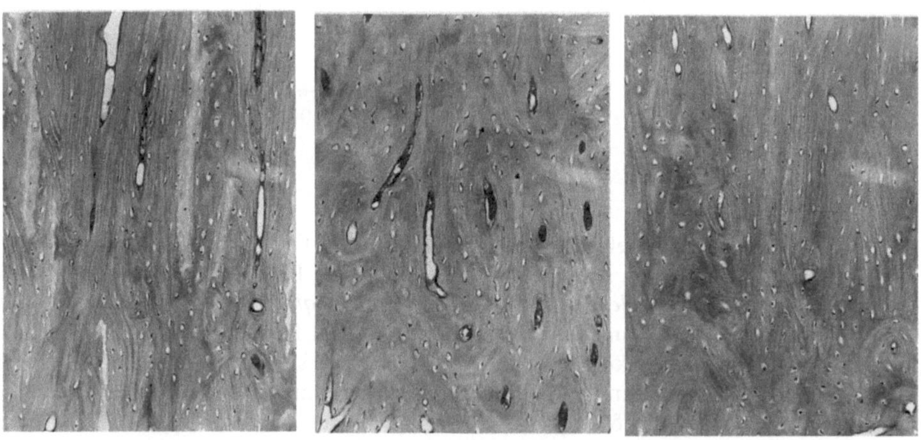
c

Fig. 8a–c. Changes in the structure of cortical bone. **a** Group A. **b** Group B. **c** Group C. Group A was constructed mostly of immature bone. In all groups, the structure of the osteon or immature bone was parallel to the bone axis. There was no evidence of osteoporosis in group C. HE, ×100

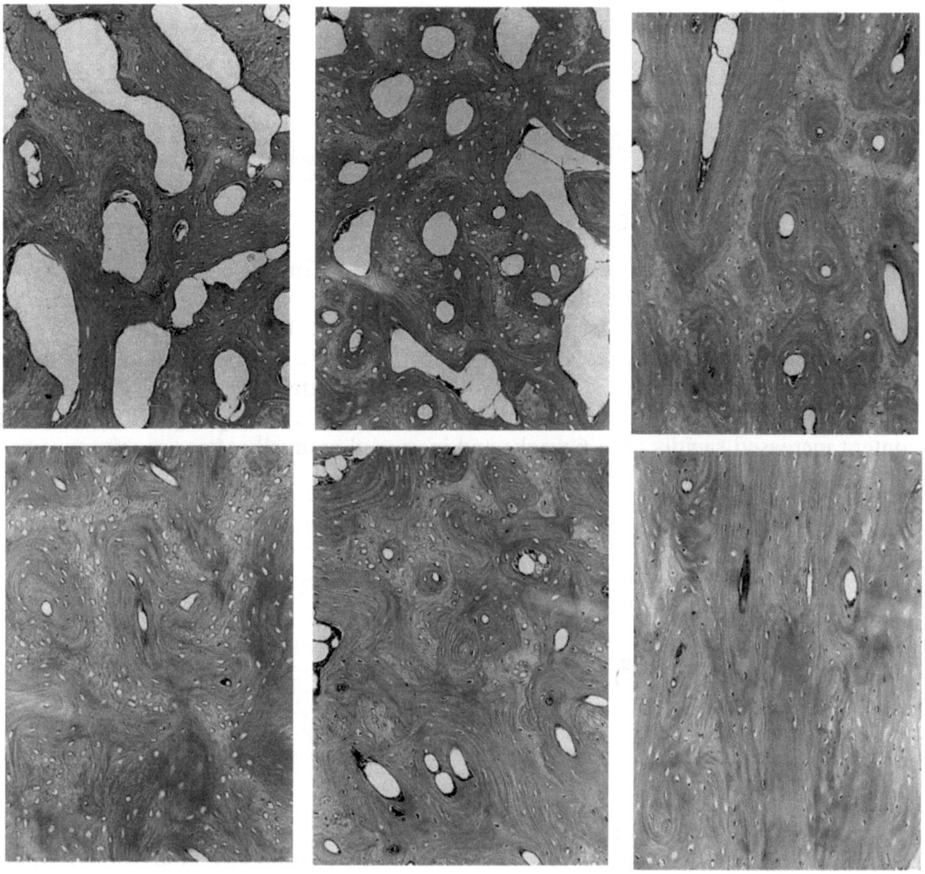

a, b c

d, e f

Fig. 9a–f. Changes in the structure of callus over time. **a** 2 weeks. **b** 4 weeks. **c** 6 weeks. **d** 9 weeks. **e** 12 weeks. **f** Cortical bone. After 2 and 4 weeks, there were low-density, immature osteons. After 6 weeks, mature osteons were observed and the density increased. However, even at 12 weeks the osteons were not oriented. HE, ×100

G′ and tan δ at 70 Hz for groups B and C differed from those for group A (Fig. 11). In callus and cortical bone, ash content correlated with G′ and with tan δ. However, the distribution after 6 weeks differed from that before 6 weeks. After 6 weeks, the effect of the change in ash content on the change in G′ or tan δ was smaller than before 6 weeks (Fig. 12).

Discussion

The dynamic viscoelastic properties of polymers have been investigated in detail. Generally, the frequency dispersion of polymers results from the micromotion of the molecules. Also, G′ changes greatly at a frequency at which both

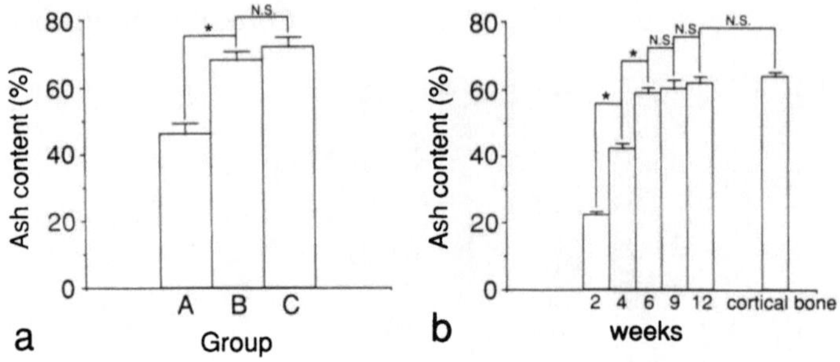

Fig. 10a,b. Changes in ash content (mean ± SEM). **a** Cortical bone specimens: group A, $n = 10$; group B, $n = 20$; group C, $n = 10$; one-way ANOVA, $P < 0.001$; *, $P < 0.05$ (Scheffe F test); N.S., not significant (Scheffe F test). **b** Callus specimens: 2 weeks, $n = 8$; 4 weeks, $n = 8$; 6 weeks, $n = 6$; 9 weeks, $n = 6$; 12 weeks, $n = 6$; cortical bone, $n = 34$; one-way ANOVA, $P < 0.001$; *, $P < 0.05$ (Scheffe F test); N.S., not significant (Scheffe F test). Group A had significantly lower values than groups B and C. Ash content increased rapidly until 6 weeks and increased gradually after 6 weeks

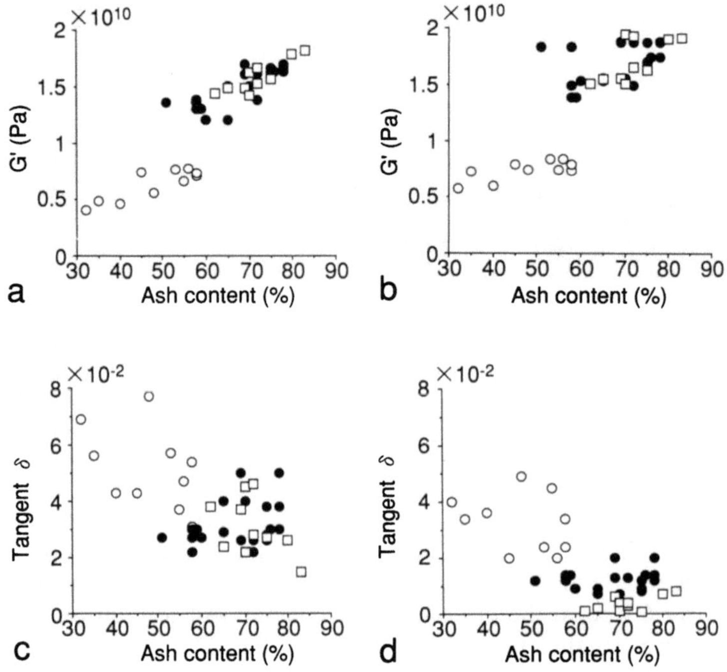

Fig. 11a–d. Relationships between ash content and dynamic viscoelastic properties in cortical bone. **a** 0.1 Hz, dynamic stroage modulus. **b** 70 Hz, dynamic storage modulus. **c** 0.1 Hz, dynamic loss tangent. **d** 70 Hz, dynamic loss tangent. Group A, *open circles*; group B, *solid circles*; group C, *open squares*. Ash content correlated with G′ (**a**, $r = 0.898$, $P < 0.001$; **b**, $r = 0.821$, $P < 0.001$) and with tan δ (**c**, $r = 0.553$, $P < 0.001$; **d**, $r = 0.679$, $P < 0.001$). However, in both G′ and tan δ at 70 Hz, the distribution of groups B and C differed from that of group A

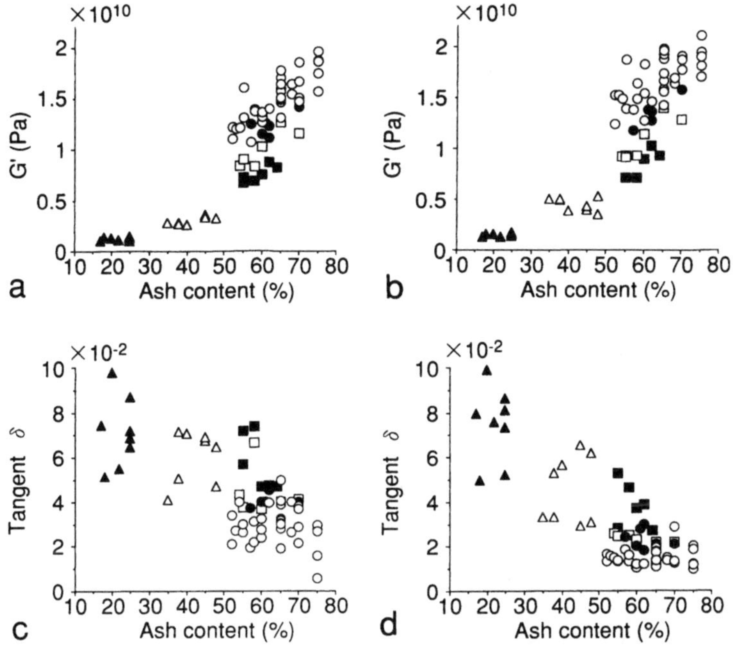

Fig. 12a–d. Relationships between ash content and dynamic viscoelastic properties in callus. **a** 0.1 Hz, dynamic storage modulus. **b** 70 Hz, dynamic storage modulus. **c** 0.1 Hz, dynamic loss tangent. **d** 70 Hz, dynamic loss tangent. Two weeks. *solid triangles*; 4 weeks, *open triargles*; 6 weeks, *solid squares*; 9 weeks, *open squares*; 12 weeks, *solid circles*; cortical bone, *open circles*. Ash content correlated with G′ (**a**, $r = 0.893$, $P < 0.001$; **b**, $r = 0.879$, $P < 0.001$) and with tan δ (**c**, $r = 0.707$, $P < 0.001$; **d**, $r = 0.826$, $P < 0.001$). However, the distribution after 6 weeks differed from that before 6 weeks

peak G″ and peak tan δ appear. Both cortical bone and callus are composite materials, and it can expected that there are various factors affecting their viscoelastic properties. Examples of factors are the ratio of various components, the density of the matrix, the rate of matrix orientation, the binding conditions between collagen and hydroxyapatite, and so on. Those factors might be combined in a complicated manner.

The dynamic viscoelastic studies of wet cortical bone by Smith and Keiper [5] showed that the dispersion of G′ was not clear at the frequency range 500–3500 Hz in tension and compression. On the other hand, Carter and Hayes [12] showed Young's modulus of compact bone increased as the strain rate increased. In addition, Black and Korostoff [7] reported that G′ of cortical bone increased as the frequency increased. Lakes et al. [8] also reported that both G′ and tan δ were dependent on the frequency in torsion. In this study, the G′, G″, and tan δ of wet cortical bone were dependent on frequency (see Fig. 3). G′ increased as the frequency increased. Both G″ and tan δ decreased as the frequency increased. The G′-frequency curves were shaped like stair-steps. However, there were no clear peaks in the shapes of the G″- or tan δ-frequency curves when G′ changed rapidly. It could be suggested that various factors were affecting the viscoelastic properties simultaneously.

Uezaki [6] concluded that there was no effect of water on the dynamic viscoelastic properties of cortical bone. However, in our current investigation, the frequency dispersion of the dry cortical specimens was significantly different from that of the wet specimens (see Fig. 5). In cortical bone, there are many canaliculi connecting osteocytes lying in bone cavities to the haversian canals. The strain applied to bone could lead to liquid motion in the canaliculus. The liquid motion should be affected by the applied strain rate. In addition, the solid surface of each canal has a electrical charge. The electrical double layer is formed on the interface between the solid surface and the electrolyte solution. Thus, the motion of the electrolyte solution should affect the mechanical properties of bone.

In mature cortical bone, there are oriented osteons with diameters of about 0.2 mm. The dynamic viscoelastic properties of 0.10-mm-thick specimens significantly differed from 0.35-mm- or 0.30-mm-thick specimens (see Fig. 6). It is suggested that the dynamic viscoelastic properties of cortical bone are influenced by the structure of each osteon.

In cortical bone, the dynamic viscoelastic properties of group A significantly differed from those of groups B and C in the frequency dispersion (see Fig. 3). In callus, properties before 4 weeks significantly differed from those after 6 weeks in the frequency dispersion (Fig. 7). Similarly, there were differences between the histological structures of group A and those of groups B and C in cortical bone (Fig. 8) and between those before 4 weeks and those after 6 weeks in callus (Fig. 9). However, it was noted that the ash content also showed the same changing patterns as the dynamic viscoelastic properties (Fig. 10). In addition, there were no significant differences among the three groups of decalcified cortical bone in the frequency dispersion aside from the magnitude of each value (see Fig. 4). It was deduced, therefore, that the age-related changes in the frequency dispersion of cortical bone or the time-related changes in the frequency dispersion of callus were affected primarily by the changes in ash content rather than changes in histological structure.

G′ is identical to Young's modulus, and tan δ is identical, in a sense, to the efficiency of energy absorption. Carter and Hayes [12] showed that the Young's modulus of compact bone correlated with apparent density at various strain rates. Currey [13] also reported that the energy absorption of bone depended on the rate of calcification. However, Watanabe et al. [14] showed that the Young's modulus of aged cortical bone from rabbits older than 45 months was the same as that of adult cortical bone from rabbits aged 25 months. Also, the energy absorption of the aged cortical bone was less than that of the adult cortical bone. Watanabe [15] also showed that the tensile strength and tensile stiffness of the callus of rats increased without an increase in ash content after the period when the strength of callus increased to about half that of cortical bone. In the current study, ash content on the whole correlated with G′ and with tan δ both in cortical bone and in callus. However, both in G′ and tan δ the distribution of groups B and C differed from that of group A (see Fig. 11). In callus, the distribution after 6 weeks differed from that before 6 weeks (see Fig. 12). These data matched the results by Watanabe. In addition, the fact that the distribution at 70 Hz differed from that at 0.1 Hz could lead to the conclusion that the change in ash content had a greater effect at the lower frequency than at the higher frequency or vice versa.

Both bone and callus change their structures, constituent elements, and mechanical properties in response to external mechanical stimulation. What kinds of stimuli are optimal for callus formation, for the transformation of callus to bone, and for bone growth? It is difficult to clarify the optimal mechanical stimulation. It should be considered whether the changes are affected by the magnitude of stress or strain, by the method by which the stress or strain is applied, by the frequency of applied stress or strain, by the total applied storage or loss energy, and so on.

Because movement at a specific frequency excites specific structures from the molecular to the histological levels, if a specific structure is related to changes in bone and callus growth, not only the magnitude but the frequency must affect the changes. Our study might contribute to an explanation of the optimal stimulation of bone and callus.

Conclusions

The dynamic viscoelastic properties of rabbit tibial cortical bone and callus were investigated. In wet cortical bone, G′, G″, and tan δ were dependent on frequency. There were clear differences between the immature cortical bone and the mature cortical bone in frequency dispersion. However, these differences were eliminated by decalcification of specimens. This frequency dispersion also changed in dry cortical bone and in thin cortical bone in which the osteon structure was destroyed. In callus, G′, G″, and tan δ were also dependent on frequency. The frequency dispersion after 6 weeks significantly differed from that before 4 weeks. The changes in frequency dispersion, the changes in histological structure, and the changes in ash content coincided over the periods that were studied. However, other factors, such as changes in water content, collagen content, and binding conditions between collagen and hydroxyapatite, should not be neglected.

References

1. Wertheim G (1847) Mémoire sur l'élasticité et la cohésion des principaux tissus du corps humain. Ann Chim Phys 21:385–414
2. Yamada H (1970) Mechanical properties of locomotor organs and tissues. In: Evans FG (ed) Strength of biological materials. Williams and Wilkins, Baltimore, pp 19–105

3. Yasuda I (1956) Mechanische Schwingung des Knochens (in Japanese). Nisshin Igaku 43(3):132–146
4. Narabayashi Y (1969) On a bone vibration (in Japanese). J Jpn Orthop Assoc 43(6):475–480
5. Smith R, Keiper D (1965) Dynamic measurement of viscoelasticity of bone. Am J Med Electron 4:156–160
6. Uezaki N (1971) Dynamic viscoelastic properties of human long bone (in Japanese). Igaku Kenkyu 41:204–213
7. Black J, Korostoff E (1973) Dynamic mechanical properties of viable human cortical bone. J Biomech 6:435–438
8. Lakes RS, Katz JL, Sternstein SS (1979) Viscoelastic properties of wet cortical bone: I. Torsional and biaxial studies. J Biomech 12:657–678
9. Shirasaki Y, Tateishi T (1985) Mechanical properties and internal structure of bone samples from horses with and without suffering fracture (in Japanese). Proc Annu Meet Jpn Soc Orthop Biomech 7:181–184
10. Shirasaki Y, Tateishi T (1986) Mechanical properties and internal structure of horse bone II (in Japanese). Proc Annu Meet Jpn Soc Orthop Biomech 8:79–82
11. Kobayashi A (1987) Study of the biomechanical characteristics of fracture calls (in Japanese). J Jpn Orthop Assoc 61:1273–1284
12. Carter DR, Hayes WC (1976) Bone compressive strength: the influence of density and strain rate. Science 194:1174–1176
13. Currey JD (1979) Change in the impact energy absorption of bone with age. J Biomech 12:459–469
14. Watanabe Y, Kusaka Y, Takai S, Inoue N, Hirasawa Y, Hirai T, Katayama H (1991) The effects of maturation on mechanical properties of cortical bone under tensile and torsional loading in rabbit (in Japanese). Proc Annu Meet Jpn Soc Orthop Biomech 13:269–273
15. Watanabe Y (1993) Evaluation of mechanical properties of fracture healing with acoustic emission technique (in Japanese). J Kyoto Pref Univ Med 102(10):1101–1116

The Mechanical Adaptive Remodeling Theory: Its Application to Bone Formation and a Comparison with Experiments on Rabbit Tibia

JIRO SAKAMOTO[1], JUHACHI ODA[1], KAZUHIRO AOYAMA[2], YASUNOBU SUEYOSHI[2], KATSURO TOMITA[2], and TAKESHI SAWAGUCHI[3]

Summary. The authors have investigated the influence of mechanical stimuli on bone formation by in vivo experiments on rabbit tibias. In the experiments, an intermittent compressive or tensile load of 1 h/day during a period of 1, 2, or 4 weeks was applied to a rabbit tibia in which a circular defect had been made at the center, along the axial direction, during 1, 2, and 4 weeks. The experimental results confirmed that bone formation is promoted by mechanical stimuli. It is necessary to establish a theoretical model of bone formation caused by mechanical stress stimuli before applying the experimental results to a clinical situation. We attempted to apply the theory of time-dependent bone remodeling proposed by Beaupré et al. to bone formation, including defect repair. Computer simulation of bone formation around the circular defect corresponded to experiments using an algorithm based on the theory, and the effectiveness is compared with the experimental results.

Key words: Bone formation—Mechanical stimulation—Theoretical model—Computer simulation

Introduction

Many researchers have tried to estimate experimentally the quantitative relationship between mechanical stress stimuli and both bone formation and bone remodeling. For example, the influence of static force on bone formation of callus was studied by Yasuda et al. [1], and Rubin and Lanyon [2] provided good reports on adaptive bone remodeling to mechanical stimuli. Mabuchi et

---

[1] Department of Mechanical Systems Engineering, Kanazawa University, 2-40-20 Kodatsuno, Kanazawa, 920 Japan
[2] Department of Orthopedic Surgery, School of Medicine, Kanazawa University, 13-1 Takara, Kanazawa, Ishikawa, 920 Japan
[3] Department of Orthopedic Surgery, Toyama Municipal Hospital, Toyama, Japan

al. [3] experimented on fracture healing of bone in various mechanical environments. The authors have also experimented with bone formation in rabbit tibia during intermittent mechanical stress stimuli in vivo [4,5]. In the experiments, a compressive or tensile load is applied along the axial direction over a few weeks to a rabbit tibia in which a circular defect had been made at the center. The quantity of bone formation around this circular defect was estimated in comparison to the degree of stress stimuli. Basic relationships between bone formation and mechanical stimuli have been clarified gradually from the aforementioned studies. It is believed that the general theoretical model of bone formation based on experimental results is necessary to apply experimental data to more complicated problems in a clinical situation such as healing of bone fractures.

Various theoretical models have been proposed from the viewpoint of bone remodeling because various common factors exist between bone remodeling and mechanical stimuli. For example, the piezoelastic model of Gjelsvik [6], the electromechanical model of Guzelsu and Saha [7], the cell biological model of Hart et al. [8], and the bone fluid model of Takakuda [9] were proposed. These models are useful for examination of the hypotheses but not for application to clinical problems. The phenomenological model called the adaptive elasticity model, in which the relationship of bone formation to stress stimuli is determined directly, is adequate for clinical problems. In fact, Weinans et al. [10] applied their phenomenological model to bone remodeling around femoral hip components and discussed the effect of material properties of the components. The most effective theory among the phenomenological models is the time-dependent model proposed by Beaupré et al. [11,12]. In this remodeling theory, the number of stimulations as well as the influence of stimulus magnitude is considered.

It is known that bone formation at injury sites occurs as an adaptation to mechanical stimuli, resembling bone remodeling at uninjured sites. Investigation of the theoretical model of adaptive bone formation at an injury site is rare compared with the bone remodeling model just described despite its relevance to clinical problems. Thus, we investigated the theoretical model of adaptive bone formation with repair at an injured bone site by applying the phenomenological remodeling theory. The time-dependent theory of Beaupré is used in our model because of its efficiency. Computer simulation corresponding to our experiments on rabbit tibia was carried out, and the effectiveness of the theoretical model is discussed by comparing the results.

Theoretical Model and Simulation Technique

Bone Remodeling Theory

Mechanical stress stimulus per day, ψ_b, is defined by Beaupré as follows:

$$\psi_b = \left(\sum_{day} n_i \sigma_i^m\right)^{1/m} \qquad \sigma_i = \sqrt{2EU} \tag{1}$$

where n_i and σ_i are the number of stimuli per day and effective stress occurring at the bone, respectively. The subscript i corresponds to various types of load applied to the bone, i.e., tensile, compression, bending, and torsional load; m is a empirical constant used as a weighting factor for the relative importance of stress magnitude and the number of stimuli. The effective stress is defined as equation 1 where E is elastic modulus and U is the strain energy density. By using the equations, we can treat the mechanical stress stimuli per day numerically in the bone remodeling theory.

It is believed that the bone apposition or resorption occurs according to the value of the stress stimuli at each point. The bone apposition and resorption are equilibrated with each other when the values of stress stimuli are normal, but bone apposition will occur when the stimuli are greater than the daily stress stimulus values. For values of stress stimuli that are less than normal, bone resorption will occur. The normal stress stimulus is called as attracter state stimulus, ψ_{bAS}, that is constant depending on the genotype and metabolic status. Considering the most simple case, it is assumed that bone apposition or resorption occur in proportion to the difference between the applied stress stimulus ψ_b and the attracter state stimulus ψ_{bAS}. The idealizer relationship between the stress stimulus and bone apposition or resorption per day assumed here is shown in Fig. 1. The relationship is defined by the following equations:

$$\dot{r} = \begin{cases} c(\psi_b - \psi_{bAS}) + cw & (\psi_b - \psi_{bAS} < -w) \\ 0 & (-w \leqslant \psi_b - \psi_{bAS} \leqslant w) \\ c(\psi_b - \psi_{bAS}) - cw & (\psi_b - \psi_{bAS} > w) \end{cases} \tag{2}$$

where $\dot{r}$ is the bone apposition or resorption per day at the bone surface, c is an empirical rate constant, and w is a half-width of the normal stress stimulus region. c is represented as a gradient of the remodeling line shown in Fig. 1.

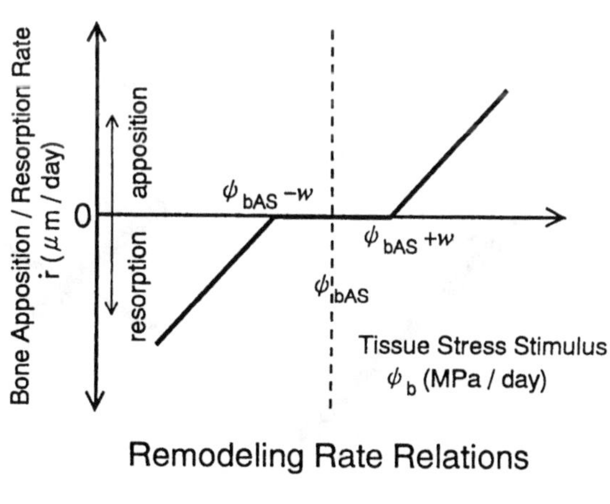

Fig. 1. Relationship between remodeling rate and stress stimulus

Theoretical Model for Experiments

The adaptive remodeling theory just described is applied to computer simulation of the bone formation with bone repair. The simulation corresponding to the bone-forming experiment on rabbit tibia in vivo examined by us is carried out by using the remodeling theory. In the experiment, bone apposition around the circular defect that was drilled at the center of the rabbit tibia in advance was estimated on a stimulation group and a no-stimulation group. An intermittent square wave axial load of 49 N at 0.5 Hz was applied 1 h/day to the stimulation group through the two pins shown in Fig. 2 for the duration of the experimental period. The no-stimulation group animals did not receive any load and were used as controls. Comparing bone apposition around the defect between stimulation and control groups, it is obvious that the bone apposition of the stimulation group is larger than in the other group. Mechanical adaptive bone formation is considered in this situation to be similar to the case of bone remodeling. Thus, computer simulation of bone formation is focused on the circular defect boundary, and results obtained by the calculation are compared with the experimental results.

The computer simulation is described as follows.

1. Calculate strain energy density around the circular defect using the finite-element method and obtain the effective stresses.
2. Calculate the stress stimuli per day, substituting the effective stress value and the number of stimulation per day into equation 1.
3. Obtain the quantity of bone apposition or resorption per day using the relationships in equation 2.
4. Change the defect shape according to the quantity of bone apposition or resorption as calculated.

The process from step 1. to step 4. corresponds to the bone formation process during a day, so the process is iterated for the duration of the experimental period in this simulation. This calculation requires great computational effort

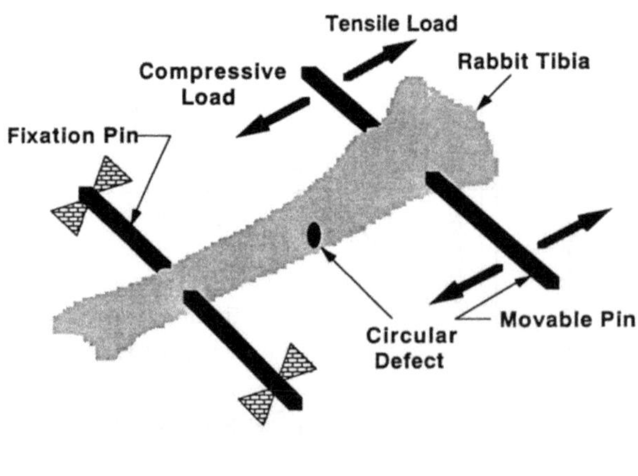

Fig. 2. Schema of rabbit tibia with circular defect in mechanical stimulation experiment

because the three-dimensional finite-element analyses are needed for each iteration. To improve calculation efficiency, an elliptic cylindrical model that approximates the rabbit's tibia (Fig. 3) is used here, and the load applied is equivalent to the experimental load. Finite-element subdivision of this model is shown in Fig. 4. Elastic modulus and Poisson's ratio are defined as 11.8 GPa and 0.5, respectively, which correspond to the material properties of cortical bone.

When the remodeling theory of Beaupré is applied to the simulation of our experiment, the relationship between bone apposition or resorption and the stress stimulus is kept as is and only the values of the parameters used in equations 1 and 2 are changed. That is, ψ_{bAS}, c, m, and w are the parameters required before doing the calculation. Beaupré et al. gave the values as $\psi_{bAS} =$ 50 MPa/day, $c = 0.02$ (μm/day)/(MPa/day), and $m = 4$; w is 10% of ψ_{bAS} in their report. Because the values of ψ_{bAS} and c were determined for a normal human condition, it is considered that the parameters take different values for the rabbit when bone formation occurs with bone repair at the defect.

The values are determined in our calculation as follows. The attracter state stimulus ψ_{bAS} is calculated from average effective stresses occurring at the rabbit tibia daily and the number of stimulation per day. Assuming that 150% of the average rabbit weight (25 N) is applied to the tibia along an axial direction in daily stimuli occurring because of the normal movements of rabbits and that the cross-sectional area of the center of the tibia is 18 mm^2, the average effective stress is approximately 2.1 MPa. Further, assuming the number of stimulations per day is 2000, the value of ψ_{bAS} is given as 14 MPa/day. On the other hand, it is difficult to determine the value of c because it is just an empirical parameter. Thus we assume the value of c from an experimental

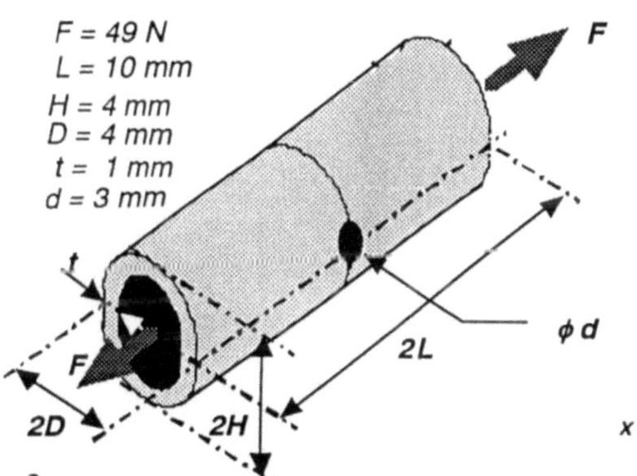

3

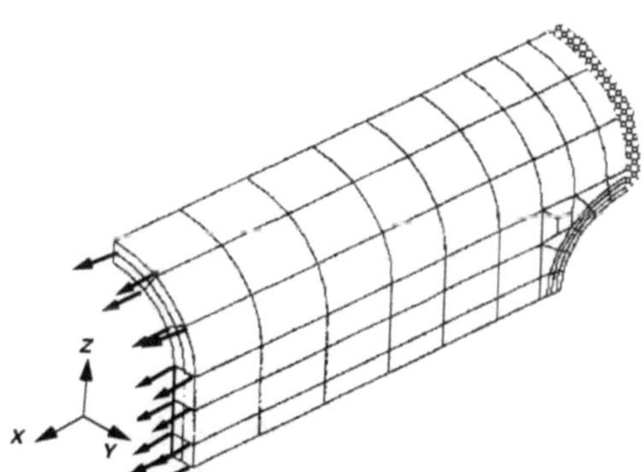

4

Fig. 3. Schema of rabbit tibia with circular defect in mechanical stimulation computer simulation

Fig. 4. Finite-element subdivision of computer simulation model of rabbit tibia result in the 4-week stimulation group, $c = 0.8\,(\mu m/day)/(MPa/day)$. Values of the other parameters m and w are the same as in analyses by Beaupré.

Results

Changes of defect shape in the rabbit tibia model obtained by computer simulation are shown in Fig. 5. Only one-fourth of the region is shown; the loading direction is along the horizontal axis, and the solid, dotted, and broken lines correspond to the shape of middle, outer, and inner surfaces, respectively. Defect shapes were estimated after 1, 2, and 4 weeks of stimulation in the experiment, so that the calculation is iterated 28 times, corresponding to 4 weeks. Figure 5 shows the defect shape at intervals of 1 week. The effective stresses around the upper-side boundary in Fig. 5 are larger in comparison with the stresses around the right-side boundary on the initial shape. Therefore, bone appositions around the upper boundary are larger than the right boundary in the bone formation process. At the lowest stimulation point of initial shape, the bone resorption can also be seen temporarily. To compare these numerical results, the bone formation after 2 weeks and 4 weeks of tensile stimulation in the experiment is shown as Fig. 6 and Fig. 7, respectively, which were evaluated by hematoxyline-eosin stain. From the computational and experimental results, the amount of bone formation that occurred inside the initial defect is evaluated. The time histories of the amount bone formation that normalized the initial defect area are shown in Fig. 8.

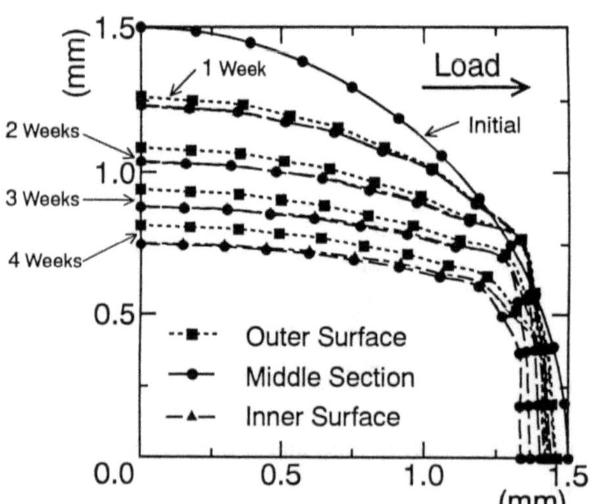

Fig. 5. Changes of defect shape of rabbit tibia model in computer simulation

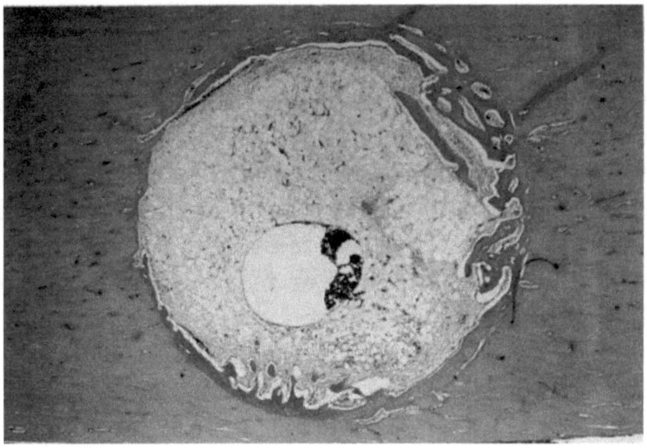

Fig. 6. Bone formation around defect of rabbit tibia after 2 weeks of experimental stress stimuli

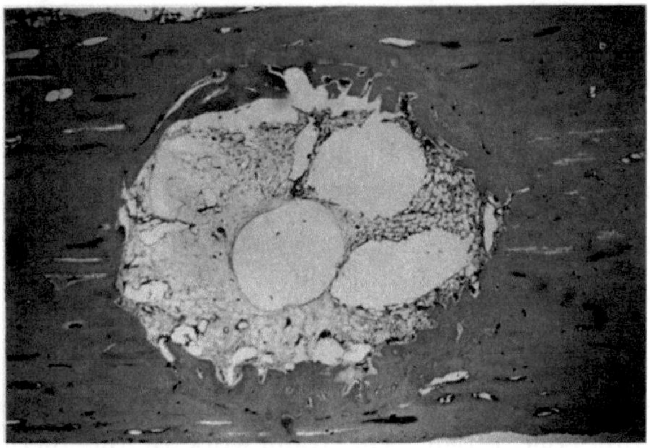

Fig. 7. Bone formation around defect of rabbit tibia after 4 weeks of experimental stress stimuli

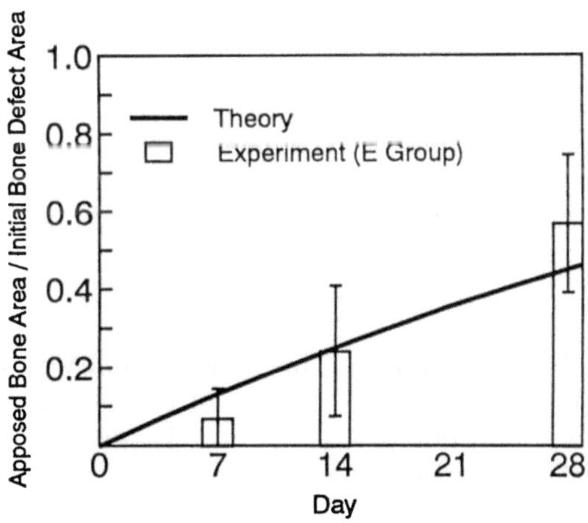

Fig. 8. Time history of bone formation quantity on computer simulation and in the experiment

Considerations

Comparing Fig. 5 with Figs. 6 and 7, locations where particular bone apposition occurred are approximately the same on both, and the defect shapes after 4 weeks of stimulation are similar. The tendency of the bone apposition increase in the calculation is very similar to the tendency in the experiment (see Fig. 8). In the results, the bone remodeling theory of Beaupré is applicable to bone formation with bone repair by tuning the parameters. Some serious problems, however, can be mentioned when examining the results in detail. Bone resorption is not found in the experimental results though it occurred transiently at the low stimulation point in the calculations. In the experiment, very little bone apposition was seen after 1 week of stimulation, and the apposition after 4 weeks of stimulation tends to be larger in comparison with the calculation. We believe that the nonlinearity of bone apposition changes and the large dispersion of the results in the experiment are influenced by the bone repair that occurred around circumference of the defect. The major cause of the difference between calculation and experiment is the effect of the bone repair at the circumference of the defect, which is not taken into account in the computation. To solve problems such as this experiment, an improved theoretical model that considers the bone repair proper is necessary.

Conclusion

In this study, the theoretical model of bone remodeling adapted to a mechanical stress stimulus provided by Beaupré is applied to a bone-forming simulation with bone repair. The framework of the theory is not changed, and some parameters related to bone formation are tuned so that the theory can be adopted to our problems. The bone formation process around the defect in the rabbit tibia while an intermittent axial load is applied is estimated by using a computer simulation based on the theory and compared with the experimental result obtained previously to determine the efficiency of the theoretical model. In the results, the bone formation process tendencies as obtained by calculation and experiment are approximately similar, and the possibility of practical use of the theoretical model is confirmed. Further, it is made clear that improvements of the theory considering the bone repair effect are necessary to apply to our problem.

Acknowledgment. This investigation was supported by the scientific research foundation of the Ministry of Education of Japan, "Biomechanics of Structure and Function of Living Cells, Tissue and Organs," in 1992 and 1993. We thank the authorities concerned.

References

1. Yasuda Y, Noguchi K, Sata T (1955) Dynamic callus and electric callus. J Bone Joint Surg 37A:1292
2. Rubin CT, Lanyon LE (1987) Osteoregulatory nature of mechanical stimuli: function as a determinant for adaptive remodeling in bone. J Orthop Res 5:300–310
3. Mabuchi K, Fujie H, Yamatoku Y, Yamamoto M, Sasada T (1992) A new methodology with an application of robotics to control the mechanical environment around experimentally fracture bone. In: Niwa S, Perren SM, Hattori T (eds) Biomechanics in orthopedics. Springer, Berlin Heidelberg New York Tokyo, pp 183–193
4. Sakai K, Tomita K, Sawaguchi T, Akagawa S, Oda J (1990) An experimental study on quantitative relation between the mechanical stimuli and bone formation. Proc Annu Meet JSOB 12:99–103
5. Aoyama K, Sueyoshi Y, Tomita K, Sawaguchi T, Sakai K, Oda J, Sakamoto J (1992) Innovation of loading apparatus for experimental studies of bone formation by mechanical stimulation. Proc Annu Meet JSOB 14:181–185
6. Gjelsvik A (1973) Bone remodeling and piezoelectricity. J Biomech 6:69
7. Guzelsu N, Saha S (1984) Electromechanical behavior of wet bone. J Biomech Eng 106:249–261
8. Hart RT, Davy DT, Heiple KG (1984) Mathematical modeling and numerical solution for functionally dependent bone remodeling. Calcif Tissue Int 36:S104
9. Takakuda K (1992) A hypothetical mechanism for adaptive bone remodeling (transportation of growth factor by mechanical loads). Trans J Soc Mech Eng 58(511A):1015–1021
10. Weinans H, Huiskes R, Grootenboer HJ (1992) Effect of material properties of femoral hip components on bone remodeling. J Orthop Res 10:845–853
11. Beaupré GS, Orr TE, Carter DR (1990) An approach for time-dependent bone modeling and remodeling—theoretical development. J Orthop Res 8:651–661
12. Beaupré GS, Orr TE, Carter DR (1990) An approach for time-dependent bone modeling and remodeling—application: a preliminary remodeling simulation. J Orthop Res 8:662–670

Effect of Mechanical Stimuli on Healing of Bone Defects in the Rabbit Tibia

Kazuhiro Aoyama[1], Yasunobu Sueyoshi[1], Katsuro Tomita[1], Takeshi Sawaguchi[2], Yasuichiro Sakai[3], Jiro Sakamoto[4], and Juhachi Oda[4]

Summary. Although it is well known that bone responds to mechanical stimuli, details of this phenomenon are still uncertain. To study the quantitative relation between bone formation and mechanical stimuli, an experiment using rabbits and three-dimensional finite-element analysis was conducted. Two pins (ϕ, 2 mm) were inserted into the rabbit right tibia 70 mm apart, and a circular bone defect (ϕ, 3 mm) was made halfway between the two pins. The 3-mm circular bone defect produced a wide range of stress distribution. An intermittent compressive or tensile axial load of 49 N at 0.5 Hz was applied for 1 h/day to the rabbit tibia through the two pins.

Bone formation at the circular bone defect was evaluated histologically. The strain energy density of the tibia was also analyzed by the three-dimensional finite-element method. The histological evidence of bone formation and strain energy density were compared. After 2 weeks of stimulation, both compressive and tensile stress generated bone formation, with more bone formation observed in the area subjected to greater stress than in that subjected to less stress. After 4 weeks, woven bone was also observed in the control specimen, but mature bone formation was observed in both groups loaded with compressive or tensile load.

Key words: Bone formation—Compressive stress—Tensile stress—Three-dimensional finite-element method

[1] Department of Orthopedic Surgery, School of Medicine, Kanazawa University, 13-1 Takara, Kanazawa, 920 Japan
[2] Department of Orthopedic Surgery, Toyama Municipal Hospital, 292 Imaizumi, Toyama, 939 Japan
[3] Department of Orthopedic Surgery, National Kanazawa Hospital, 1-1 Ishibiki 3, Kanazawa, 920 Japan
[4] Institute of Technology, Kanazawa University, 40-20 Kodatsuno 2, Kanazawa, 920 Japan

Introduction

It is well known that bone responds to mechanical stimuli [1–5]. We previously conducted an experiment focusing on the relation between compressive load and bone formation [6]. This time we applied not only a compressive but also a tensile load to the rabbit tibia to further clarify the relationship between the quantity and quality of mechanical stimuli and bone formation. We investigated this relationship by comparing histological evaluation and the analysis of mechanical stimuli by the three-dimensional finite-element method (3D-FEM).

Materials and Methods

Forty-five female Japanese white rabbits were used in this investigation. Each pin (2 mm in diameter) was inserted into the proximal part and distal part of the right tibia 70 mm apart, and a 3-mm circular bone defect was made halfway between the two pins (Fig. 1a). The right leg was fixed with external fixators through the two pins and cast postoperatively (Fig. 1b). In the experimental groups, an intermittent compressive or tensile axial load of 49 N with 0.5 Hz was applied for 1 h/day to the rabbit tibia through the two pins by the self-monitoring loading apparatus devised by us [7] (Fig. 1c). In the control group, no load was applied and the rabbit tibia remained fixed by the external fixators and the casts. The tibia was resected at 1, 2, or 4 weeks after operation, and bone formation at the circular bone defect was evaluated histologically. Compression, tension, and control groups were established, and each group was divided into three subgroups: 1-week, 2-week, and 4-week groups. Each subgroup contained five rabbits. Image-analyzing software (ULTIMAGE, Graftek, 19, bis rue Duguay-Trouin, 78960 Voisins-Le-Bretonnex, France) was used for quantitative analysis of bone formation. Quantity of mechanical stimuli was calculated by 3D-FEM. Strain energy density was used as a representative quantity of mechanical stimuli [8].

Results

Histology

No bone formation at the circular bone defect was observed in the compression group, tension group, or control group after 1 week. After 2 weeks of stimulation, considerable immature bone formation was observed at the area parallel to the loading axis in both the compression and tension groups. In contrast, there was no or only slight bone formation at the area perpendicular to the loading axis. There were two cases in which bone formation was observed and three cases in which bone formation was not observed in the control group. In the former two cases, however, the amount of bone formation was slight and the site of bone formation was random. After 4 weeks of stimulation,

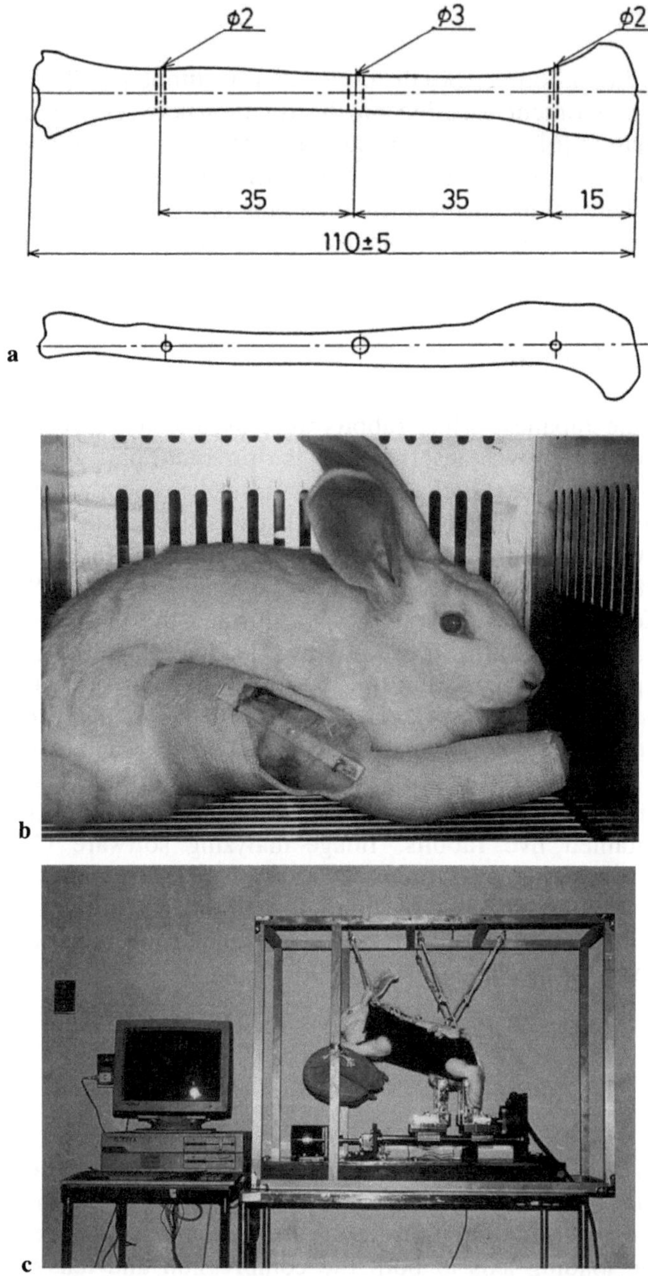

Fig. 1. **a** Schema of rabbit tibia. Fibula was excised. Two pins (ϕ, 2 mm) were inserted 70 mm apart; hole 3 mm in diameter was drilled halfway between the two pins. **b** Nonstimulating status. Tibia was fixed with external fixator and cast. **c** Setup of experiment. Rabbit tibia was fixed to the stimulator. Intermittent compressive or tensile axial load of 49 N at 0.5 Hz was applied 1 h/day

the woven bone that had been observed in both the compression group and tension group of 2 weeks had become mature, and the bone formation extended to the center of the circular bone defect. In the control group of 4 weeks, bone formation was observed but it was immature and was much less in amount than that of the compression and tension groups (Fig. 2).

Quantitative Analysis of Bone Formation by Image-Analyzing Software

The amount of bone formation within the circular bone defect was compared in the compression, tension, and control groups quantitatively by means of image-analyzing software. The quantity of bone formation in both the compression group and tension group was observed to be larger than that in the control group. In comparing the compression group and tension group in the 2-week groups, the quantity of bone formation in the tension group tended to be larger, while at 4 weeks that in the compression group tended to be larger.

Three-Dimensional Finite-Element Method

Strain energy density around the circular bone defect was analyzed by means of 3D-FEM to relate the quantity of mechanical stimuli to the quantity of bone formation. Analysis was carried out by approximating a part of the rabbit tibia near the circular bone defect to an elliptical cylinder model (Fig. 3, preceding chapter). A quarter of the model was used as an object of analysis because the shape of the model was symmetrical (Fig. 3a). As shown in Fig. 3a, we assumed the direction of nine o'clock to be 0° and the direction of twelve o'clock to be 90°. The distribution of strain energy density at a border of the circular bone defect that was generated by the compressive or tensile load of 49 N along the axial direction is shown in Fig. 3b. The quarter of circular bone defect was divided into two areas; one was a high-strain-energy density area (45°–90°) and the other was a low-strain-energy density area (0°–45°). The strain energy density between 0.3 and 4.9×10^{-3} Mpa is generated in the higher area and the strain energy density between 0.0 and 0.7×10^{-3} Mpa is generated in the lower area. Quantitative evaluation of bone formation was carried out separately and the two areas were compared.

Relationship Between Quantity of Mechanical Stimuli and Bone Formation

The circular bone defect was divided by the lines of 45°, 135°, 225°, and 315° as shown in Fig. 4a, and we assumed the higher strain energy density area to be area I and the lower strain energy density area to be area II. The control group was divided into the same two areas to be compared. The strain energy density at the border of the bone defect changes as the bone is generated and the shape of the bone defect changes. Thus, the quantitative evaluation of bone formation was performed in the 2-week specimen, the change of shape of

COMPRESSION 2W

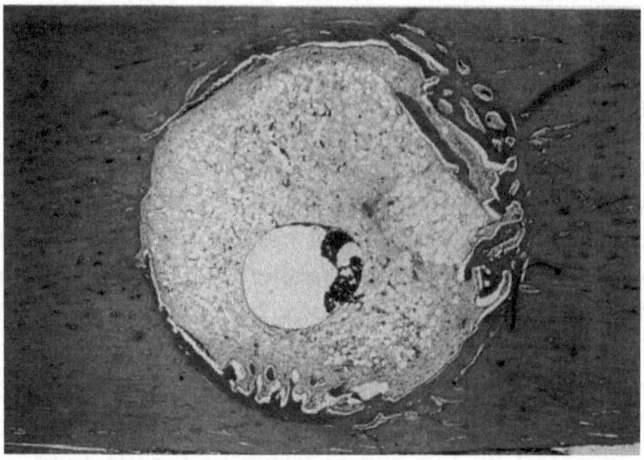

TENSION 2W

CONTROL 2W

Fig. 2. Sections perpendicular to the axis

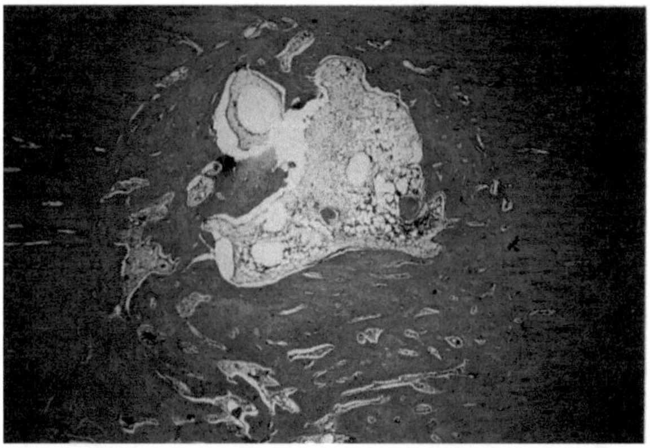

COMPRESSION 4W

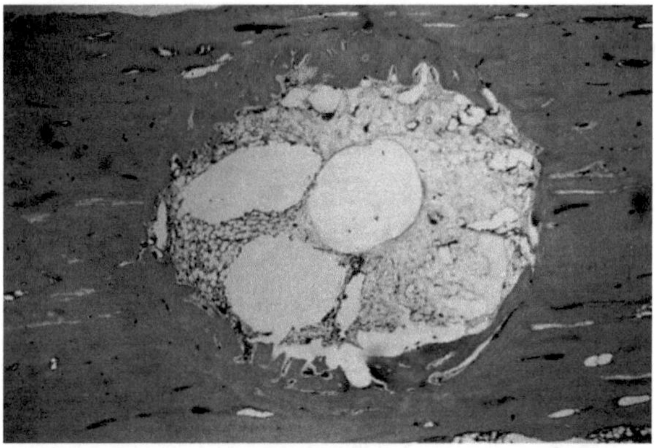

TENSION 4W

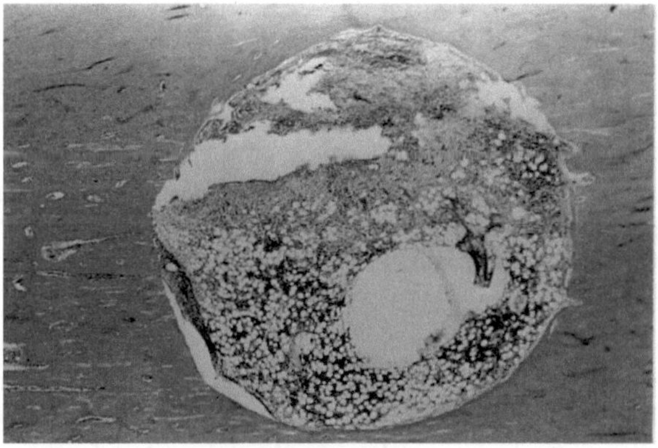

CONTROL 4W

Fig. 2. (*Continued*)

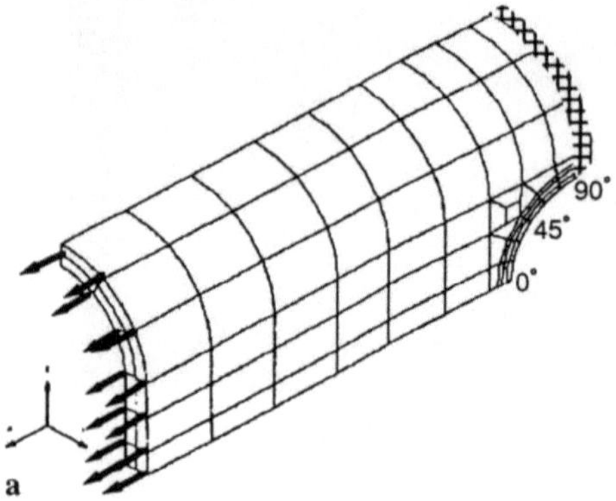

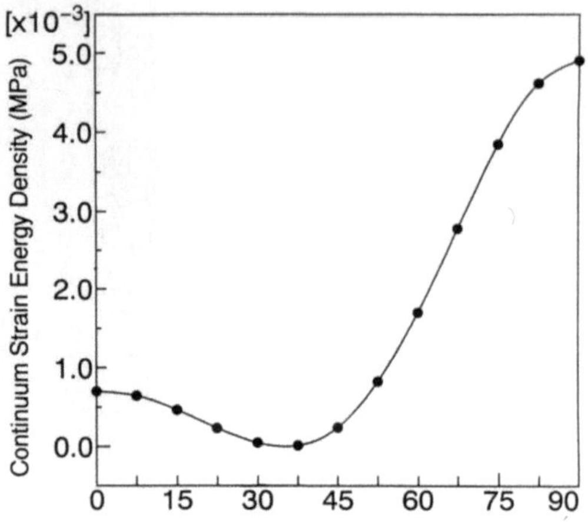

Fig. 3. **a** Finite-element subdivision. **b** Strain energy density around circular bone defect

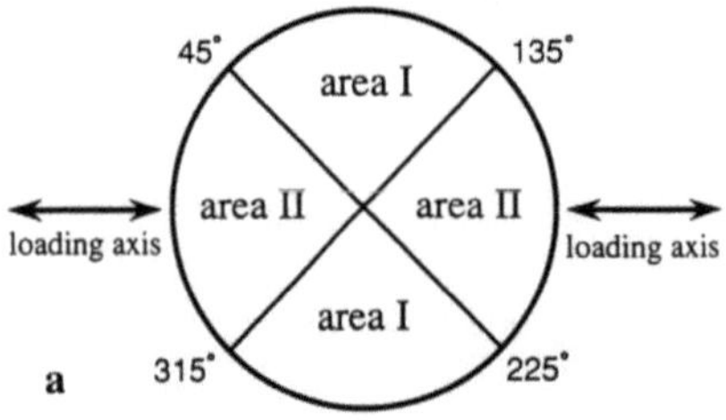

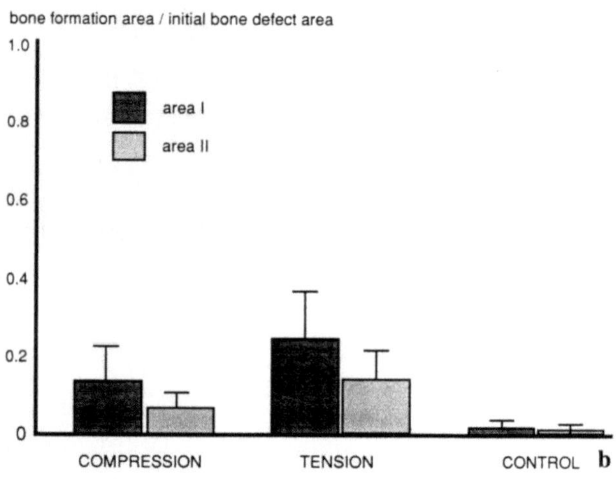

Fig. 4. **a** Division of the circular bone defect into two areas. Strain energy density generated in area I was relatively higher than in area II. **b** Quantitative analysis of bone formation compared between area I and area II

which was the least. The quantity of bone formation was statistically tested between area I and area II in each group. There was no significant difference in the control group whereas the quantity of bone formation in area I was larger than that in area II in both the compression group and the tension group ($P < 0.05$) (Fig. 4b).

Discussion

Since the formulation of Wolff's law it has been recognized that the structure of bone adapts itself to the mechanical environment modifying its morphology and density. Factors involved in bone formation include electrical factors [9], bone morphological factors [10], circulatory factors, and systemic factors such as hormones and metabolism in addition to mechanical stimuli; these factors are thought to interact. Because many varied interactions, not all of which are understood, occur among these factors, the mechanisms involved in bone formation are extremely complex and require much additional study. We have aimed at clarifying the quantitative relationship between mechanical stimuli and bone formation in this research. As regards the type of mechanical stimulus, it has been clarified by past research [11] including ours [6] that compressive stress promotes bone formation. It is widely considered that tensile stress acts as a bone absorption effect, as it has hitherto been explained by the example of self-correcting of malunion in long bone, but Yasuda et al. [12], Okada [13], and Matsushita and Kurokawa [14] concluded that tensile stress promotes bone formation. However, this issue is still controversial. In this study, we performed an experiment in which tensile load was applied to the rabbit tibia to clarify the influence of tensile stress on bone formation. Also, the loading apparatus used in the former experiment was devised for the purpose of comparing tension with compression under the same conditions [7].

Strain energy density on the border of the circular bone defect analyzed by 3D-FEM was compared with histological data and the results of image analysis after 2 weeks. More bone formation was observed in area I in which a strain energy density of $0.3–4.9 \times 10^{-3}$ MPa was produced during the initial period than in the area II in which a strain energy density of $0.0–0.7 \times 10^{-3}$ MPa was produced. Although bone formation was observed in two of the control rabbits at 2 weeks, the site of bone formation was erratic and the quantity formed was less than that of the experimental group. This bone formation is thought to be caused not by mechanical stimuli but because of the natural repair mechanism, because the bone defect, namely damage to the bone, was deliberately made in this experiment. More, and also more mature, bone was generated after 4 weeks than after 2 weeks in both the compression group and the tension group, while less and more immature bone was generated in the control group. Our experiment demonstrated that tensile stress as well as compressive stress promotes both the amount and maturation of bone, and that the promotion of bone formation depends on the quantity of the stress. After 2 weeks the quantity of bone formation was greater in the tension group than in the compression group, but after 4 weeks the result was reversed. It is unclear which kind of stress exerts a greater promoting action on bone formation, but it is clear that these two types of stress differ in some way in their bone growth-promoting properties.

Conclusions

1. It was clarified that not only compressive stress but also tensile stress promotes bone formation.
2. The quantity of bone formation in the high-energy-density area was greater than that in the low-energy-density area when both compressive stress and tensile stress were applied.

Acknowledgments. This study was supported by funds provided by the Ministry of Education, Japan, Science and Culture, a Grant-in-Aid for Scientific Research on Priority Area, 1992 and 1993. We are grateful to Mr. Shigeyuki Tabata, Mr. Hiroaki Kobayashi, Mr. Yaoki Nakano, and Mr. Satoru Kenmochi.

References

1. Rubin CT, Lanyon LE (1987) Osteoregulatory nature of mechanical stimuli: function as a determinant for adaptive remodeling in bone. J Orthop Res 5:300–310
2. Pead MJ, Skerry TM, Lanyon LE (1988) Direct transformation from quiescence to bone formation in the adult periosteum following a single brief period of bone loading. J Bone Miner Res 3:647–656
3. Pead MJ, Suswillo R, Skerry TM, Vedi S, Lanyon LE (1988) Increased ^{3}H-uridine levels in osteocytes following a single short period of dynamic bone loading in vivo. Calcif Tissue Int 43:92–96
4. Mabuchi K, Fujie H, Yamatoku Y, Yamamoto M, Sanou Y, Sasada T (1992) Utility of robotics to control the mechanical environment on the fracture healing of bone and to assess the healing. Trans Jpn Soc Mech Eng 58(551A):1030–1035
5. Turner CH, Akhter MP, Raab DM, Kimmel DB, Recker RR (1991) A noninvasive, in vivo model for studying strain adaptive bone modeling. Bone 12:73–79
6. Sakai K, Tomita K, Sawaguchi T, Akagawa S, Oda J (1990) An experimental study on quantitative relation between the mechanical stimuli and the bone formation. Orthop Biomech 12:99–103
7. Aoyama K, Sueyoshi Y, Tomita K, Sawaguchi T, Sakai K, Oda J, Sakamoto J (1992) Innovation of the loading apparatus for the experimental studies of bone formation by mechanical stimulation. Orthop Biomech 14:181–185
8. Huiskes R, Weinans H, Grootenboer HJ, Dalstra M, Fudala B, Slooff TJ (1987) Adaptive bone-modeling theory applied to prosthetic-design analysis. J Biomech 20:1135–1150
9. Hirasawa Y, Okada K (1989) Application of electric phenomena to fracture treatment: recent trend and future. Sogo Rehab 17(1):5–11
10. Takakuda K (1992) A hypothetical mechanism for adaptive bone remodeling (transportation of growth factors by mechanical loads). Trans Jpn Soc Mech Eng 58(511A):1015–1021
11. Nakabayashi Y, Shiba R (1987) An experimental study of bone remodeling influenced by mechanical stress. J Jpn Orthop Assoc 61:1429–1436

12. Yasuda I, Nagayama H, Kato T, Hara O, Okada K, Noguchi K, Sata T (1953) Fundamental problems in the treatment of fracture. J Kyoto Med Soc 4:395–406
13. Okada K (1956) Dynamic studies on bone fractures. J Jpn Orthop Assoc 30(2): 105–133
14. Matsushita T, Kurokawa T (1991) Mechanical condition promoting bone healing (comparison between compression and tension). J Jpn Orthop Assoc 65(7):S1216

Biomechanical Study of the Human Mandible on Mechanical Response of Its Shape and Structure

NORIO INOU[1], HIROSHI FUJIWARA[2], and KOUTAROU MAKI[3]

Summary. This paper describes a biomechanical study of the human mandible based on individual X-ray computed tomography (CT) data. The purpose of this study is to estimate the mechanical response in the human mandible during biting and to discuss the mechanical rationality of the mandible. It is indispensable for any biomechanical study to establish reliable mathematical models. We constructed two types of three-dimensional finite-element models based on individual X-ray CT data. One is symmetrical in shape, and the other one is nonsymmetrical. For stress analysis, we estimated the masticatory forces and directions from the CT images, and also estimated reaction forces acting on the teeth using a pressure sheet. The computational results showed that stress is distributed in the entire area of these mandibles, but stress around the condyle is less than in other regions. Bone density distributions of the mandibles were measured by the same X-ray CT. These results showed a strong relationship between stress and bone density distribution for both mandibles.

Key words: Biomechanics—Stress analysis—Three-dimensional finite-element model—Mandible—Bone density—Functional adaptation

Introduction

Mandibles are essential to the mastication of foods. We can consider them to be mechanical elements that receive outer forces from teeth and masticatory muscles during biting. It is said that bone is well-designed in a mechanical sense. Mandibles are also expected to be well-designed structures because they

[1] Department of Mechanical Engineering and Science, Tokyo Institute of Technology, O-okayama, Meguro-ku, Tokyo, 152 Japan
[2] Ricoh Corporation, Ohta-ku, Tokyo, 143 Japan
[3] Department of Orthodontics, Showa University, Ohta-ku, Tokyo, 145 Japan

must support large loads during biting. However, there are not many studies on this theme.

On the other hand, bone has an adaptive function to change the shape or the bone density for outer forces. The portion where high stress is set up is strengthened, but the portion where stress is absent is weakened. As a result of the continuous processes, the bone structure adapts to outer circumstances. The phenomenon is called functional adaptation or remodeling. Human mandibles are also expected to have this functional adaptation.

Recent food processing technologies supply us plenty of easy-to-chew foods, and we do not need strong mastication. However, some researchers are concerned about these comfortable eating habits because such soft foods may decrease the ability of mastication or obstruct the normal growth of the mandible at young ages.

To discuss these points, we must know the exact mechanical characteristics of mandibles. This chapter examines human mandibles from the biomechanical viewpoint. First, we explain a simulation method to execute stress analysis based on X-ray computed tomography (CT). Next, we state the computational results. Finally, we discuss the functional adaptation in the mandible comparing stress distribution with bone density distribution.

Individual Modeling of the Mandible

The finite-element method is a useful tool for biomechanical studies to estimate the stresses produced by applied loads. It is important for executing reliable stress analysis to provide individual models because each mandible differs from every other mandible. There are few reports concerned with the individual simulation because of difficulty of individual modeling. In this article, we explain our method of developing an individual finite-element model.

First, we provided a standard type of mandibular finite element model based on a plastic model, as in Fig. 1. This model is an original form modified into an individual model. It consists of 1596 isoparametric elements including 2380 nodes. As the corpus mandibulae (front part of mandible) takes a kind of shell structure, we selected the elasticity of inside elements as 1/100 compared with outside elements.

Second, for the individual modeling, we provided X-ray CT data taken around the mandibles from two persons. We call the two persons and the corresponding models "A" and "B" for convenience. Figure 2a,b provides images of the X-ray CT for persons "A" and "B", respectively. Mandible A is approximately symmetrical in shape, and B is nonsymmetrical. These images were taken at intervals of 2 mm for each person. The upper circles in the images are phantoms that are necessary for estimation of bone density (the details are described later).

Third, we extracted contour lines of mandibular shape from each CT image and made three-dimensional contour images by stacking them at the same intervals as shown in Fig. 3a,b.

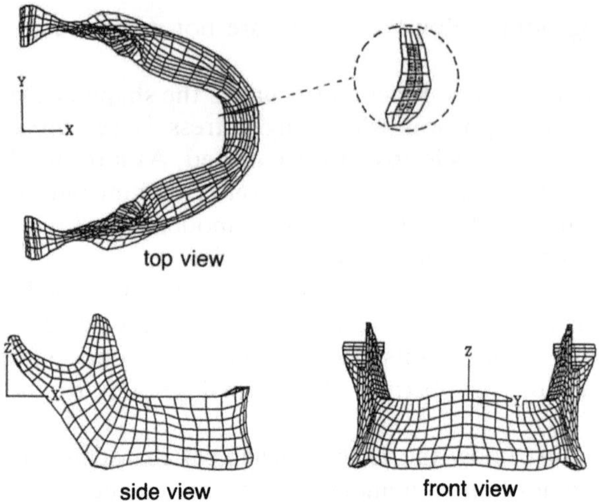

Fig. 1. Three-dimensional finite-element model (original form): *shaded*, 9.8 × 10 MPa; *unshaded*, 9.8 × 10^3 MPa

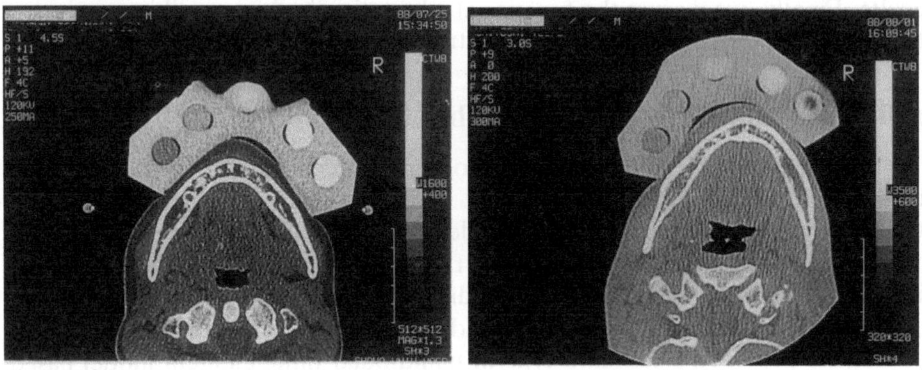

Fig. 2. X-Ray CT image around mandible: **a** person "A"; **b** person "B"

Fourth, we modified the standard model by moving the nodal positions or making geometrical transformation so that they become the same shape as the contour images. The final models are shown in Fig. 4a,b.

Estimations of Biting Condition

To execute stress analysis we must determine the mechanical condition during biting. The necessary data are reaction forces on the teeth and masticatory forces and the directions on the mandible.

The reaction forces were measured by a pressure sheet made with polyethylene telephtharate that changes the color where the forces are exerted. We asked the persons to bite the sheet in the bilateral condition. Figure 5a,b shows

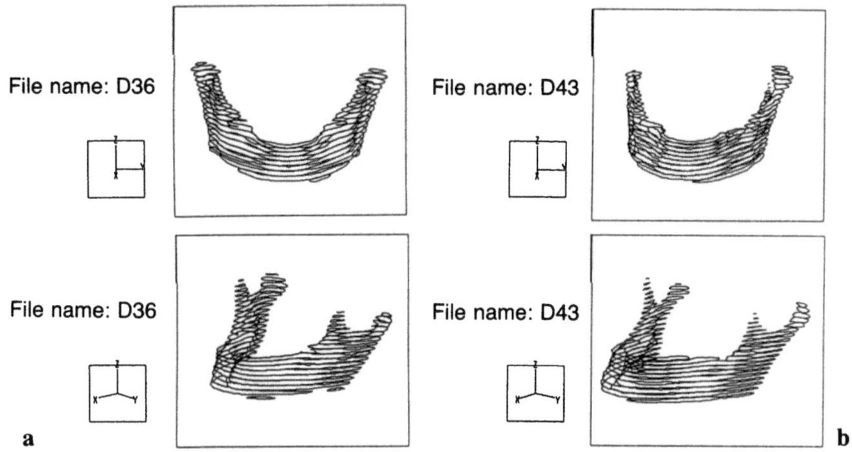

Fig. 3. Three-dimensional contour image extracted from X-ray CT: **a** person "A"; **b** person "B"

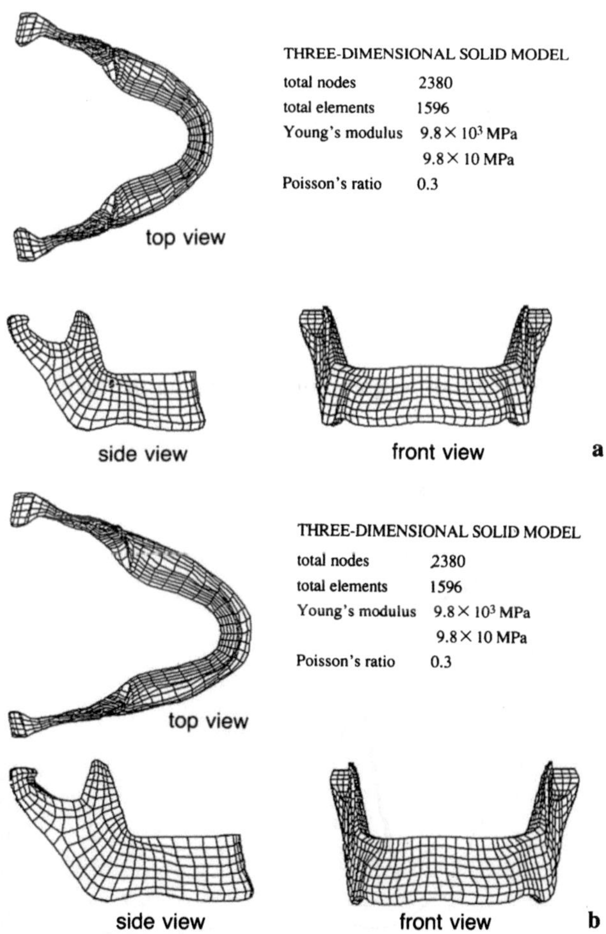

Fig. 4. Individual finite-element model: **a** model "A"; **b** model "B"

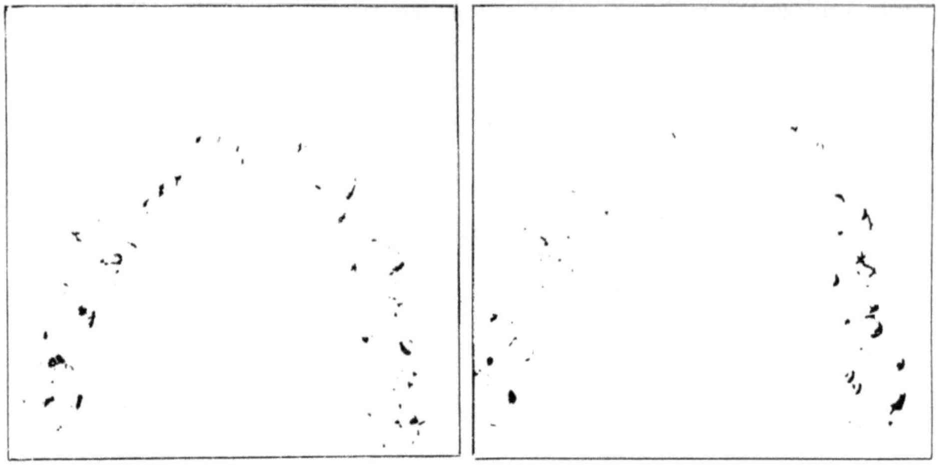

a b

Fig. 5. Developed spots on a pressure sheet after biting: **a** person "A"; **b** person "B"

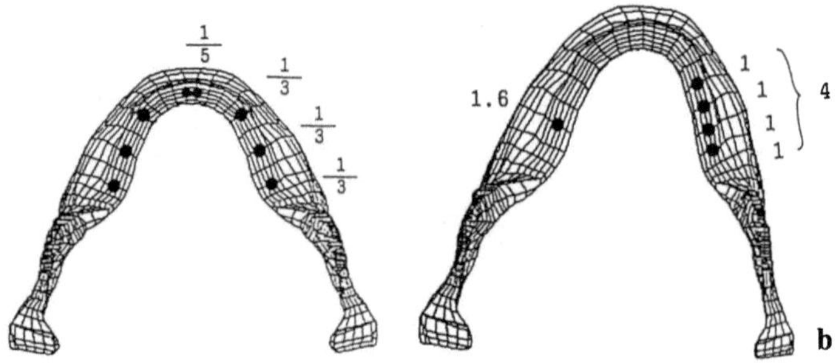

a b

Fig. 6. Ratio of reaction forces acting on the teeth: **a** person "A"; **b** person "B"

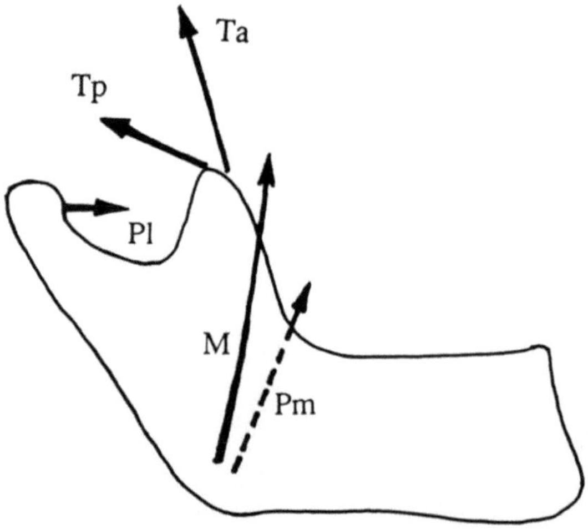

Fig. 7. Masticatory muscles and the direction of acting on mandible. *Ta*, anterior temporalis; *Tp*, posterior temporalis; *M*, masseter; *Pm*, pterygoideus medialis; *Pl*, pterygoideus lateralis

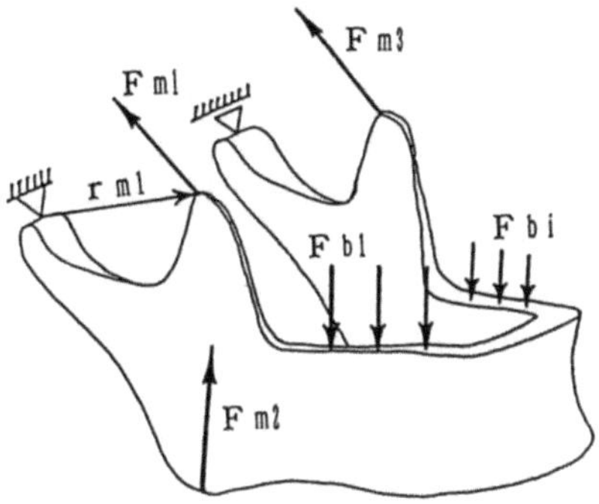

$$T_m = \Sigma\, r_{mi} \times F_{mi}$$
$$T_b = \Sigma\, r_{bi} \times F_{bi}$$
$$T_m + T_b = 0$$

Fig. 8. Supporting system for stress analysis and balance of the moment

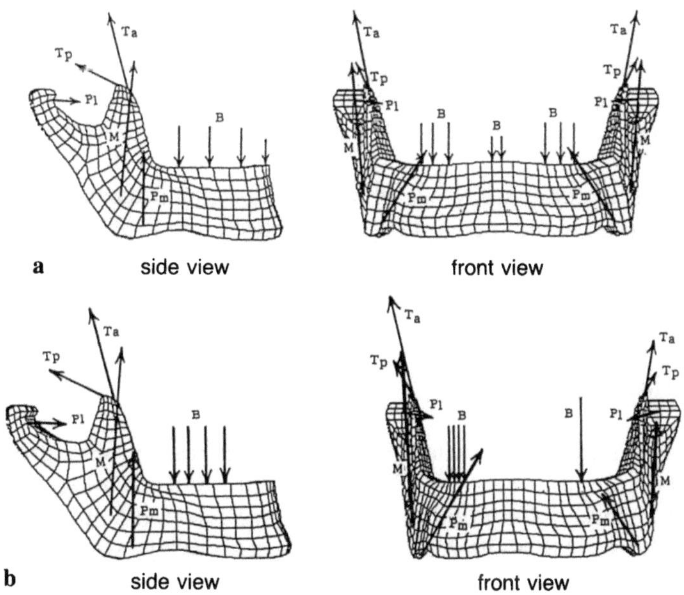

Fig. 9. Biting condition: **a** model "A"; **b** model "B"

developed distributions for persons A and B. We estimated the ratio of reaction forces and the locations acting on the teeth from the spots using an image analyzer as shown in Fig. 6a,b. In this study, we assumed that masticatory forces act on the teeth in the vertical direction.

There are many kinds of muscles on the mandible to control the movement. For stress analysis in the biting condition, we consider four kinds of masticatory muscles as shown in Fig. 7. The muscular directions were obtained from the X-ray CT while examining the position of muscular attachment to the mandible with a specimen of skull. As we cannot directly measure muscular forces with any current technique, we measured the effective muscular areas using a three-dimensional (3-D) image instrument (Voxel Flinger, Reality Imaging Corp.) and estimated the muscular forces as the ratio of the values assuming that muscular forces are roughly proportional to the muscular areas. We also referred to other data reported by Hatcher et al. [1] and Faulkner et al. [2] to estimate the muscular forces and the directions.

The supporting system for stress analysis (or boundary condition) was set as shown in Fig. 8. A condyle can turn in any direction like a pivot and the other condyle can also turn in any direction and move freely along the Y axis though it is restricted to move freely in the X and Z directions. This support system does not produce abnormal moment on both condyles. Under the mechanical condition, we adjust the reaction forces on the teeth as a whole so that it keeps the moment balance around Y axis produced by the muscular forces and the reaction forces.

For executing stress analysis by the finite-element model (FEM), because the entire rotation of the model must be restricted, only one location of loading points on the teeth was fixed in the Z direction though it is allowed to move freely in the X–Y plane. Even this fixation does not produce any new forces or moments because the model completely keeps moment balance around the Y axis.

Through above these steps, we obtained muscular forces and the directions on each individual mandible as shown in Tables 1 and 2. Figure 9a,b illustrates the biting conditions. These pictures tell us that a way of biting differs with the mandibular shape.

Computational Results

For the computational execution, we used a general-purpose program called SAP6 on the computer center at Tokyo Institute of Technology and also used NISA II on a workstation at our laboratory. We selected Young's modulus and Poisson's ratio as 9.8×10^3 MPa and 0.3, respectively, which are commonly used for stress analysis of bone. We have now finished providing all necessary

Table 1. Masticatory muscular directions

Muscle	Muscular direction (left side)
Ta	−0.255i + 0.174j + 0.951k
Tp	−0.893i + 0.174j + 0.416k
M	0.087i + 0.087j + 0.992k
Pm	0.000i − 0.500j + 0.866k
Pl	0.763i − 0.643j − 0.067k

i, j, and k denote the vectors of x, y, and z directions, respectively.

Table 2. Ratio of muscular forces

Muscle	Model A	Model B	
		right	left
Ta	3.75	5.00	2.50
Tp	2.25	3.00	1.50
M	6.75	9.00	4.50
Pm	3.23	4.30	2.15
Pl	1.05	1.05	1.05

data for executing the finite-element analysis and can execute the individual simulation. In this simulation, the important thing is not to know the absolute stress values but to know the way of stress distribution because we estimated the muscular forces as relative values.

Figure 10a,b shows maximum stress distribution (tensile stress) and minimum stress (compressive stress) distribution for the model A. Overlapping the tensile and compressive stress distributions, the stresses almost cover in the entire area of the mandible. Figure 11a,b shows the equivalent stress distributions related to the shearing strain energy for the models A and B, respectively.

Looking at these results, we notice that there is not a highly stressed portion around condyles (joint portions) compared with other portions. It may be explained that the portions around condyles look slightly built. For the non-symmetrical model, high stress distributes on the working side (stronger biting side), but also distributes considerably on the balancing side. It is common to bite foods with each side of teeth alternatively in actual mastication. We showed in our previous paper that even symmetrical mandibles cause considerable stress on a balancing side during unilateral biting [3].

There may be an adjustable mechanism so that the biting condition is controlled, thus distributing the stress in the entire area. In other words, we can say that mandibles are rational mechanical structures as a fully stressed design.

Comparison of Bone Density Distribution

Compact bones are generally rigid, and have high bone density. Cancellous bones are fragile and have low bone density compared with compact bones. We can see the difference easily from X-ray CT because the X-ray attenuation is related to the bone density. Maki proposed a quantitative method of the density distribution by placing a calibration phantom when X-ray CT is taken [4]. The calibration phantom is made by hydroxyapatite [$Ca_{10}(PO_4)_6OH_2$], which is a main component of bone. As the phantom consists of several rods that contain different densities of hydroxyapatite, we can make a quantitative estimation of the bone mineral.

Figure 12a,b shows dot images of the mandibular bone density where the locations of high bone density (more than 90% of maximum value) are plotted. Comparing these images with stress distributions during biting, as in Fig. 11, we can see a strong relationship between them. It should be noticeable that even the nonsymmetrical mandible has this correlation. This means that there is a close relationship between biting condition and mandibular shape and structure.

Discussion

It is natural to consider that the mandibular shape is determined by genes as are other structures. However, it is difficult to consider that the nonsymmetrical shape or bone density distribution is also determined inherently. As we stated, there is a strong relationship between stress distribution during biting and bone density distribution, even in the nonsymmetrical mandible. This fact shows that interactions between the mandible and the masticatory muscular forces can change the bony shape and the structure. Therefore, we cannot neglect circumstance factors.

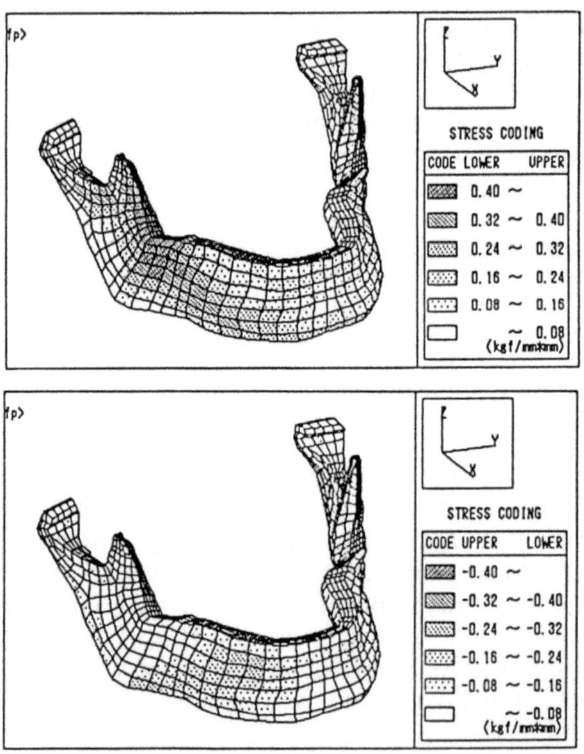

Fig. 10. Stress distribution of mandible for model "A": **a** tensile stress; **b** compressive stress

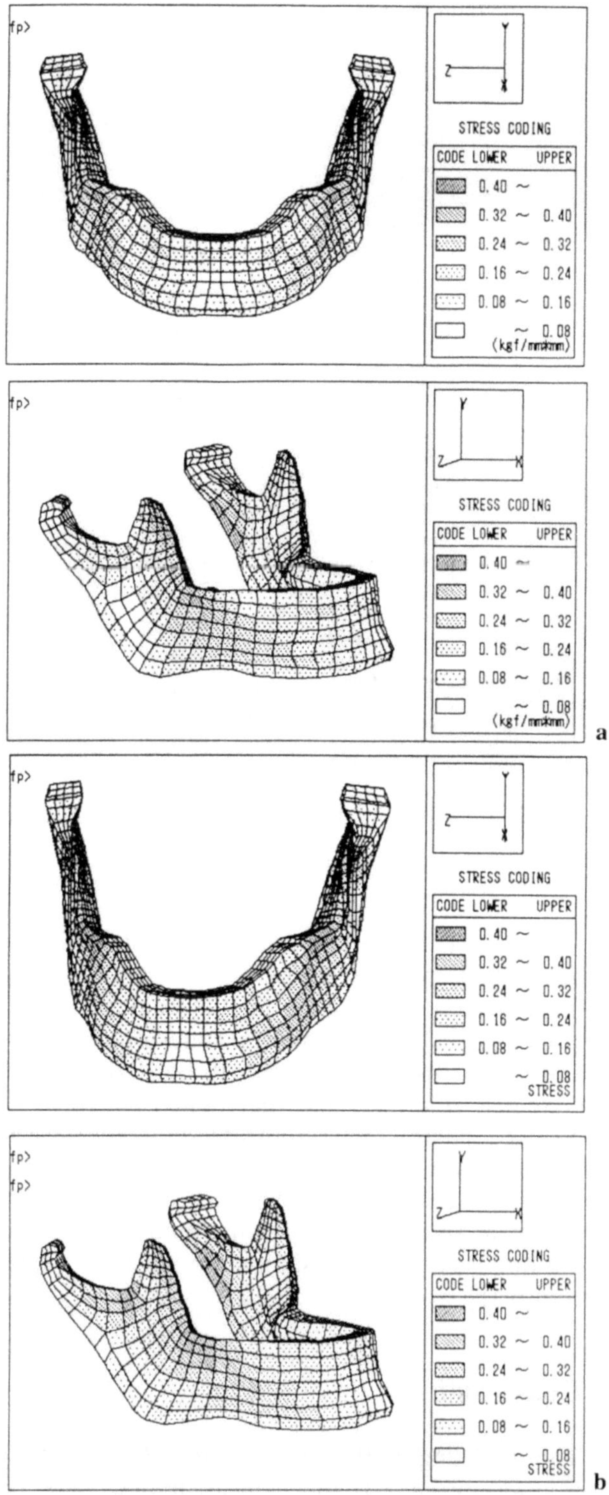

Fig. 11. Equivalent stress distribution: **a** model "A"; **b** model "B"

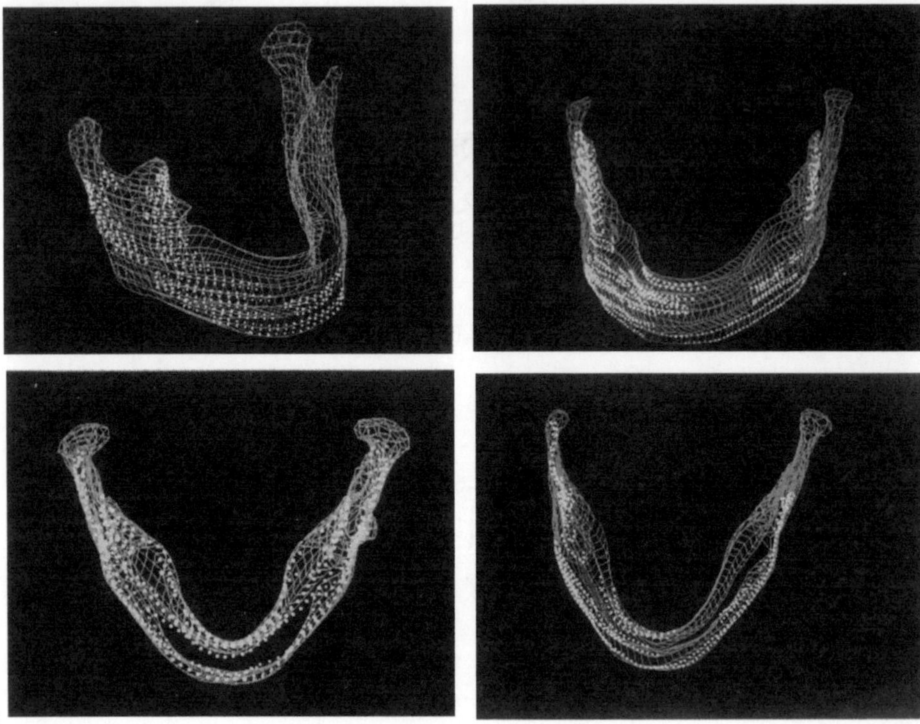

a b

Fig. 12. Bone density distribution: **a** person "A"; **b** person "B"

The relationship between stress distribution and bone density distribution shows that mandibles also have functional adaptation (or remodeling). The phenomenon is expressed in the engineering sense as follows. Each part of bone "senses" the mechanical states produced by outer forces at each position and adjusts its material strength by itself. As a result, the order of the whole system is reformed and maintained properly. If we simulate the successive process, making the Young's modulus increase for the highly stressed portion and decrease for the less stressed portion while calculating the stress of each part produced by the applied load by every changing process, we will finally obtain a Young's modulus distribution that is similar to bone density distribution. This computer simulation has been executed and showed the expected results [5].

Conclusion

The modeling method for an individual finite-element model based on X-ray CT was proposed. Mechanical analysis was performed under the bilateral biting condition. The bone density distribution also obtained from the same X-ray CT. The computational results showed that a mandible has a mechanical rationality and can change its shape and structure to adapt the mechanical circumstances.

Our final goal in this study is to comprehend the adaptive functions of mandible and to apply this knowledge to medical examinations. To realize our goal, we must have as reliable biomechanical information as possible. The stress analysis of the mandible is considerably developed, but further improvements are necessary for this research.

References

1. Hatcher DC, Faulkner MG, Hay A (1986) Development of mechanical and mathematic models to study temporomandibular joint loading. J Prosthet Dent 55:377–384
2. Faulkner MG, Hatcher DC, Hay A (1987) A three-dimensional investigation of temporomandibular joint loading. J Biomech 20:997–1002
3. Umetani Y, Inou N, Fujiwara H (1989) Stress analysis of mandibular shape during unilateral biting. In: Kubota K (ed) Mechanobiological research on the masticatory system. VEB Verlage für Medizin und Biologie, Berlin, pp 50–54
4. Maki K (1988) A study about the X-ray CT cephalometrics providing functions as the standard for three-dimensional reconstruction and bone mineral measurement. J Orthop Soc 47:380–390
5. Inou N, Fujiwara H, Umetani Y (1993) A computational method generating three-dimensional structure based on the adaptive function of bone. Trans Soc Instrum Control Eng 29:1221–1226

Biomechanical Simulation of the Osteoporotic Trabecular Pattern Around a Tooth

MOTOFUMI SOGO[1], SADAMI TSUTSUMI[2], YOSHINOBU MAEDA[1], and MASANORI OKA[2]

Summary. The purpose of this study was to elucidate the biomechanical mechanisms controlling the relationship between the mechanical loading environment on a tooth and its internal architecture, or so-called trabecular pattern, in the mandible. A computer-assisted simulation of the biomechanical rearrangement process of trabeculae around a tooth in the mandible was developed using the two-dimensional finite element method. This simulation predicted the stable trabecular pattern around a tooth, which was then compared with a human physiological pattern. The principal stress distribution of the simulated configuration showed a close similarity to the real trabecular pattern. In addition, the simulation procedure was used to model the trabecular pattern in the osteoporotic situation. The resulting patterns contained a pattern of compressive trabeculae that were thick and tensile trabeculae that were thin. Further, the resulting trabeculae had a high level of stress. The results suggest a high clinical risk of microfracture in an osteoporotic mandible.

Key words: Biomechanics—Osteoporosis—Finite element method—Trabecular pattern—Mandible—Stress distribution

Introduction

The basic work of Julius Wolff [1] described as "Wolff's law," performed a century ago, showed that the arrangement of cancellous bone was in accordance with the direction of the applied loads. This theory must also apply to the trabecular pattern around a tooth in the mandible.

---

[1] Department of Removable Prosthodontics, Faculty of Dentistry, Osaka University, 1-8 Yamadaoka, Suita, Osaka, 565 Japan
[2] Research Center for Biomedical Engineering, Kyoto University, 53 Kawahara, Shogoin Sakyo-ku, Kyoto, 606-01 Japan

The principal stress was previously determined using finite element analysis and compared with the trabecular pattern in the human specimens reported by Masuda [2] (Figs. 1, 2). When loaded by a vertical bite force, the vector representation of the principal stress was similar to the human trabecular pattern. Minimum principal stress spreads in a fan-shaped manner, and transverse maximum principal stress runs almost horizontally in the cancellous bone (Fig. 2a). The similarity between the stress distribution and the actual trabecular pattern in the molar region of the human mandible suggests that the trabecular pattern coincides with a stress distribution [3].

In the usual conventional finite element models, uniform material properties are assigned to the cancellous bone area. However, this approach does not consider trabecular bones and voids in the real cancellous bone. We improved the computer simulation method, and this report describes the introduction of the trabecular pattern simulation in normal situation and mostly the application on osteoporotic situation.

Materials and Methods

Normal Situation

A two dimensional finite-element model of the cross section of the molar region of the mandible was created that was composed of small square elements (0.2 × 0.2 mm) (Fig. 1, Fig. 3). With this model the cancellous bone area could

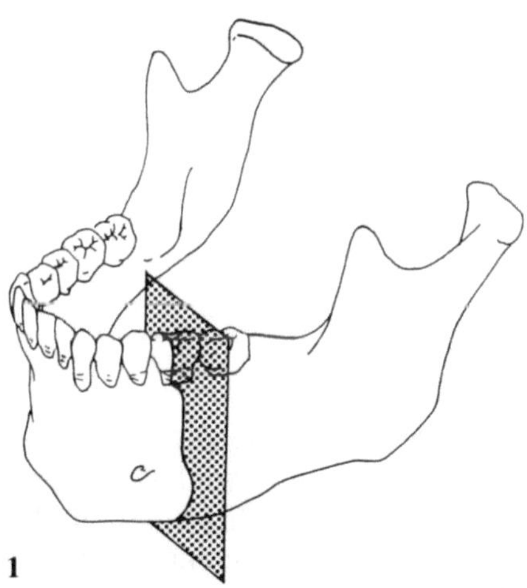

1

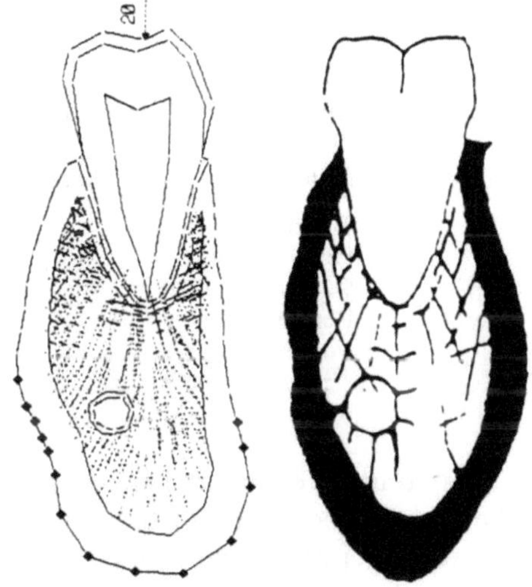

a b 2

Fig. 1. Cross section of the mandible modeled for finite element analysis

Fig. 2. **a** Distribution of principal stress from finite-element analysis. **b** Trabecular pattern at molar region on mandible in human specimen with high frequency. *Solid line*, tensile trabeculae; *dotted line*, compressive trabeculae

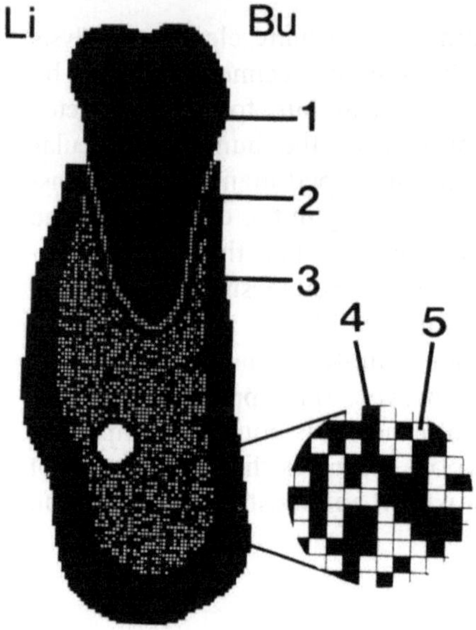

3

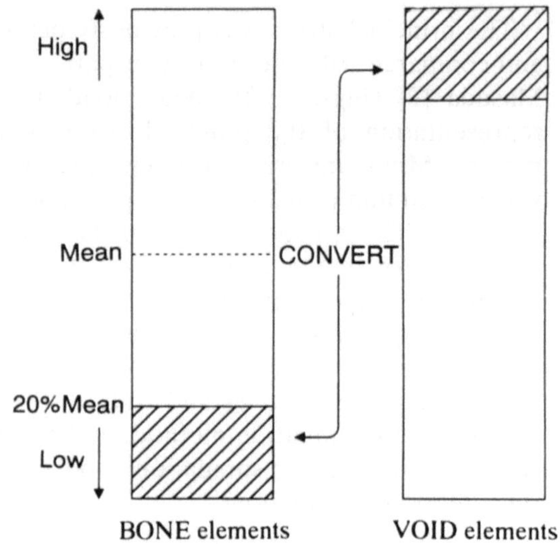

4

Fig. 3. Finite element model of the mandible cross section: *1*, Tooth; *2*, periodontium; *3*, cortical bone; *4*, bone element; *5*, void element; *Li*, lingual site; *Bu*, buccal site

Fig. 4. Rearrangement algorithm of bone and void elements based on their equivalent stress levels

Table 1. Material properties used in the finite element model

Material	Young's modulus (MPa)	Poisson's ratio
Tooth	14 000	0.15
Periodontium	10	0.49
Cortical bone	10 000	0.15
Bone element	10 000	0.15
Void element	1	0.49

be created using elements whose material properties were assigned to those of either cortical bone or soft tissue (void). The ratio of bone to void element or bone volume ratio, bone/(bone + void space), was 50% for this situation [4]. Initially, the elements in the cancellous bone region were randomly assigned one of the two material properties. Table 1 shows the material properties used in the model.

We previously reported an algorithm altering the material properties of elements [5,6]. For this study we further improved the algorithm. That algorithm, altering locations of bone elements and void elements in the cancellous bone area in this study, is as follows. Initially, a vertical bite force of 20 N was applied to the occlusal fossa, and the equivalent stress (von Mises stress) level in each element was determined. The mean stress in all the bone elements was determined, and those bone elements with a stress level less than 20% of the mean were converted into void elements by changing the material properties assigned to the element. To maintain the bone: volume ratio, an equal number of the highest stressed void elements were similarly converted to the bone elements (Fig. 4). The cutoff criteria of 20% was chosen so that remodeling would progress slowly. The stress distribution was then recalculated and the conversion process repeated. This process was continued until the stress in the bone elements were relatively uniform.

Osteoporotic Situation

Osteoporosis is defined as a disease characterized by low bone mass [7]. Initially, the osteoporotic condition was considered to have a bone volume ratio of 35%. To attain this ratio, the lowest stressed bone elements in the converged 50% bone volume ratio model were converted into void elements. Then, the bone–void interactive conversion process was repeated. Once the model had converged, the bone volume ratio was further reduced to 20% and the interactive conversion process repeated until convergence was again achieved.

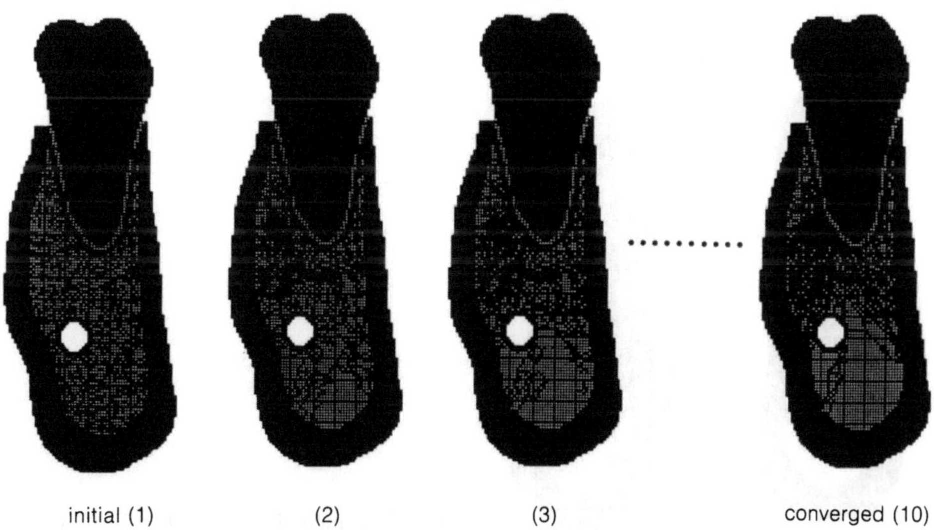

Fig. 5. Changes in the trabecular pattern drive in the simulation process

Results

Normal Situation

Figure 5 shows the initial condition, second stage (first interaction), third stage (second interaction), and converged stage (tenth interaction) of the trabeculation process. A dense accumulation of bone elements was found around the tooth root as the simulation steps progressed.

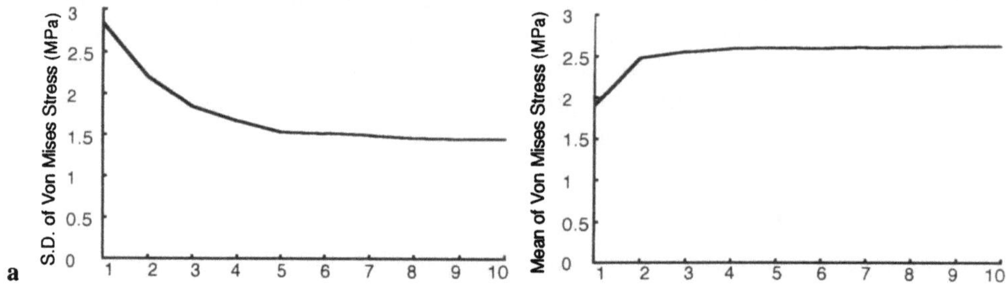

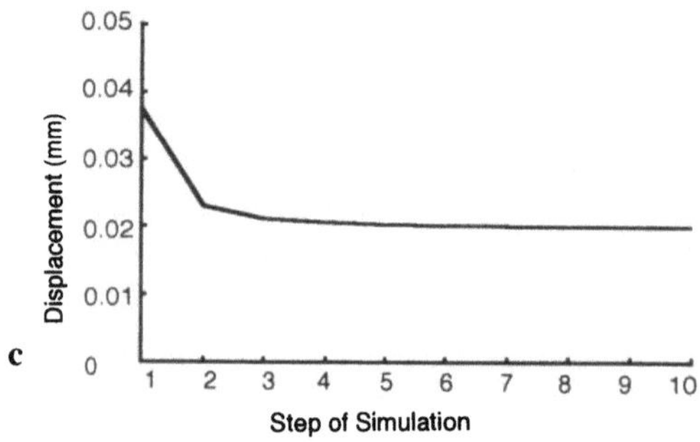

Fig. 6. Variation of the SD (**a**) and mean (**b**) of the equivalent stress in the bone elements and the displacement (**c**) for the 50% bone volume model with simulation stage

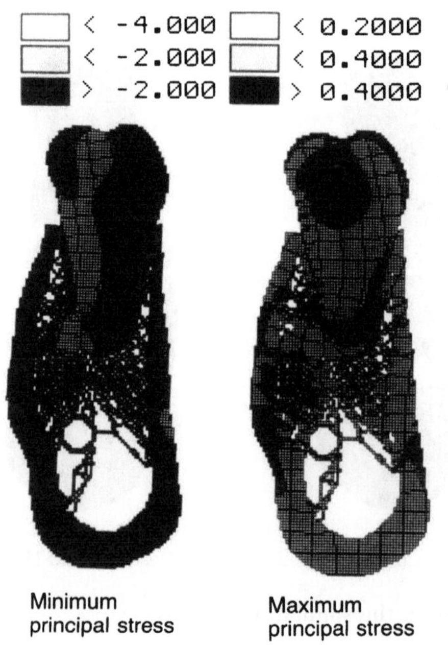

Fig. 7. Principal stress distribution on converged trabecular pattern in normal model

The mean and standard deviation (SD) of the equivalent stress in the bone elements and the displacement of the model are shown in Fig. 6. Figure 7 presents the principal stress distribution superimposed on the converged trabecular pattern.

Osteoporotic Situation

Figure 8 shows the osteoporotic simulation steps from the initial 50% bone volume ratio configuration to the converged configuration of the 20% bone volume ratio.

Figure 9 presents the principal stress distribution on the converged trabecular pattern in osteoporosis, and Figure 10 shows the variation of the mean equivalent stress in bone elements with iteration.

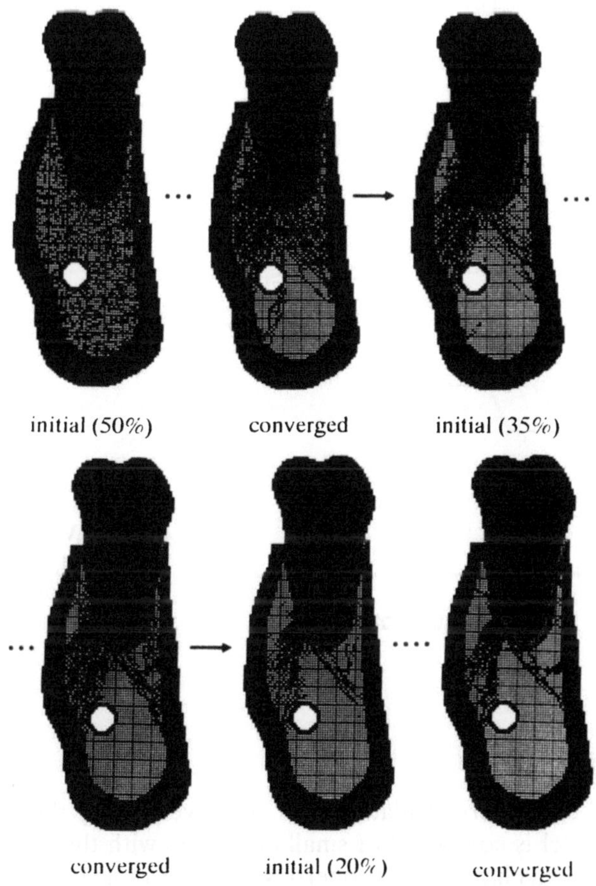

Fig. 8. Changes in the trabecular pattern for the osteoporotic case during the simulation process

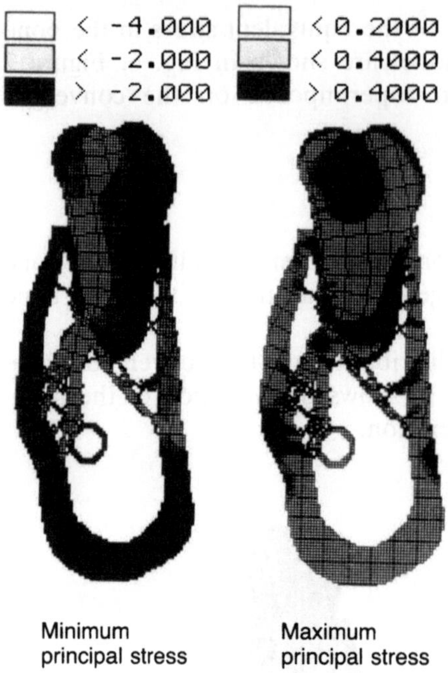

Fig. 9. Principal stress distribution on converged trabecular pattern in osteoporosis

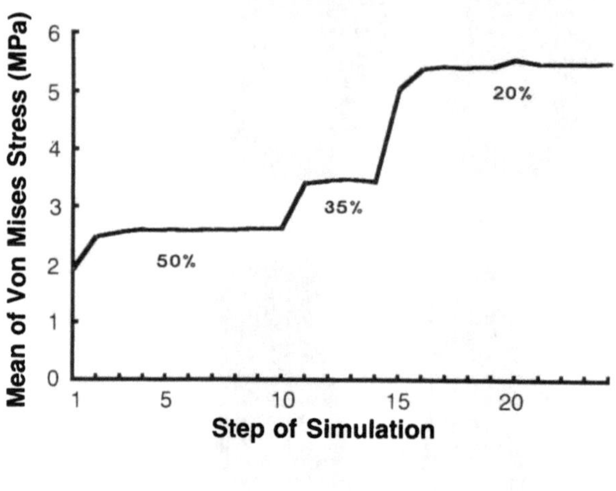

Fig. 10. Variation in the mean equivalent stress in the bone elements for the osteoporotic case with simulation stage

Discussion

Normal Situation

The interactive computer simulation was improved using the finite element method. The model is composed of small elements, with the cancellous region consisting of elements having the properties of either bone or soft tissue (void). The simulation procedure constructed the trabecular pattern around a tooth using the criterion that the bone elements will remodel until they can support similar stress. In this way bone elements were located where mechanically necessary. The criterion that bone elements whose stress was less than 20% of the mean stress was the cutoff used to limit the speed of the remodeling process. The simulation was repeated until bone elements with extreme low stress were missing. As the simulation progressed, the SD of the equivalent stress in the bone elements decreased, and the mean equivalent stress increased until it reached a constant level (Fig. 6a,b). This implies that each bone element was supporting an equal portion of the load. As displacement of the model decreased, the stiffness of the model increased (see Fig. 6c).

The principal stress distribution on the converged trabecular pattern showed that minimum principal stress spreads in a fanshaped manner and that transverse maximum principal stress runs almost horizontally (see Fig. 7). This stress distribution is similar to the actual trabecular pattern in the molar region (see Fig. 2b). This suggests that the trabecular pattern around the root is biomechanically arranged to support the bite load. The trabeculae that spread in a fanshape could be called compressive trabeculae, and the other trabeculae that run horizontally could be called tensile trabeculae.

Osteoporotic Situation

The trabecular pattern for the osteoporotic case was simulated with this method. The principal stress distribution for the converged osteoporotic configuration is as follows. Minimum principal stress spreads thickly in a fan shape. Maximum principal stress runs horizontally, and the trabeculae are very thin (see Fig. 9). This configuration resembles the internal structure of the osteoporotic vertebra, and also resembles the trabecular pattern with thick vertical trabeculae and thin horizontal trabeculae as described by Masuda [3].

The mean equivalent stress in the bone elements increased with the degree of osteoporosis such that all bone elements were highly stressed in the converged osteoporotic case in comparison with the normal case (50%). These results suggest that a incidence of trabecular bone resorption and bone microfractures can be anticipated around a natural tooth root in the osteoporotic situation.

Conclusions

1. The trabecular pattern around a human molar can be simulated using a finite element model consisting of bone and void elements that mimic cancellous bone.
2. The pattern of human trabeculae was found to be arranged so as to optimize the load transfer.
3. Trabeculae can be separated into two groups, compressive trabeculae and tensile trabeculae.

4. In osteoporosis, the trabecular pattern around a tooth is remodeled to support load as in other bone architecture; that is, compressive trabeculae are thick and tensile trabeculae are thin.
5. For the osteoporotic condition, trabeculae have a high risk of microfracture or resorption because of the high stress level.

References

1. Wolff J (1892) The law of bone remodeling (Das Gesetz der Transformation der Knochen). Translated by Msquet P, Furlong R, 1986, Springer, Berlin, pp 3–74
2. Takao Masada (1960) Architecture and biomechanical study in a mandible. (Kagaku-kotu no kouzo narabini rikigakuteki kenkyu, in Japanese.) Achievements of Department of Anatomy in Tokyo Dental College, pp 498–506
3. Sogo M, Maeda Y, Tsutsumi S, Okada M, Enomoto K, Nokubi T, Okuno Y (1993) Biomechanical studies on trabeculation of mandible: Part 1. Trabecular pattern around natural tooth and load transmission of bite force. J Jpn Prosthodont Soc 37:505–513
4. Yanagisawa S (1991) Dental implant modality vs alveolar ridge augmentation with bone implant. In Reconsidering a mastication impediment: 3. ("Soshaku-shogai" wo Saikou-suru 3, in Japanese.) Dental Outlook 78:115
5. Ueo T, Tsutsumi S, Yamamuro T, Okumura H, Shimizu A, Nakamura T (1985) Biomechanical aspects of the development of aseptic necrosis of the femoral head Arch Orthop Trauma Surg 104:145–149
6. Tsutsumi S, Maruyama T, Okuno Y, Ida K (1988) Computer simulation for design of dental prostheses with biomechanical compatibility. The Organizing Committee of the Third World Biomaterial Congress, Business Center for Academic Societies of Japan, Tokyo, p 116
7. Consensus development conference on osteoporosis—diagnosis, prevention and treatment, October 19–20, Copenhagen, 1990

Influence of Hypergravity on Osteoprogenitor Cells

TAKASHI USHIDA and TETSUYA TATEISHI

Summary. We loaded cultured osteoprogenitor cells (MC3T3-E1) with hypergravity (150 g) by using centrifugal forces. The cultured cells were loaded with the hypergravity 1 h/day during 4 days. The cell numbers in test dishes and control dishes were measured every day, and alkaline phosphatase activities of the cultured cells in the two kinds of dishes were also measured on the fourth day of incubation. The result showed that the cell growth rates in test dishes were almost the same as those in control dishes, suggesting that the hypergravity used in this study did not influence the cell growth of osteoprogenitor cells. Measurement of alkaline phosphatase activity showed that the cells loaded with hypergravity had higher activity of alkaline phosphatase than the cells in control dishes, indicating that the hypergravity of this study had a promotive influence on the bone formation ability of osteoprogenitor cells. In other words, it is possible that the hypergravity promoted the differentiation of osteoprogenitor cells to osteoblast-like cells.

Key words: Bone formation—Hypergravity—Osteoblast—Osteoprogenitor—Alkaline phosphatase

Introduction

The influence of mechanical forces on osteoblasts has been studied at cellular and molecular levels, and it is now probable that mechanical forces have an important role in bone remodeling. The question is to elucidate the mechanism of transforming mechanical stress, or gravity into cellular response at the molecular level to determine the reason that bone rigidity and bone mass

Bionic Design Group, National Institute for Advanced Interdisciplinary Research, 1-1-4 Higashi, Tsukuba, Ibaraki, 305 Japan decrease when one remains in bed while ill or undergoes long periods of no gravity, such as in the space shuttle [1].

The classical theory, the trajectory theory of Culman and Wolff, states that the bone loss is related to the fact that the trabeculae of cancellous bone at the proximal end of the human femur are oriented along the lines of principal stress, corresponding with the lines of maximum and minimum loading. Recent studies reveal that mechanical stress stimulates the production or release of second messengers such as cyclic AMP (cAMP), inositol triphosphate, and intracellular Ca^{2+} [2,3]. Although inositol triphophate and cAMP are produced in response to prostaglandin E2 (PGE2) and PTH in osteoblasts [4], mechanical stress also stimulates the production of inositol triphosphate as well as cAMP [5]. In studies of cytokines as a mediator of bone remodeling in response to mechanical stress, the hypothesis states that whether formation or resorption of bone occurs depends on the cytokines produced locally by mechanically activated cells [6]. On the other hand, some researchers suggest that microfilaments as a cytoskeleton could transduce mechanical force that is loaded on the cell itself and on the extracellular matrix because these form a junctional complex with extracellular matrix on the plasma membrane. Banes et al. [7] showed that mechanical stress reduced the amount of tublin and indicated that the cytoskeleton would be a mediator of transduction of mechanical stress.

As an animal model for studying mechanical stress on bone, Hert et al. developed an experimental model in which the tibiae of rabbits were transfixed by Kirschner wires. It was found by using this model that continuous loading to bone did not stimulate bone remodeling [8,9]. Moreover, they tried to apply intermittent forces by using a electromagnetic device and showed that the intermittent forces promoted bone formation on the endosteal and periosteal surfaces of the cortex in cases of both compressive and tensile stress [10]. Lanyon and Rubin also showed by using a similar model that daily loading of 100 consecutive 1-Hz load cycles was sufficient to maintain normal osseous architecture, provided peak stress was greater than 0.001 [11].

To provide an in vitro model, models were developed of petri dishes with flexible bases on which cultured cells were spread. Increase in DNA synthesis was observed with these models [12]. Recent studies with these models suggested that a number of factors of different molecular weights capable of stimulating bone resorption were produced by osteoblasts in response to mechanical stress [13].

In this study, we loaded cultured osteoprogenitor cells with hypergravity (150 *g*) by using centrifugal forces, and measured cell growth rates and alkaline phosphatase activities as an indicator osoteoblastic activity for bone formation.

Materials and Methods

Osteoprogenitor cells (MC-3T3-E1) derived from murine calvaria and established by Kodama et al. [14] were used (Fig. 1). The cells were cultured in α-modified Eagle's minimal essential medium (α-MEM) with 10% fetal

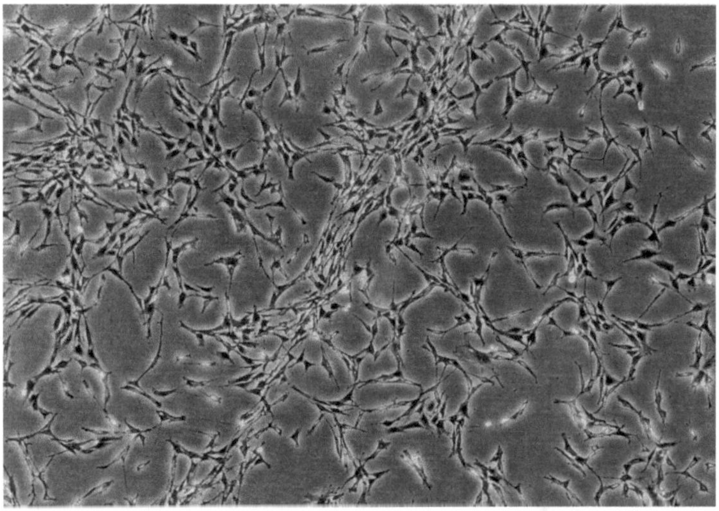

Fig. 1. Cultured osteoprogenitor cells (MC-3T3-E1) derived from murine calvaria

bovine serum (Flow Laboratory, Australia), and maintained at 37°C in a fully humidified atmosphere of 5% CO_2 in air. The cells (0.5×10^4 cells/ml, 3 ml per dish) were seeded into 35-mm-diameter petri dishes (25000, Corning, USA) for the experiments.

An apparatus made of aluminum alloy was used for loading the cells with hypergravity (Fig. 2). The petri dish was set into the apparatus, which was suspended in low- and medium-speed centrifuge (CR5B2, Hitachi). The cells were loaded with hypergravity 1 h/day during 4 days. The number of cells was measured every day using an electrical cell counter (Model D, Coulter Electronics, Bedfordshire, UK). Alkaline phosphatase activity was measured on the fourth day of loading according to the Bessey-Lowry method [15].

Results

The result of counting cell numbers during 4 days in both control and test petri dishes showed that (1) the cell number in both categories decreased on the first day in comparison with initial seeding cell number because cell adhesion rates were normally not 100% and the cell division cycle was not recovered in 1 day; (2) the cells in both sets of dishes were in the logarithmic growth phase from the second to the fourth day of loading; and (3) the cell growth rates in the test dishes were almost the same as those in the control dishes (Fig. 3). These data suggest that the hypergravity used in this study did not influence the cell growth of osteoprogenitor cells.

Measurement of alkaline phosphatase activity showed that the alkaline phosphatase values for the cells loaded with hypergravity were 0.134 ± 0.006

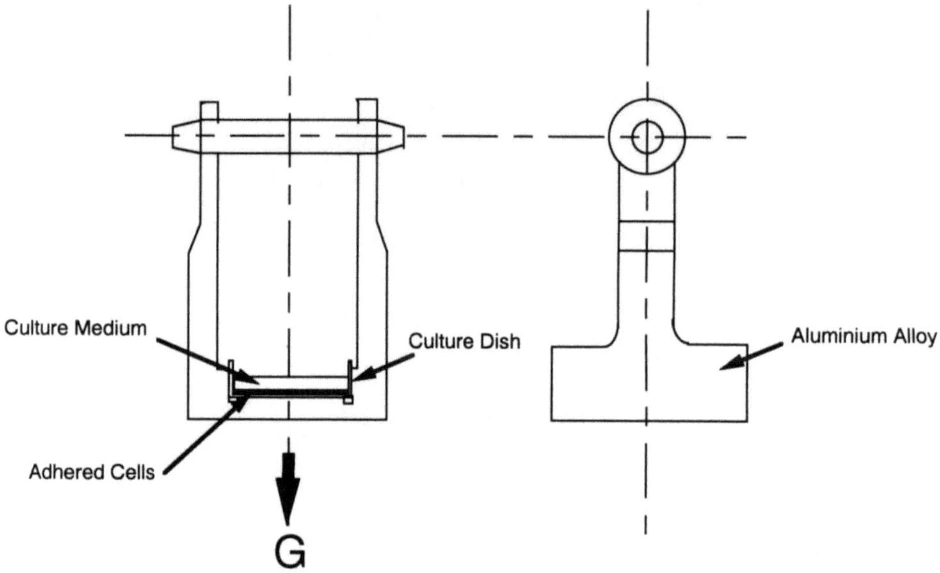

Fig. 2. Apparatus for loading cultured cells with hypergravity, made of aluminium alloy, is suspended from a centrifuge with an attached culture flask

($M \pm$ SD, $n = 4$), and those of the cells in control dishes were 0.116 ± 0.003 (Fig. 4). According to the *t* test, the activity of alkaline phosphatase in cells loaded with hypergravity was significantly higher than that in control dishes. These data show that hypergravity in this study had a promotive influence on the bone formation ability of osteoprogenitor cells. In other words, it is possible that the hypergravity promoted the differentiation of osteoprogenitor cells to osteoblast-like cells.

Discussion

It is difficult to evaluate physical deformation of the cells that hypergravity could cause in this system in comparison with a petri dish model which deforms the bottom of dishes where the cells adhere. In our model, the cells also adhered and spread on the bottom of dishes. The influence of apparent static pressure would be slight, because the density of the cells is about $1.07 \times 10^3\,Kg/m^3$, that of the culture medium is about $1.01 \times 10^3\,Kg/m^3$, and the difference between these densities is small. On the other hand, the influence of hypergravity itself on organelles, the cytoskeleton, and other cytoplasmic proteins seems also minor, because several hundred thousand times the force of gravity (g) is normally necessary to separate these substances by using a ultracentrifuge. Therefore, the rearrangement of organelles, the cytoskeleton, and other cytoplasmic proteins would be negligible under the conditions in this study.

3

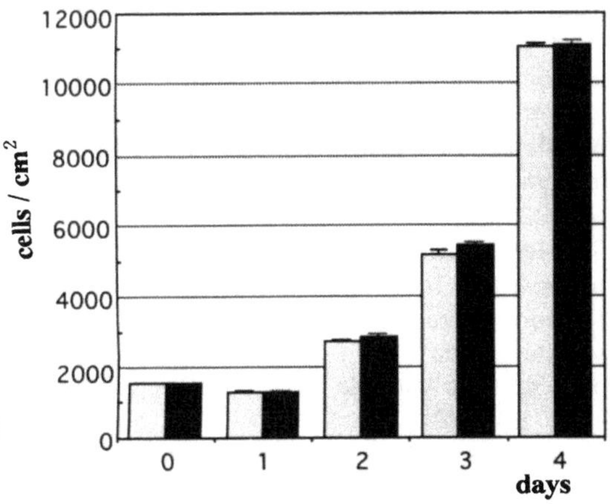

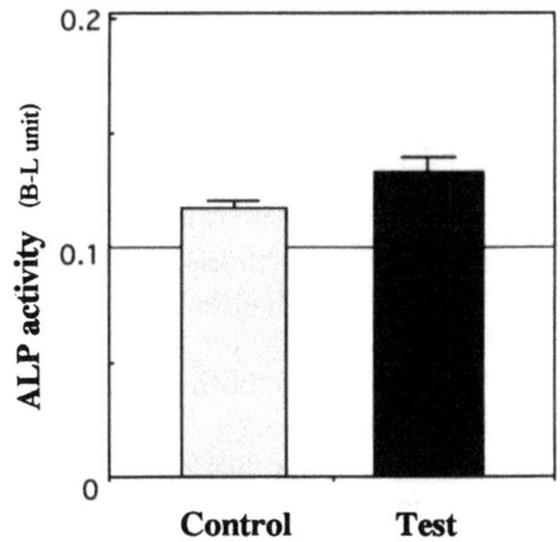

Fig. 3. Average ($M \pm$ SD) of cell numbers on 1, 2, 3, and 4 days after seeding under conditions of 370°C and 5% CO_2 atmosphere *Light shaded bars*, controls; *dark shaded bars*, tests

Fig. 4. Average ($M \pm$ SD) of alkaline phosphatase (ALP) activities on 4 days after seeding. Alkaline phosphatase activity of hypergravity-loaded cells (*dark shaded bars*) was significantly higher than that of control cells (*light shaded bars*) at $P < 0.05$

The hypergravity used in this study is equivalent with that needed to cause the cells to precipitate. Although the influence of hypergravity in this study is physically small, the experimental results showed that the loading of hypergravity acts promotively on bone formation because of the augmented alkaline phosphatase activity but has neither a promotive nor suppressive effect on cell proliferation. Considering that the cells used in this study were initially at the osteoprogenitor stage and differentiated to osteoblast-like cells with high alkaline phosphatase activity, if the cells become confluent and resemble epithelial cells it is possible that hypergravity loading promoted the differentiation of osteoprogenitor cells to osteoblast-like cells without contact inhibition. In any case, it would be very likely that a unknown mediator transformed hypergravity into cellular events such as the promotion of alkaline phosphatase activity. We wish to determine this mediator, as well as that of mechanical stress, from the molecular biology point of view.

References

1. Wu AH, Taylor GR, Graham GA, McKinley BA (1993) The clinical chemistry and immunology of long-duration space missions. Clin Chem 39(1):22–36
2. Harell A, Dekel S, Binderman I (1977) Biochemical effect of mechanical stress on cultured bone cells. Calcif Tissue Res [Suppl] 22:202–207
3. Somjen D, Binderman I, Berger E, Harell A (1980) Bone remodelling induced by physical stress in prostaglandin E2 mediated. Biochim Biophys Acta 627:91–100

4. Farendale RW, Sandy JR, Atkinson SJ, Pennington SR, Meghji S, Meikle MC (1988) Parathyroid hormone and prostaglandin E2 stimulate both inositol phosphates and cyclic AMP accumulation in mouse osteoblast cultures. Biochem J 252:263–268
5. Sandy JR, Meghji S, Farndale RW, Meikle MC (1989) Dual elevation of cyclic AMP and inositol phosphates in response to mechanical deformation of murine osteoblasts. Biochim Biophys Acta 1010:265–269
6. Meikle MC, Health JK, Atkinson SJ, Hembry RM, Reynolds JJ (1989) Molecular biology of stress on connective tissues in vitro. In: Norton LA, Burstone CJ (eds) The biology of tooth movement. CRC Press, Boca Raton, pp 71–86
7. Banes AJ, Gilbert J, Taylor D, Monbureau O (1985) A new vacuum-operated stress-providing instrument that applies static or variable duration cyclic tension or compression to cells in vitro. J Cell Sci 75:35–42
8. Hert J, Liskova M, Lanrgot B (1969) Influence of the long-term continuous bending on the bone. An experimental study on the tibia of the rabbit. Folia Morphol (Praha) 17:389–399
9. Hert J, Liskova M, Landa J (1971) Reaction of bone to mechanical stimuli. Part 1. Continuous and intermittent loading of tibia in rabbit. Folia Morphol (Praha) 19:290–300
10. Liskova M, Hert J (1971) Reaction of bone to mechanical stimuli: Part 2. Periosteal and endosteal reaction of tibial diaphysis in rabbit to intermittent loading. Folia Morphol (Praha) 19:301–317
11. Rubin CT, Lanyon LE (1985) Regulation of bone mass by mechanical strain magnitude. Calcif Tissue Int 37:411–417
12. Hasegawa S, Sato S, Saito S, Suzuki Y, Brunette DM (1985) Mechanical stretching increases the number of cultured cells synthesising DNA and alters their pattern of protein synthesis. Calcif Tissue Int 37:431–436
13. Sandy JR, Meghji S, Scutt AM, Harvey W, Harris M, Meikle MC (1989) Murine osteoblasts release bone-resorbing factors of high and low molecular weights: stimulation by mechanical deformation. Bone Miner 5:155–169
14. Kodama H, Amagai Y, Sudo H, Kasai S, Yamamoto S (1981) Establishment of a clonal osteogenic cell linc from newborn calvaria. Jpn J Oral Biol 23:899–901
15. Bessey OA, Lowry OH, Brock MJ (1946) J Biol Chem 164:321–329

Monitoring of the Mechanical Strength of the Callus in Fracture Treatment and Callotasis by Acoustic Emission

Yasusuke Hirasawa, Yoshinobu Watanabe, Yoshiyuki Arai, and Nozomu Inoue

Summary. The acoustic emission (AE) technique was applied clincially to monitor the mechanical strength of the callus in fracture treatment, callotasis and arthrodesis. Twenty patients treated using various external fixators were the subjects of this study. AE signals were detected through the half-pins driven into the fractured bone fragment under partial weight bearing of the involved limb. All cases demonstrated that the load at initiation of the AE signal increased with the time of healing. Intramedullary nailing was applied for three tibia fractures and two femur fractures which were originally treated by external fixator. In the other 15 cases in this series, the external fixator was removed after AE testing. There were no cases of recurrent fracture after removal of the external fixator when the strength of the healing bone was estimated by AE testing to be higher than the body weight of the patient. Thus AE testing can accurately estimate the mechanical strength of healing fractures and callotasis.

Key words: Fracture healing—Acoustic emission—External fixator—Non-invasive test—Biomechanics—Callotasis—Arthrodesis

Introduction

The callus in the healing of fractures and the osteotomy of callotasis are important contributors to the maintenance of the strength of healing bone. There are many studies on the biomechanical monitoring of the the strength of healing bone [1,2]. Acoustic emission (AE) is autogenously produced sound from crystalline structures under loading, and AE testing is categorized as

---

Department of Orthopaedic Surgery, Kyoto Prefectural University of Medicine, Kawara, Hirokoji, Kamigyo-ku, Kyoto, 602 Japan a noninvasive test because the basic structural integrity of the tested bone remains intact [3]. It is belived that Kaiser was the first investigator to use AE testing for metallic materials as early as the 1950s. AE properties of bio-materials have been studied since the 1970s. Previous reports have shown that bone has marked AE properties [4,5]. The possibility of Clinical AE testing for monitoring fracture healing was first reported by Nicholls and Berg in 1981 [3]. However, the AE properties of healing fractures remained a question of debate and there were few reports of the clinical application of this technique. The objective of the present work was to estimate the mechanical strength of healing fractures based on AE characteristics.

Experimental Approach

Experimental femur fractures in rats were tested in tension to determine their mechanical strength. AE signals were registered during mechanical testing using two piezoelectric transducers attached to the fractured femur during mechanical testing. Figure 1 shows typical load-strain and cumulative AE signal curves. The AE signal can be detected before the complete failure of the specimen, and the total AE event count increases exponentially as the load increases. As the AE characteristics of the callus are very similar to those of composite materials such as fiber reinforced plastics, we belive that the first AE event indicates the local yielding of the fracture callus. Watanabe defined the load at the first AE event the as AE the initiating load (AEIL), and reported that AEIL was highly correlated with tensile strength (Fig. 2) [6]. Thus AEIL

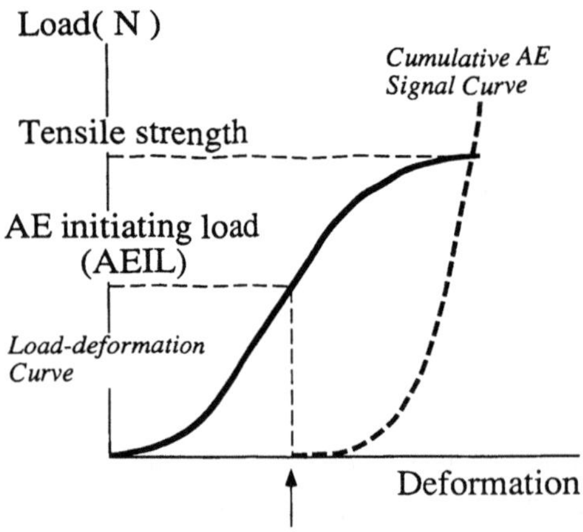

1

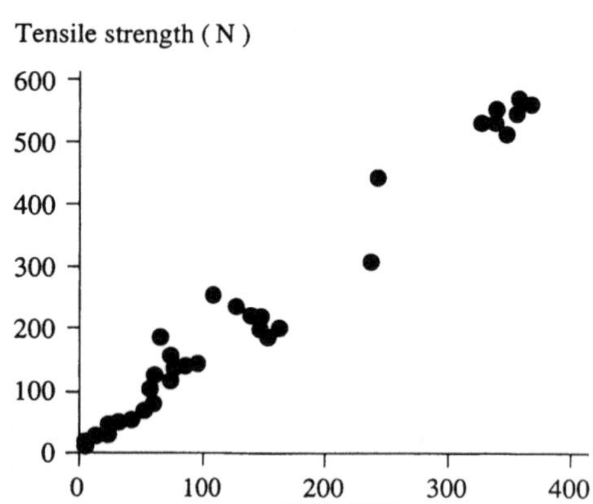

2

Fig. 1. Acoustic emission (AE) initating load (AEIL) is defined as load at initiation of AE signal

Fig. 2. Relationship between AEIL and tensile strength. AEIL is highly correlated with ultimate strength ($r = 0.98$, $P < 0.0001$)

can be used as a new parameter for predicting the mechanical strength of healing fractures.

Materials and Methods

Twenty patients treated by external fixator were repeatedly tested by the AE technique after surgery (Table 1). Piezoelectric transducers were placed on the half-pins driven into the bone fragment. AE signals generated in the fractured bone were detected through the half-pins under weight-bearing conditions, and the load-AE event count relationship was obtained using the AE measuring system. The load applied to the fracture site was measured by the digital load analyzer in the partial weight bearing rehabilitation machine (Fig. 3).

Results

External fixators were applied in the treatment of 11 cases of bone fracture, 5 of callotasis, 2 of pseudoarthrosis, and 2 of arthrodesis. AE testing was done in all cases at least twice during treatment as well as general radiographic

Table 1. Patient summary

Case	Diagnosis	Surgery	External fixator
1	Lower limb length discrepancy	Callotasis (tibia)	Orthofix
2	Lower limb length discrepancy	Callotasis (femur)	Orthofix
3	Lower limb length discrepancy	Callotasis (femur)	Orthofix
4	Lower limb length discrepancy	Callotasis (femur)	Orthofix
5	Pseudoarthrosis (tibia)	Papinaue	Orthofix
6	Pseudoarthrosis (femur)	Application of external fixator	Hifixator[a]
7	Tibial fracture	Application of external fixator	Orthofix
8	Tibial fracture	Application of external fixator	Stuhler Heise[b]
9	Tibial fracture (open)	Application of external fixator	Orthofix
10	Femur fracture	Application of external fixator	Orthofix
11	Femur fracture	Application of external fixator	Hoffmann[c]
12	Femur fracture (open)	Application of external fixator	Orthofix
13	Supracondylar fracture	Application of external fixator	Stuhler Heise
14	Forearm length discrepancy	Callotasis of radius	Orthofix
15	RA	Arthrodesis (knee)	Stuhler Heise
16	Humeral fracture	Application of external fixator	Hoffmann
17	Humeral fracture	Application of external fixator	Hoffmann
18	Humeral fracture	Application of external fixator	Stuhler Heise
19	Radius fracture	Application of external fixator	Hoffmann
20	RA	Arthrodesis (ankle)	Charnley[d]

RA, Rheumatoid arthritis.
[a] Hifixator, Nagano Keiki Seisakusho, Ltd., Tokyo, Japan.
[b] External Fixation System Stuhler-Heise, Aesculap, Tuttlingen, Germany.
[c] Hoffmann External Fixation System, Jaquet Orthopédie S.A., Geneva, Switzerland.
[d] Shearer External Fixation System, Charnley Type, Thackray, Ltd., Leeds, England.

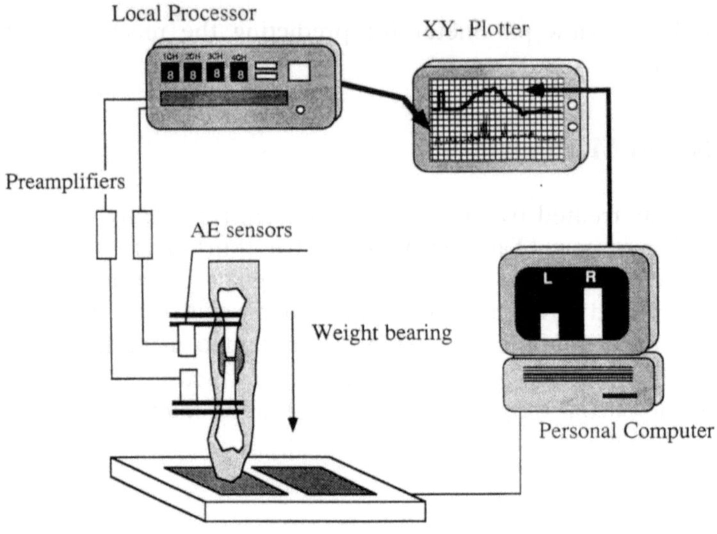

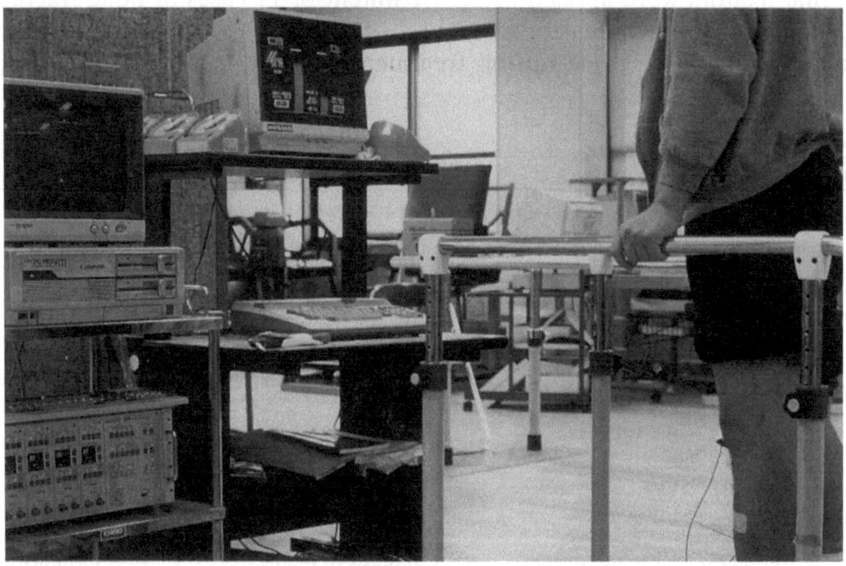

Fig. 3. AE measuring system for a patient suffering from bone fracture

examination. Intramedullary nailing was done for three tibia fractures and two femur fractures a few weeks after the original external fixations. External fixators were removed after AE testing in the other 15 cases. The AEIL depended mainly on the length of the interval after application of the external fixator. AEIL mostly increased with the time of healing, but the increase in AEIL did not necessarily terminate with the maturation of the callus as eval-

4 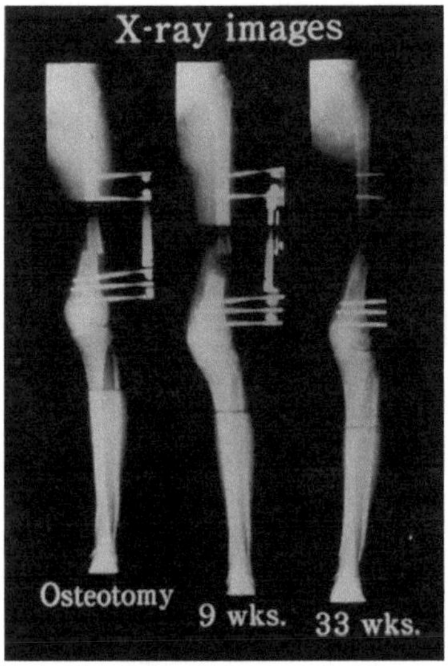

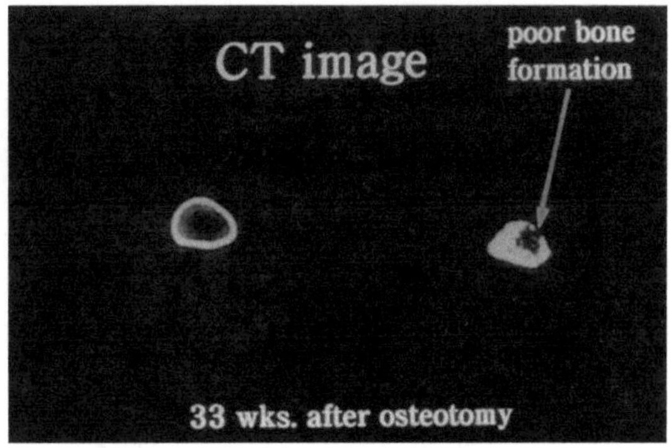 5

Fig. 4. Case 2, plane X-ray photogram

Fig. 5. Case 2, computed tomography (CT) scan at 33 weeks after surgery

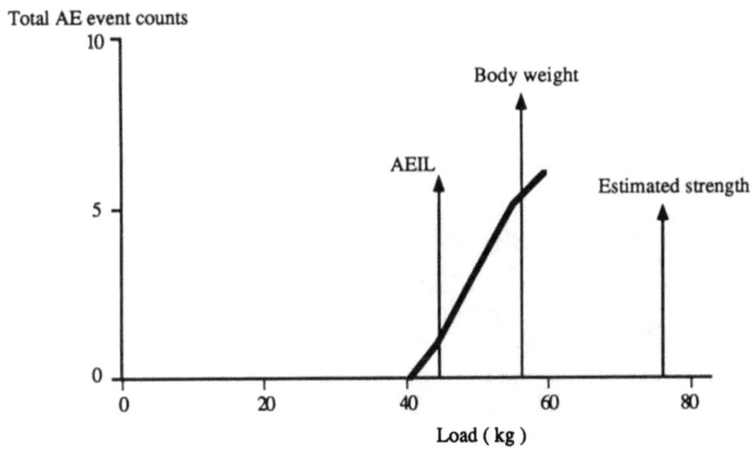

Fig. 6. Case 2, AEIL and estimated failure load

uated from radiographic examination. There were no recurring fractures after removal of external fixators when the estimated strength of the healing bone was greater than the body weight of the patient.

Case 2. A 15-year-old female patient with lower limb length discrepancy was treated with callotasis and correctional osteotomy of the femur using the Orthofix external fixator (Orthofix Modulsystem, Orthofix, Verona, Italia). Radiography showed sufficient callus formation at 33 weeks after surgery (Fig. 4). However, a computed tomography (CT) scan demonstrated poor bone formation on the lateral side (Fig. 5). AEIL was about 45 kg, indicating the

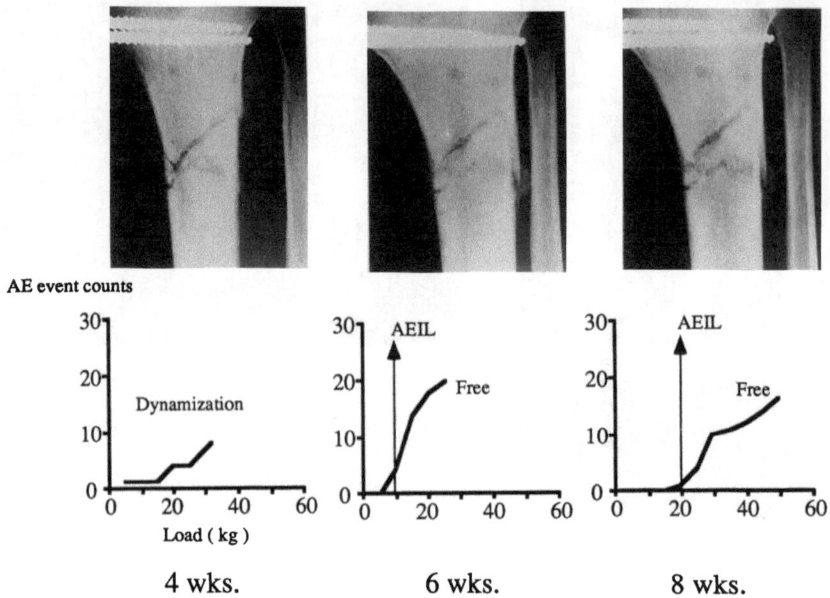

Fig. 7. Case 9, change in AEIL and X-ray image for fracture site

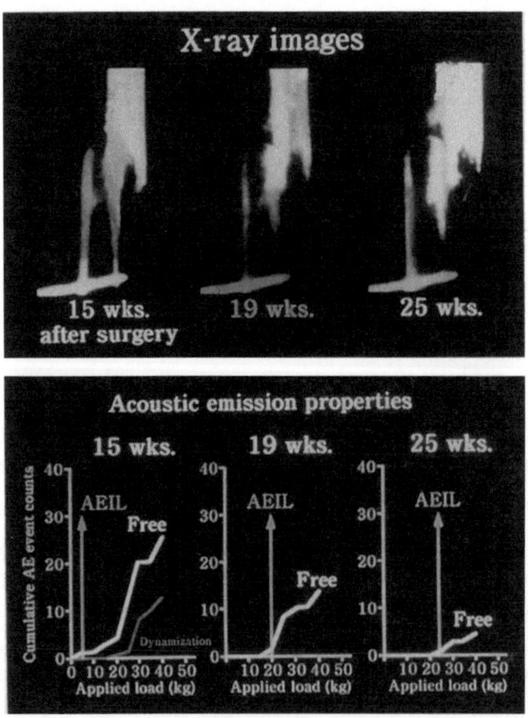

Fig. 8. Case 10, change in AEIL and X-ray image of fracture site

ultimate strength of the callus as about 75 kg (Fig. 6). Her body weight was 55 kg, so the external fixator was removed at 34 weeks after surgery. Full weight bearing was permitted 2 weeks after removal of the fixator.

Case 9. A 16-year old male patient who sustained open proximal tibia fracture in a motor vehicle accident was originally treated with the Orthofix external fixator. AE signals under weight bearing were measured at 4, 6, and 8 weeks after surgery. Only minimum changes in radiograph were observed during these periods. At 6 weeks after surgery, AEIL was about 10 kg; however, at 8 weeks, it increased to 25 kg. This result was interpreted as indicating that the mechanical strength of the fracture site had at least doubled in two weeks without significant change in radiograph (Fig. 7).

Case 10. A ten-year-old boy who sustained a shaft fracture of the right femur with the Orthofix external fixator. AE signals under weight bearing were measured at 15, 19, and 25 weeks after surgery. Radiography revealed an increase in mineral density during this period. AEIL increased from 5 kg to 25 kg is (Fig. 8). The external fixator was removed at 26 weeks after surgery.

Discussion

Currently, the external fixation method has progressed in the repair of fractured bones and for callotasis. External fixation has many biological and mechanical advantages for the treatment of bone fractures, however, the external fixator and half-pins should be removed as early as possible after the callus has developed sufficient strength because pin tract infections remain a serious problem. So it is necessary to monitor the mechanical strength of healing fractures. There have been trials to monitor the mechanical status of healing fractures nondestructively [7–9]. These include measurement of callus mineral density by dual energy X-ray absorptiometry (DEXA), ultrasonic examination, strain measurement of the external fixator, and stress wave transmission through the fracture site. These methods are likely to be effective for estimation of the elasticity or stiffness of the fracture callus, especially in the early stage of fracture healing. However, it is likely to be difficult to estimate its strength after the callus has partially matured. Previous studies showed that the intiation of AE signals under stress corresponded to the local yielding of bone. The advantage of AE methods for monitoring healing fractures it the possibility of estimating the yielding load of the callus. Nicholls and Berg reported that the fractured tibia in rabbits produced no AE at less than 50% of shearing strength [3]. Watanabe also showed that the fractured femur in rats also produced no AE at less than 60% of tensile strength and torsional strength, and that there was a good correlation between mechanical strength and the load or torque at initiation of AE signals [6]. This study was the first report which shows the clinical usefulness of the AE technique for monitoring the healing of fractures, callotasis, and arthrodesis. There are some problems to be solved before applying this method routinely, but we belive the AE technique will contribute greatly to these fields in the near future.

References

1. Macey LR Kana SM Jigushi S, Terrek RM Borretos J, Bolander ME (1989) Defects of early fracture-healing in experimental diabetes. J Bone Joint Surg (Am) 71-A: 722–733
2. Northmore-Ball MD, Wood MR, Meggitt BF (1980) A biomechanical study of the effects of growth hormone in experimental fracture healing. J Bone Joint Surg (Br) 62-B:391–396
3. Nicholls PJ, Berg E (1981) Acoustic emission properties of callus, Med Biol Eng Comput 19:416–418
4. Hanagud S, Clinton R, Lopez JP (1973) Acoustc emission in bone substance. Proceedings of Biomechanics Symposium of the American Society of Mechanical Engineers. ASME, New York, pp 74
5. Wright AA Vosburgh F, Burstein AH (1981) Permanent deformation of compact bone monitored by acoustic emission. J Biomech 14:405–409
6. Watanabe Y (1993) Evaluation of mechanical properties of fracture healing with acoustic emission techique—An experimental study in rats (in Japanese). J Kyoto Pref Univ Med 102:1101–1116
7. Markel MD, Wikenheiser MA, Chao EYS (1990) A study of fracture callus material properties. Relationship to the torsional strength of bone. J Orthop Res 8:843–850
8. Nishimura N. (1984) Serial strain gauge measurement of bone healing in Hoffmann external fixation. Orthopaedics 8:677–684
9. Nikiforidis G, Bezerianos A, Diarogonas A, Sutherland C (1990) Monitoring of fracture healing by lateral and axial vibration analysis. J Biomech 23:323–330

Part 2. Joint

Application of Robotics to Studies of Joint Biomechanics

Hiromichi Fujie[1], Savio L.-Y. Woo[2], Glen A. Livesay[2], and Kiyoshi Mabuchi[1]

Summary. Joint biomechanical studies have been successfully performed using robotics. In those studies, robotics was found to be very appropriate for various reasons. First, the mathematics originally developed in the field of robotics and kinematics can be applied to describe joint motion and joint forces and moments. This allows for systematic treatment of joint motion as well as joint force and moment. Second, using a manipulator, it is possible to reconstruct the three-dimensional joint motion while simulating a physiological situation. Joint motion can be applied along a degree of freedom (DOF) of the joint while controlling force and moment along other DOF of the joint. Third, sensor instrumentation and techniques such as the universal force-moment sensor can be utilized for biomechanical measurements. It was found through these applications that robotics methodology can provide a revolutionary step for studies of joint biomechanics.

Key words : Robotics—Control—Human knee joint—Three-dimensional biomechanical test—Force and moment—In situ force—Ligaments—Diagnosis of ligament rupture

Introduction

The first application of robotics to the study of biomechanics in our laboratories was designed to determine the effect of mechanical stimulation on fracture healing [1,2]. A robot manipulator was utilized to simulate cyclic motion as well as cyclic loading around the fracture site of rabbit tibia. We found from

[1]Department of Biomedical Engineering, School of Medicine, Kitasato University, 1-15-1 Kitasato, Sagamihara, Kanagawa, 228 Japan
[2]Department of Orthopedic Surgery, University of Pittsburgh, E1641 Biomedical Science Tower Pittsburgh, PA 15213, USA

the study that robotics makes it possible to simulate a mechanical environment in multiple degrees of freedom (DOF) for biomechanical studies. Since then, robotics has been employed in other biomechanical applications. To determine the in vitro mechanical behavior of synovial joints, a robot system has been developed combining a 6-DOF manipulator and the mathematics for describing joint motion as well as joint force and moment [3–7]. Under the position and force control of the system, which permitted the natural motion of the joints, the three-dimensional mechanical behavior of knee joints was determined. Based on the success of this approach, in vivo stability testing of the knee has been clinically performed to diagnose knee ligament rupture applying robotics technology [8].

A methodology has also been developed to determine in situ forces in knee ligaments using a universal force-moment sensor (UFS) and a force transformation scheme [9–11]. This approach allows the determination of the magnitude, direction, and point of application of in situ forces in ligaments. A methodology to determine lubrication behavior of synovial joints has been also developed using robotics [8].

These biomechanical studies have been successfully performed through the use of robotics technology to obtain valuable data. It was found that not only application of the manipulator but also the mathematics and the methodology originally developed in the field of robotics are applicable to biomechanical studies. From these studies, in vitro as well as in vivo studies to determine the mechanical behavior of knee joint and an in vitro study to determine the in situ force in knee ligaments are the focus of this chapter.

In Vitro Studies

Three-Dimensional Biomechanical Testing System

The human knee joint experiences a great variety of mechanical loads in normal as well as sports activity. External forces or internal forces, applied to human body as a whole, influence the three-dimensional motion of the knee by producing forces and moments about the joint. Soft tissues around the knee function to resist these forces and moments by generating in situ forces that maintain knee stability. To obtain the baseline data needed to study the mechanical process of knee ligament rupture, to understand the methodology for the diagnosis of rupture, and to improve the surgical treatment of ligament reconstruction it is important to perform mechanical testing on human knee joints in three dimensions.

In previous studies [12–14], a material testing machine with multi-DOF clamps was utilized for the mechanical testing to experimentally simulate a mechanical condition physiologically experienced by the knee. Force or moment was applied to the knee while the natural motion of the knee was maintained via the smooth link motion of the clamps. We have proposed a new methodology to accomplish this by the application of robotics. A commercially available 5-DOF articulated manipulator (RV-P2, Mitsubishi, Nagoya, Aichi, Japan) was modified through the addition of a sixth DOF and gearing down of each original axis (Table 1) [7]. This modification decreased the speed of the manipulator, increased its load capacity from 20 to 200 N, and improved its position repeatability from 0.5 to 0.05 mm. The manipulator was rigidly fixed to an aluminum base plate with a thickness of 25 mm. A universal force-moment sensor (UFS) (UFS I, Nitta, UFS-4015A-100, Yamatokohriyama, Nara, Japan) (Table 1), capable of measuring three forces and three moments at the sensor, was attached to the hand of the manipulator. The clamp for the tibia was mounted on the UFS, such that the manipulator held the tibia through the UFS. With the femoral clamp rigidly fixed to the base of the manipulator, motion as well as forces and moments were applied to the joint by moving the tibia (Fig. 1).

To describe the motion as well as forces and moments, sets of coordinate systems were devised for various parts of the setup. The reference coordinate system was fixed to the base plate. The femoral coordinate system was fixed to femur, and its position and orientation with respect to the reference coordinate system were described by a $[4 \times 4]$ constant homogeneous transformation P. The manipulator coordinate system was fixed to the hand of the manipulator, and its position and orientation to the reference coordinate system were mathematically described by a $[4 \times 4]$ homogeneous transformation Y, consisting of the 6-DOF displacement along the manipulator. This matrix was served as input to a control computer, allowing the manipulator to be moved to the position and orientation described by the matrix. The sensor coordinate system was fixed to the UFS.

A knee joint coordinate system, originally developed by Grood and Suntay [15] and by Chao [16], was applied to this test setup for the description of the 6-DOF knee kinematics. This coordinate system is commonly used in the field

Table 1. Specifications of the manipulator and universal force-moment sensors used in biomechanical studies of the human knee

Manipulator (custom-made)		
Degrees of freedom		6
Load capacity		200 N
Position repeatability		0.05 mm
Universal force-moment sensors (UFS)		
	UFSI	UFS II
Company	Nitta (Japan)	JR^3 (USA)
Model no.	4015A-100	4015A-50
Degree of freedom	6	6
Force capacity	440 N (880 N for axial direction)	220 N (440 N for axial direction)
Moment capacity	450 Nm	225 Nm

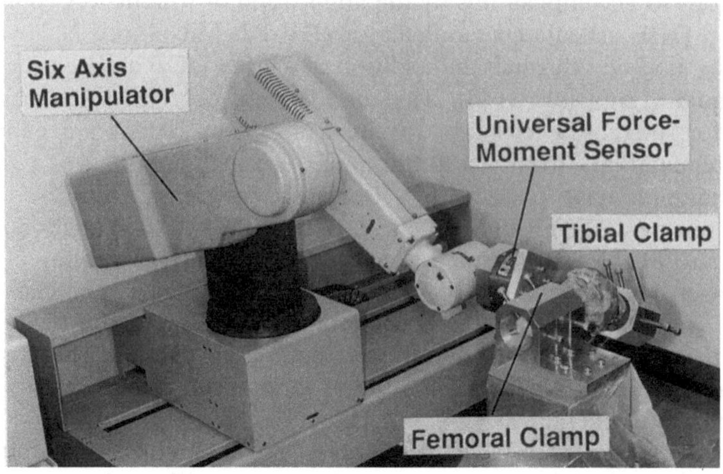

Fig. 1. Robot system used to determine the three-dimensional mechanics of knee joints. Motions as well as forces and moments are applied to a joint by the movement of the manipulator, while the forces and moments are measured with a universal force-moment sensor mounted between the tibial clamp and the manipulator hand

of biomechanics because it allows for a rigorous description of the 6-DOF motion of the knee as a sequence of six serial links between the femur and tibia with clinical relevance (Fig. 2). The 6-DOF knee motion was then defined, from femur to tibia, as extension-flexion (E-F) rotation, Θ_{EF}, lateral-medial (L-M) translation, d_{LM}, varus-valgus (V-V) rotation, Θ_{VV}, anterior-posterior (A-P) translation, d_{AP}, internal-external (I-E) rotation, Θ_{IE}, and proximal-distal (P-D) translation, d_{PD}, respectively, with respect to the knee joint coordinate system. It was required to relate these displacements to the position and orientation of the manipulator coordinate system so that the 6-DOF displacement of the knee could be controlled by the movement of the manipulator. To do this, the position and orientation of the manipulator coordinate system with respect to the femur was mathematically described by a homogeneous transformation, X, consisting of the 6-DOF knee displacement (Appendix A).

Now there are two paths to reach the manipulator coordinate system from the reference coordinate system in a kinematic relationship, and Equation 1 is obtained (Fig. 3).

$$Y = PX \tag{1}$$

When a target 6-DOF displacement of the knee joint is given, the position and orientation of the manipulator must be controlled as described by Equation 1. Conversely, when the position and orientation of the manipulator coordinate system are known, the current 6-DOF displacement of the knee can be determined by the inverse of Equation 1 [6,7].

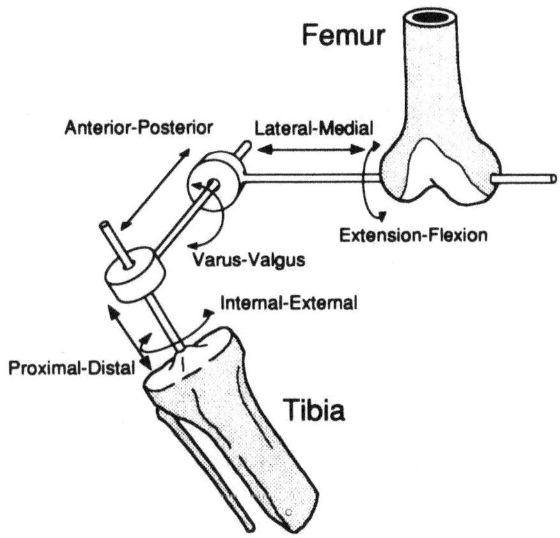

Fig. 2. Knee joint coordinate system for the description of the 6 degree of freedom (*DOF*) of the human knee joint (Adapted from [15])

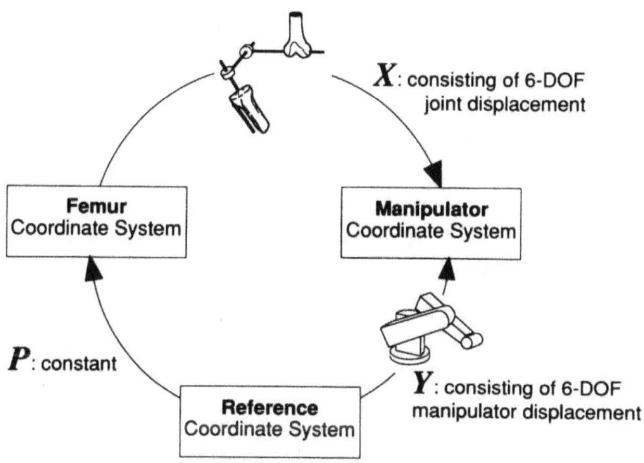

Fig. 3. Kinematic relationship between the reference, manipulator, and femur coordinate systems in relation to 6-DOF knee displacement as well as 6-DOF manipulator displacement

In the same manner, the knee forces and moments were defined as E-F moment, M_{EF}, L-M force, F_{LM}, V-V moment, M_{VV}, A-P force, F_{AP}, I-E moment, M_{IE}, and P-D force, F_{PD}, respectively, with respect to the knee joint coordinate system. To control the 6-DOF forces and moments at the knee, a relationship between the femoral and the sensor coordinate systems in terms of differential motion was mathematically described by a [6 × 6] Jacobian transformation matrix, J, consisting of the 6-DOF displacement of the knee except E-F and L-M, respectively (Appendix B) [17]. By applying the principle of virtual work to the relationship, the forces and moments at the knee joint, F, were related to those at the UFS, F_s, as follows:

$$F = J^T F_s \tag{2}$$

where the superscript T is transpose. From the force and moment measured by the UFS, 6-DOF force and moment applied to the knee can be calculated by Equation 2. Conversely, to apply a target 6-DOF force and moment to the knee, the manipulator was controlled through a feedback loop so that the force and moment, calculated by the inverse of Equation 2, could be applied to the sensor. Finally, position control as well as force control were incorporated in the control program of the system, in which each DOF of the knee would be either position-controlled or force-controlled using Equations 1 and 2 (Fig. 4). The basic function of the control system was position control; however, force control was possible by mathematically transforming the forces and moments to motion at the knee joint using a virtual compliance.

To verify the developed system, anterior-posterior translation tests as well as varus-valgus rotation tests of a human cadaveric knee joint were performed. The knee was dissected down to the joint capsule and fixed to the system at 0° of flexion. The knee was placed in its neutral position by controlling the forces and moments at the knee and requiring them to be 0. Anterior and posterior forces up to ±80 N were then applied to the knee at a rate of 0.5 mm/s of translation, while forces and moments at all other DOF of the knee, except E-F and A-P, were maintained at 0 with force control. In the same manner, varus and valgus moments up to ±6 Nm were applied to the knee at a rate of 0.5°/s of rotation, with force control. Both of the tests were performed on the knee at 0°, 30°, 60°, and 90° of flexion. Joint forces and moments as well as joint motion in 6 DOF were acquired in each test.

The forces and moments at the knee were well controlled during the anterior-posterior and varus-valgus tests. Using force control as well as position control,

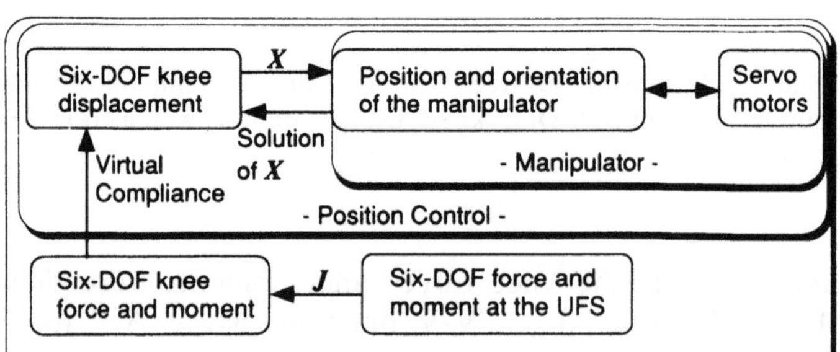

Fig. 4. Hybrid control algorithm for both position and force control. *UFS*, universal force-moment sensor

A-P force or V-V moment was applied to the knee while allowing the natural motion of the knee. A typical force-displacement curve during A-P testing as well as a typical moment-displacement curve during V-V testing are represented in Fig. 5. When the knee was placed in an unconstrained condition under force control and position control, the forces or moments applied to the knee to achieve a given displacement were smaller than those with the knee constrained without force control. The A-P laxity, under ±50 N of anterior and posterior forces, increased as flexion angle increased from 0° to 30° (Fig. 6). Conversely, the A-P laxity slowly decreased as flexion angle increased from 30° to 90°. The V-V laxity, under ±4 Nm of varus and valgus moments, rapidly increased as flexion angle increased from 0° to 30°. As flexion angle increased from 30° to 90°, the V-V laxity slowly increased. These results corresponded well with clinical observation as well as a previous experimental data [18].

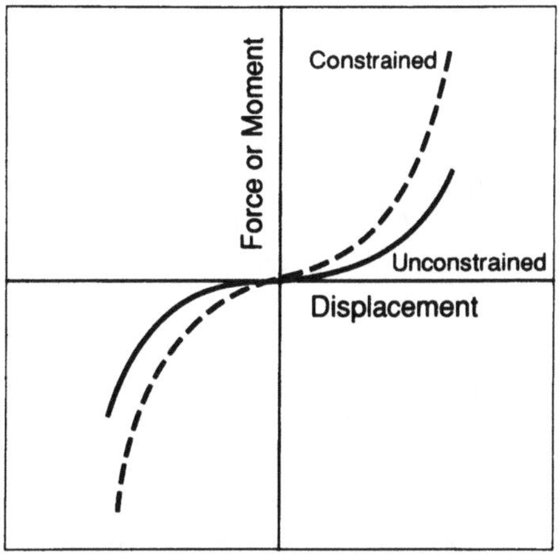

Fig. 5. Typical force-displacement curve during an A-P translation test or moment-displacement curve during a V-V rotation test of the human knee. The knee was placed under unconstrained conditions using both position and force controls

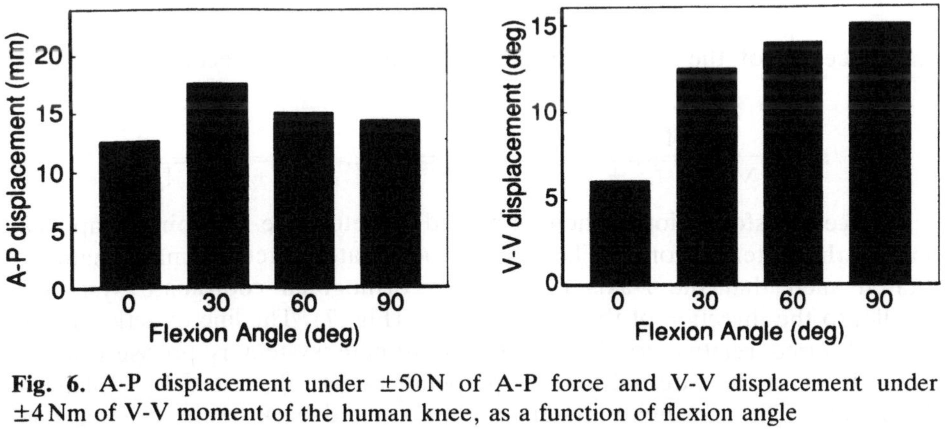

Fig. 6. A-P displacement under ±50 N of A-P force and V-V displacement under ±4 Nm of V-V moment of the human knee, as a function of flexion angle

In comparison to the setup previously utilized with the material testing machine and the multi-DOF clamp, the presented robotics method has several advantages. In the previous method, it was impossible to reconstruct the same path of knee motion after soft tissues were transected; this ability is needed for such investigations as cutting studies. Also, the weight of the clamp was not negligible and affected the experimental outcome. The presented robotics methodology can solve these problems using force control and position control. We believe this robotics methodology will provide a revolutionary step for mechanical testing of biological joints.

In Situ Forces in Ligaments

Better understanding of ligament function and contribution to joint biomechanics depends on an accurate measurement of the in situ forces generated in the ligament in response to motion or external loading of the intact joint. For this reason, previous approaches have employed the use of buckle transducers [19–21], strain gauges affixed near the bony insertions [22], instrumentation of the tibial insertion [23], kinematic linkage methods [24], and implantable transducers [25] to examine the in situ forces in ligaments of the human body, in particular those experienced by the human anterior cruciate ligament (ACL). Although providing valuable data, these approaches are invasive to the knee, and require the attachment of mechanical devices on or near the ACL.

To solve these problems, we developed a noncontact, direct method to determine in situ forces in ligaments using robotics whereby not only the magnitude but also the direction and point of application of in situ forces can be determined using the UFS in conjunction with a force transformation scheme [9–11]. When a rigid body is fixed to the UFS, any external force applied to the body will be transmitted to and measured by the UFS as a force and moment. Using the three force components f_x, f_y, and f_z, the magnitude of the external force can be determined:

$$\text{Magnitude} = \sqrt{f_x^2 + f_y^2 + f_z^2} \tag{3}$$

The direction of the external force is described by a direction vector **a** as follows:

$$\mathbf{a} = \left(\frac{f_x}{\sqrt{f_x^2 + f_y^2 + f_z^2}}, \frac{f_y}{\sqrt{f_x^2 + f_y^2 + f_z^2}}, \frac{f_z}{\sqrt{f_x^2 + f_y^2 + f_z^2}} \right)^T \tag{4}$$

A force transformation is then employed to determine the point of application of the external force. The sensor coordinate system is mathematically rotated such that the z-axis of the new orthogonal coordinate system is parallel to the direction of the external force (Fig. 7). The line of action of the external force relative to the rotated coordinate system is precisely determined because the external force is perpendicular to the rotated x-y plane of the system. The coordinate system is then rotated back to the original con-

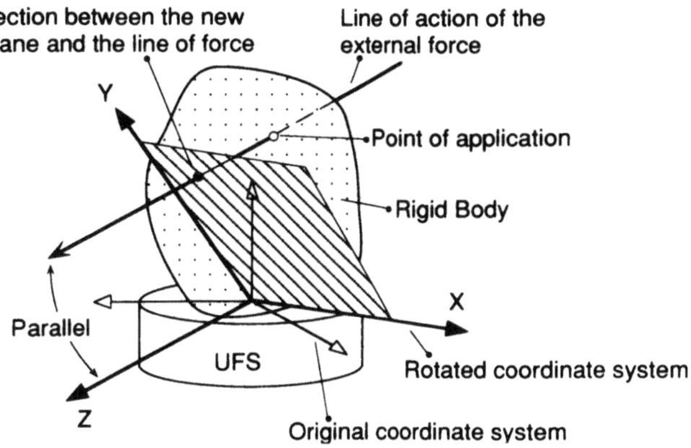

Fig. 7. Mathematical rotation of the sensor coordinate system for the precise determination of the line of action of the external force applied to a rigid body

figuration to describe the line of action of the applied force with respect to the original coordinate system as follows:

$$P_x = \frac{-f_z m_y + f_y m_z + f_x \sqrt{f_x^2 + f_y^2 + f_z^2}\, p}{f_x^2 + f_y^2 + f_z^2}$$

$$P_y = \frac{-f_x m_z + f_z m_x + f_y \sqrt{f_x^2 + f_y^2 + f_z^2}\, p}{f_x^2 + f_y^2 + f_z^2}$$

$$P_z = \frac{-f_y m_x + f_x m_y + f_z \sqrt{f_x^2 + f_y^2 + f_z^2}\, p}{f_x^2 + f_y^2 + f_z^2} \tag{5}$$

where p is a parameter that indicates a location along the line of action of the external force with respect to the sensor coordinate system. Please note that the developed Equation 5 is symmetric with respect to the x-, y-, and z-axes of the sensor. This feature makes it particularly useful, because the line of action of the in situ force in the ligament may change significantly with knee motion or external loading. With information about the geometry of the surface of the rigid body, the point of application of the external force can be determined by sequentially cutting ligaments of the knee.

To verify the methodology, seven intact human cadaveric knee joints were dissected down to the joint capsule and rigidly fixed to femoral and tibial clamps on a material testing machine (Model 4502, Instron, Canton, Massachusetts, USA) at 30° of flexion. The femoral clamp was fixed to the base of the machine, while the tibial clamp was fixed to the crosshead through a UFS (UFS II, JR^3, UFS-4015-A-50, USA) (Table 1). After the knee was manually placed in a stable position and orientation based on force and moment data from the UFS, an anterior displacement and posterior displacement were applied to the knee up to ±100 N of anterior and posterior force, at a rate of 20 mm/min. During the test, force and moment applied to the UFS as well as

A-P displacement were recorded. The lateral collateral ligament (LCL), medial collateral ligament (MCL), posterior cruciate ligament (PCL), and ACL were sequentially transected near their tibial insertions. Every time a ligament was transected, the same A-P displacement pattern was applied to the knee, while measuring forces and moments. Using the principle of superposition, the in situ force in each ligament was determined as a function of applied A-P force from Equations 3, 4, and 5.

The magnitude of the in situ forces in the LCL, MCL, and PCL were found to be less than 5 N during application of anterior force up to 100 N. The magnitude of the in situ force in the ACL was found to be slightly larger than the applied anterior force under 100 N of anterior force, which corresponded well with previous data (Fig. 8) [23,24]. However, it was found that the in situ force was smaller than the applied force when the applied force was smaller than 60 N. This indicates that the ratio of the in situ force in the ACL to an externally applied load is dependent not only on knee orientation, as found by previous investigators [23,24], but also on the magnitude of the externally applied loads. The direction of the in situ force in the ACL was approximately identical to its anatomical orientation. However, the point of application of the in situ force in the ACL was found to be located at the posterior and medial section of the anteromedial portion of the tibial insertion, which indicated a slight contrast with clinical knowledge. Further study is required to obtain more detailed data.

The approach outlined here does not require mechanical contact with the ligament, so in situ forces in intact ligaments can be accurately determined. Obtained data can provide useful information for studying the mechanism of ligament injury, improvement of the procedure of ligament reconstruction, and rehabilitation protocols.

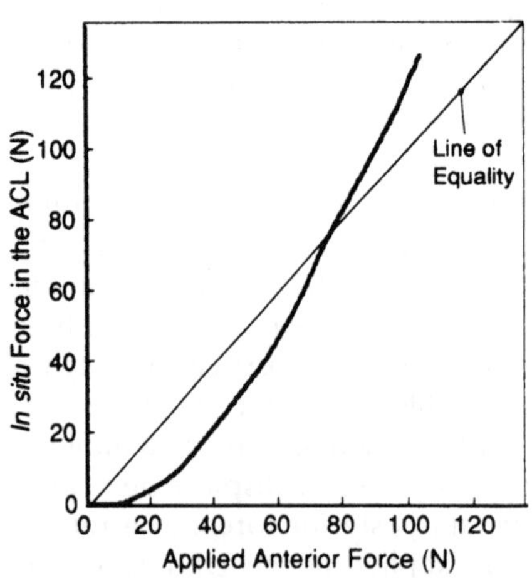

Fig. 8. Magnitude of the in situ force in the ACL as a function of the applied anterior force

In Vivo Study

Diagnosis of Ligament Rupture

Based on the data obtained in previously described in vitro studies, ACL rupture has been diagnosed using a robotic system (Fig. 9) [8]. This system was modified from the system for the in vitro study, mainly by the addition of a special tibial clamp using ski boots for tight fixation. In vivo anterior translation testing was performed on 26 intact knees, 12 partially ruptured ACL knees, and 10 completely ruptured ACL knees at 30° of flexion. The status of the tear was confirmed by arthroscopic examination. Stiffness as well as laxity of the knees under 80 N of anterior force were determined. Stiffness values of intact, partially ruptured, and completely ruptured knees were $3.04 \pm 0.14 \times 10^4$ (mean ± SD) N/m, $1.53 \pm 0.20 \times 10^4$ N/m, and $0.84 \pm 0.07 \times 10^4$ N/m, respectively. Laxity values of intact, partially ruptured, and completely ruptured knees were 3.07 ± 0.27 (mean ± SD) mm, 5.46 ± 0.49 mm, and 8.95 ± 1.33 mm, respectively. Statistically significant differences of stiffness as well as

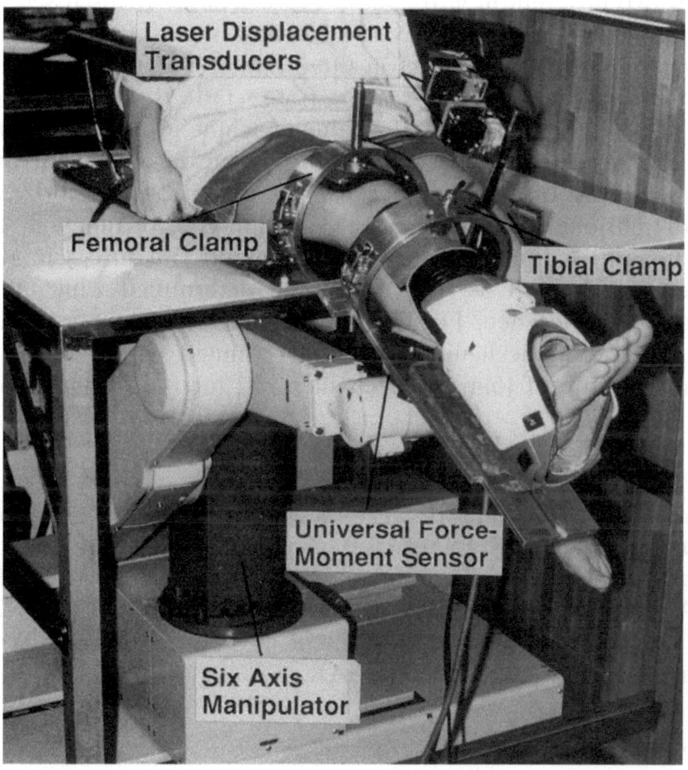

Fig. 9. Setup of the robotic system for the in vivo diagnosis of ligament rupture in clinic patients

laxity were found among the three different states of ACL rupture. These values were approximately identical to those obtained in in vitro studies. In the orthopedic clinic in Kitasato East Hospital, this robot system has been used for diagnosis of ACL as well as PCL rupture to supply fundamental data for surgical and postoperative treatment of ligament reconstruction.

Discussion

In biomechanical studies of the human knee joint, robotics has been found to be very appropriate. The mathematics originally developed in the field of robotics and kinematics can be applied to describe joint motion and joint forces and moments. This allows for systematic treatment of joint motion as well as joint force and moment in the experiments. Second, it is possible to reconstruct the three-dimensional path of joint motion using a manipulator while simulating a physiological situation. Moreover, the sensor instrumentation and techniques, such as the UFS, can be used for biomechanical measurements.

Although two different in vitro studies were separately discussed here, these will be combined to obtain more valuable data in the future. In the cutting study, it is required to reconstruct in three dimensions the same path of joint motion after a particular soft tissue is transected. It was impossible to do this using the traditional material testing machine with multi-DOF clamps. However, this is possible using the robotic system developed for the study of mechanical behavior of the knee, because the system can memorize and repeat the three-dimensional motion using the position control function. Combining the other in vitro methodology to the system, in situ forces in response to external loading as well as motion will be determined while allowing the natural motion of the knee. Finally, the function of ligaments as well as their contribution to joint mechanics will be determined to provide useful information on the study of joint biomechanics and for the diagnosis of ligament rupture.

Acknowledgments. The authors wish to acknowledge the assistance of Dr. Yukio Tsukamoto and Dr. Tadashi Sasada for their technical and clinical support and advice. These studies were supported by the Ministry of Education, Grant-in-Aid for Scientific Research on Priority Area No. 05221229 and 06213229, Grant-in-Aid for Encouragement of Young Scientists No. 06771157 (Japan) and NIH grant AR 39683 (USA).

References

1. Fujie H, Sasada T, Mabuchi K, Yamatoku Y, Tsukamoto Y, Yamamoto M (1985) The evaluation of mechanical factors influencing the bone fracture healing—the

application of robotics to an animal experiment (in Japanese). Proc Jpn Soc Orthop Biomech 7:199–206
2. Mabuchi K, Fujie H, Yamatoku Y, Yamamoto M, Sasada T (1992) A new methodology with an application of robotics to control the mechanical environment around experimentally fractured bone. In: Niwa S, Perren SM, Hattori T (eds) Biomechanics of Orthopedics. Springer-Verlag, Berlin Heidelberg New York Tokyo, pp 183–193
3. Fujie H, Mabuchi K, Tsukamoto Y, Yamamoto M, Sasada T (1987) Application of robotics to the knee instability test—preliminary experiment of canine knee joints (in Japanese). Proc Jpn Soc Orthop Biomech 9:105–110
4. Fujie H, Mabuchi K, Tsukamoto Y, Yamamoto M, Sasada T, Arai S (1989) Application of robotics to palpation of injury of ligaments—development of a new method of knee instability test. Tissue Eng BED-14:119–122
5. Fujie H, Livesay GA, Mabuchi K, Tsukamoto Y, Woo SL-Y (1991) A new methodology to perform joint kinematics studies using robotics. Adv Bioeng BED-20:407–408
6. Fujie H, Mabuchi K, Tsukamoto Y, Sasada T (1992) Six DOF hybrid control of the human knee joint using robotics: simulation of physiological situation (in Japanese). Proc Jpn Soc Orthop Biomech 14:119–122
7. Fujie H, Mabuchi K, Woo SL-Y, Livesay GA, Arai S, Tsukamoto Y (1993) The use of robotics technology to study human joint kinematics: a new methodology. ASME J Biomech Eng 115:211–217
8. Fujie H, Mabuchi K, Tsukamoto Y, Yamamoto M, Sasada T (1989) Trial making of knee mechanical diagnosis robot (in Japanese). Proc Jpn Soc Orthop Biomech 11:81–84
9. Fujie H, Livesay GA, Kashiwaguchi S, Blomstrom G, Woo SL-Y (1992) Determination of in situ force in the human anterior cruciate ligament: a new methodology. Adv Bioeng BED-22:91–94
10. Fujie H, Livesay GA, Kashiwaguchi S, Woo SL-Y (1992) In situ forces in knee soft tissues during anterior load: magnitude, direction and point of application (in Japanese). Proc Jpn Soc Orthop Biomech 14:115–118
11. Livesay GA, Fujie H, Blomstrum G, Kashiwaguchi S, Woo SL-Y (1993) Determination of in situ forces in the human anterior cruciate ligament with a universal force sensor. Transactions of the 39th annual meeting of the ORS 18, San Francisco, USA, Section 1:1
12. Fukubayashi T, Torzilli PA, Sherman MF, Warren RF (1982) An in vitro biomechanical evaluation of anterior-posterior motion of the knee. J Bone Joint Surg 64A:258–264
13. Lewis JL, Lew WD, Schmidt J (1988) Description and error evaluation of an in vitro knee joint testing system. ASME J Biomech Eng 110:238–248
14. Hollis JM, Takai S, Adams DJ, Horibe S, Woo SL-Y (1991) The effect of knee motion and external loading on the length of the anterior cruciate ligament (ACL): a kinematic study. ASME J Biomech Eng 113:208–214
15. Grood ES, Suntay WJ (1983) A joint coordinate system for the clinical description of three-dimensional joint motions: application to the knee. ASME J Biomech Eng 105:136–144
16. Chao EYS (1986) Biomechanics of the human gait. In: Schmid-Schonbein, Woo SL-Y, Zweifach BW (eds) Frontiers in Biomechanics. Springer, Berlin Heidelberg New York, pp 225–244

17. Fujie H, Livesay GA, Tsukamoto Y, Woo SL-Y, Blomstrom G, Mabuchi K, Fujita M, Sasada T (1993) Proceedings of the second international symposium on 3-D analysis of human movement, Satellite meeting of the International Society of Biomechanics, Poitiers, France, pp 39–42
18. Woo SL-Y, Adams DJ, Takai S (1992) The human anterior cruciate ligament and its replacement: biomechanical considerations. In: Niwa S, Perren SM, Hattori T (eds) Biomechanics of Orthopedics. Springer, Berlin Heidelberg New York Tokyo, pp 183–193
19. Lewis JL, Lew WD, Schmidt J (1982) Note on the application and evaluation of the buckle transducer for knee ligament force measurement. J Biomech Eng 104:125–128
20. Barry D, Ahmed AM (1986) Design and performance of a modified buckle transducer for the measurement of ligament tension. J Biomech Eng 108:149–152
21. Ahmed AM, Hyder A, Burke DL, Chan KH (1987) In vivo ligament tension patterns in the flexed knee in passive loading. J Orthop Res 5:217–230
22. France PE, Daniels AU, Goble ME, Dunn HK (1983) Simultaneous quantitation of knee ligament force. J Biomech 16:553–564
23. Markolf KL, Gorek JF, Kabo JM, Shapiro MS (1990) Direct measurement of resultant forces in the anterior cruciate ligament. J Bone Joint Surg 72A:557–567
24. Takai S, Woo SL-Y, Livesay GA, Adams DJ, Fu FH (1993) Determination of the in situ loads on the human anterior cruciate ligament. J Orthop Res 11:686–695
25. Holden JP, Korvick DL, Grood ES, Cummings JF, Bylski-Austrow DL (1992) In vivo forces in the anterior cruciate ligament during walking and trotting in a quadruped. Proceedings of the second North American congress on biomechanics, Chicago, USA, pp 133–134

Appendix A

$$X = \begin{bmatrix} C_{EF}C_{IE}-S_{EF}S_{VV}S_{IE} & -S_{EF}C_{VV} & C_{EF}S_{IE}+S_{EF}S_{VV}C_{IE} & d_{AP}C_{EF}-d_{PD}S_{EF}C_{VV} \\ S_{EF}C_{IE}+C_{EF}S_{VV}S_{IE} & C_{EF}C_{VV} & S_{EF}S_{IE}-C_{EF}S_{VV}C_{IE} & d_{AP}S_{EF}+d_{PD}C_{EF}C_{VV} \\ -C_{VV}S_{IE} & S_{VV} & C_{VV}C_{IE} & d_{PD}S_{VV}+d_{LM} \\ 0 & 0 & 0 & 1 \end{bmatrix} X'$$

where X' is a constant matrix representing the position and orientation of the manipulator coordinate system with respect to the tibial coordinate system. The X' can be numerically determined by measuring the kinematic relationship between two coordinate systems. Also, C_{EF} and S_{EF} represent cosine (θ_{EF}) and sine (θ_{EF}), respectively.

Appendix B

$$J = J' \begin{bmatrix} d_{AP}S_{VV}S_{IE}-d_{PD}C_{VV}C_{IE} & -C_{VV}S_{IE} & -d_{PD}S_{IE} & C_{IE} & 0 & 0 \\ d_{AP}C_{VV} & S_{VV} & 0 & 0 & 0 & 1 \\ -d_{AP}S_{VV}C_{IE}-d_{PD}C_{VV}S_{IE} & C_{VV}C_{IE} & d_{PD}C_{IE} & S_{IE} & 0 & 0 \\ -C_{VV}S_{IE} & 0 & C_{IE} & 0 & 0 & 0 \\ S_{VV} & 0 & 0 & 0 & 1 & 0 \\ C_{VV}C_{IE} & 0 & S_{IE} & 0 & 0 & 0 \end{bmatrix},$$

where, J′ is a constant Jacobian relating the tibial coordinate system to the sensor coordinate system in terms of the differential motion. J′ can be numerically determined by measuring the kinematic relationship between two coordinate systems.

Kinematic Analyses of the Knee Joint with Regard to the Load Transfer on the Cartilage

Wolfgang Thoma[1], Alwin Jäger[1], and Stefan Schreiber[2]

Summary. In reconstructive procedures of the anterior cruciate ligament (ACL), the exact localization of the attachment points is of great importance. The purpose of this study was to analyze the effect of small dislocations of the proximal insertion point with special regard to the ratio of rolling and sliding. Using a specimen, the insertion sites of the cruciate ligaments and the geometric shapes of the femur condyle and the tibia plateau were registered. A mathematical model was developed that allowed calculation of the range of instabilities and the quotient of rolling and sliding for eight different directions of dislocation. The poorest results could be found for deviations in the anteroproximal and in the anterodistal directions. With dislocations in the amount of 3.9 mm, instabilities between 0° and 90° of flexion occurred and the sliding component increased by as much as 200%.

Key words: Knee joint—Kinematics—Biomechanics—Cruciate ligaments—Rolling and sliding—ACL replacement—Knee surgery

Introduction

This study should be helpful for planning and evaluating reconstructive procedures of the anterior cruciate ligament (ACL). The purpose of the study was to analyze the effect of small dislocations of the proximal insertion point of the ACL. With special regard to the ratio of rolling and sliding, we wanted to describe quantitatively how an incorrect replacement of the ACL may disturb the kinematics of the knee joint and influence the load transfer between the cartilage surfaces of the femur condyles and the tibia plateau.

[1] Orthopädische Universitätsklinik Frankfurt, Marienburgstraße 2, 60528 Frankfurt am Main, Germany
[2] Institut für Mechanik der Technischen Hochschule, Hochschulstraße 1, 64289 Darmstadt, Germany

Methods

In our measurements we used a specimen with intact cruciate ligaments, collateral ligaments, and menisci. First, we registered the proximal and distal insertion sites of the cruciate ligaments. We then cut the medial femur condyle and the corresponding tibia plateau in a sagittal plane and recorded the shapes of the cartilage surfaces using a two-dimensional digitizer. On the basis of a four-bar kinematic linkage, a mathematical model was developed that allowed us to calculate the following parameters: the relative forces within the two cruciate ligaments, the range of instabilities, the contact zone between femur and tibia, the envelope curve, and the quotient of rolling and sliding.

For these calculations our model had to consider a series of interacting phenomena:

1. After the dislocation of the proximal insertion point of the ACL, a force within the ACL occurs during knee motion and a certain range of instability appears depending on the direction of the dislocation.
2. The force within the ACL will pull the femur condyle forward so that the posterior cruciate ligament (PCL) also comes under stress.
3. Depending on the ratio of stiffness between the ACL and PCL, the femur condyle will then be balanced in an intermediate position meeting the condition of minimal energy within the two cruciate ligaments.
4. Because the forces within the cruciate ligaments and the position of the femur condyle depend on the angle of flexion, the contact zone between femur and tibia and the ratio of rolling and sliding must be calculated continuously for the total range of motion.

The described parameters were calculated for eight different proximal attachment points of the ACL. Figures 1 and 2 illustrate the directions in which the point was dislocated, numbered from 0 to 7. The radius r represents one-tenth of the total length of the ACL, equivalent to 3.9 mm in our specimen.

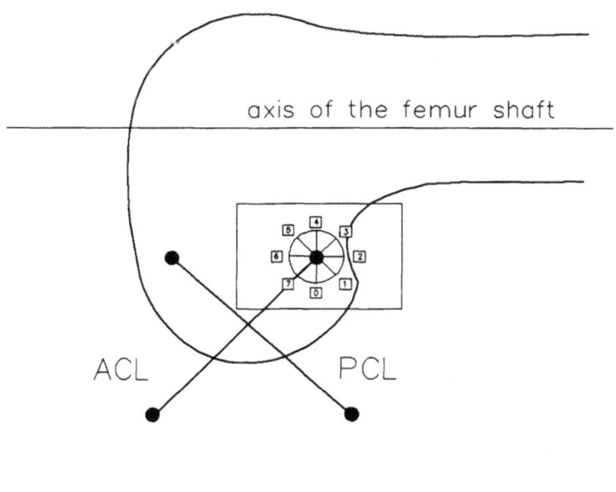

Fig. 1. Lateral femur condyle shows the geometry of the anterior cruciate ligament (*ACL*) and posterior cruciate ligament (*PCL*)

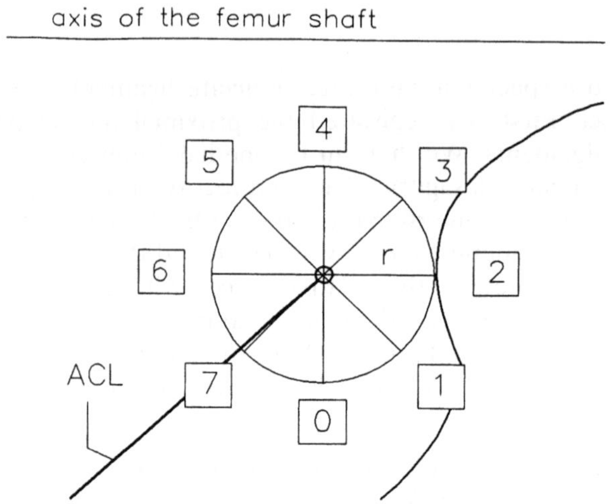

Fig. 2. Proximal attachment point of the anterior cruciate ligament (ACL) and eight directions of dislocation (1–8). *r*, radius

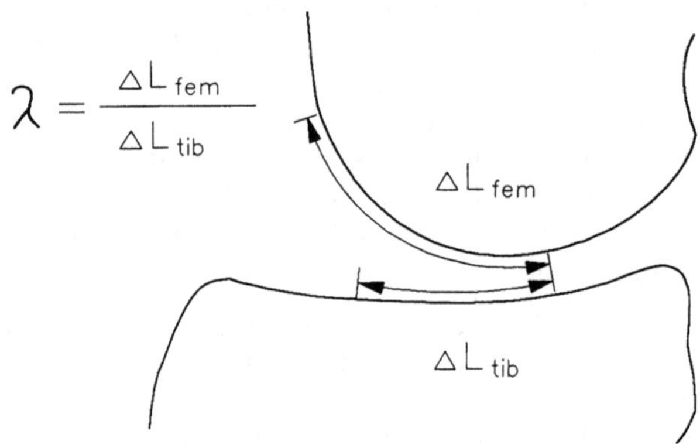

Fig. 3. Quotient of rolling and sliding. λ, ratio; ΔL_{fem}, change in femur length; ΔL_{tib}, change in tibia length

The calculation presumed a dislocation of the insertion point in a position of 90° flexion without applying any preload on the graft.

In Fig. 3, the formula ($\lambda = \Delta L_{fem}/\Delta L_{tib}$) for the quotient of rolling and sliding is specified and graphically explained: λ is the ratio, ΔL_{fem} the change in femur length, and ΔL_{tib} the change in tibia length.

Results

Figure 4a–h shows the results of our measurements and calculations. The horizontal axis represents the angle of joint flexion while the ratio of rolling and sliding is depicted in the vertical direction. The physiological ratio λ is plotted as a thin curved line in the graphs.

Within the hatched areas a negative force was calculated for at least one of the cruciate ligaments after dislocation of the proximal ACL insertion point. This is equivalent with the relaxation of the ligaments and will result in an instable knee joint. The ratio of rolling and sliding is not defined in these ranges of instability. On the other hand the light areas represent the zones in which the motion is controlled by the cruciate ligaments and the quotient λ can be calculated. The amount of the quotient is plotted as a thick curved line in Fig. 4a–h.

Discussion

In ACL reconstruction and replacement measures, the exact localization of the proximal attachment point is crucial. If the operative procedure fails to reproduce the correct geometry of the ACL, instabilities or contractions will be the result with the consequence of insufficient function and a physiologically incorrect load on the cartilage.

The measurements and calculations of our study allowed us to predict the effect of such mistakes in a quantitative way. Referring to our results, we want to point out two main factors that influence the functional outcome of ACL replacement procedures: the residual range of instability and the ratio of rolling and sliding.

If we look to the range of instability in Fig. 4a–h, the different directions of dislocation can be divided in two groups. The first group consists of dislocations in the dorsoproximal and the dorsodistal direction, represented by the numbers 0, 1, 2, and 7 (see Fig. 2). Dislocations in these directions are characterized by a stable knee joint between flexion angles of 0° to 90°. The second group consists of dislocations in the anteroproximal and the anterodistal direction, represented by the numbers 3, 4, 5, and 6 (Fig. 2). This group shows instability in the range of 0° to 90° but stability between 90° and about 140°. In relation to the geometry of the femur, the two groups are divided by a line that is almost parallel to the axis of the femur shaft. Regarding daily activities of the patient, it is obvious that those in the first group have better conditions for a good functional result than those of the second, although the absolute amount of dislocation is not different.

In addition, our results demonstrate very clearly that instabilities of the knee joint caused by small dislocations, elongations, or partial ruptures are highly dependent on the angle of flexion. Careful clinical examinations should therefore always be done in several flexion angles within a broad range.

Regarding the ratio of rolling and sliding we again can separate the dislocation directions into the same two groups. In the first group (numbers 0, 1, 2, and 7 in Fig. 2), the deviations from the physiological ratio are not very great while the second group shows an increase of the sliding component up to 200% within the range of stability.

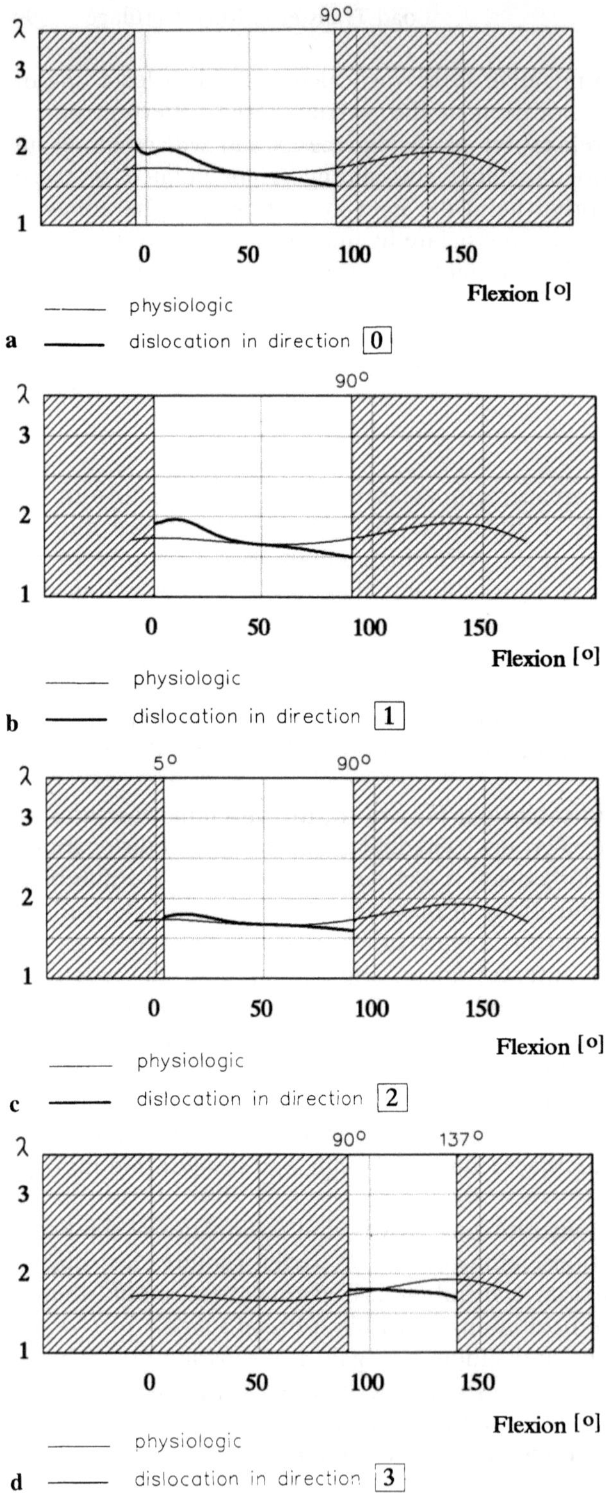

Fig. 4a–h. Quotient of rolling and sliding against the flexion of the knee joint; *hatched areas* represent range of instability. **a–h** Dislocation directions 0–7, respectively

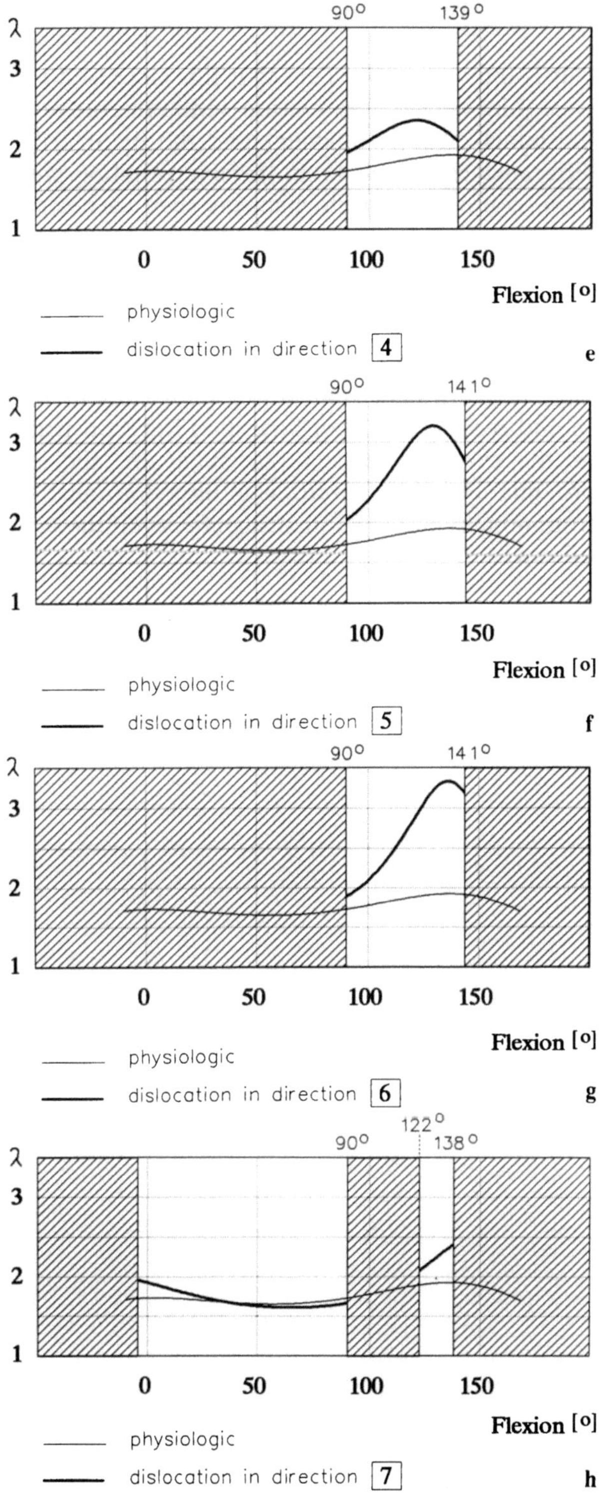

Fig. 4. (*Continued*)

It is important to reflect that this increased ratio cannot be smoothed by the knee joint because the motion within the range of stability is fully controlled by the intact cruciate ligament and the incorrectly fixated graft. This is a basic difference compared to an instable situation without ACL replacement, where some compensation mechanisms such as muscular activity may still work. In addition, one should consider that the curves in Fig. 4a–h show the effect of a dislocation of 3.9 mm. If the dislocation exceeds this value, the differences in the ratio of rolling and sliding will further increase.

In conclusion, our results show that even small dislocations of the proximal ACL attachment point will cause remarkable changes in the kinematics of the knee joint. In particular, a physiologically incorrect ratio of rolling and sliding should receive more consideration in reconstructive procedures. The aim must be the exact isometric replacement. At the least, dislocation in the anteroproximal or in the anterodistal direction must be avoided to prevent the joint from nonphysiological shearing effects on the cartilage with the risk of excessive joint wear.

Developing a Sensor for the Dynamic Measurement of Joint Pressure Distributions

STEVEN CULLIVER[1], MARK N. NABARRO[2], BRUCE MILTHORPE[1], ALBERTO CIMMINO[3], WILLIAM J. CUMMING[2], and KLAUS SCHINDHELM[1]

Summary. A system that would display pressure distributions in synovial joints in real time has many potential benefits for orthopedic surgery and research. High contact pressures have been implicated in the pathogenesis of several degenerative diseases of the joints. However, accurate contact stress distributions of joints are technically very difficult to obtain. This problem may be addressed by the development of a pressure-sensing pad that may be used in vivo as well as in vitro. Design considerations for the pressure sensor are that it must measure pressure distributions dynamically and display the distribution in real time, that it not disturb joint mechanics, and that it be biocompatible and suitable for arthroscopic insertion. A sensor sheet has been designed to measure joint pressures in vitro and in vivo. A prototype has been built and tested to determine the feasibility of the chosen technology. This prototype sensor does display in vitro pressure distributions in real time and has proved the sensor to be a workable device.

Key words: Contact pressure distribution—Force measurement—Joints: contact areas—Patellofemoral—Sensors: applications—Sensors: array—Synovial joints—Medical transducers

Introduction

Attaining knowledge of the fundamental dynamic properties of bones and joints is the primary goal in biomechanics research, but direct measurements of contact pressure distributions at articular surfaces of synovial joints continue to

[1] Centre for Biomedical Engineering, University of New South Wales, Sydney, NSW 2052, Australia
[2] Orthopaedic Department, St. George Hospital, Kogarah, NSW 2217, Australia
[3] School of Physics, University of Melbourne, Vic 3052, Australia

pose a challenge. The accurate estimation of dynamic pressure distribution within the larger diarthrodial joints is also of great benefit to orthopedic surgeons attempting reconstructive surgery of the joint to alleviate degenerative problems associated with inappropriate pressures in the joint.

A large number of different techniques have attempted to ascertain the pressure distribution in model [1–3] and cadaveric [1,4–10] studies. There are few data comparing these otherwise valuable studies to the in vivo situation. The lack of in vivo data is directly related to a lack of a suitable implantable sensor that can be tolerated by the joint and that will not affect the biomechanics of the joint.

Because researchers have experienced difficulties in quantifying joint contact forces using available methods, a prototype sensor sheet has been developed to facilitate the direct measurement of joint pressures in vitro and in vivo. This sensor has been designed to measure contact pressures in the patellofemoral joint but it may also be used for all the large diarthrodial joints.

Design considerations for the pressure sensor are as follows. The device must:

Dynamically measure pressure distributions
Display the distribution in real time
Be sufficiently thin so as not to disturb joint mechanics
Be biocompatible
Be suitable for arthroscopic insertion
Be sterilizable or reusable

The prototype sensor has fulfilled the first two design criteria and has proved to be a feasible and workable device.

Review

The properties of joints that can be measured include motion by the intrinsic architecture of the joint, the restrictions imposed by the soft tissue envelope, and the contact characteristics, which include area and pressure.

Contact stress distributions have been difficult to quantify because of inherent problems in measuring compressive forces between joint surfaces while retaining normal relationships and mechanics. The contact stress distribution of joints is an important physical and clinical measurement, but accurate values are technically very difficult to obtain. High contact pressures have been implicated in the pathogenesis of chondromalacia [11], and a plethora of reconstructive procedures have been described to decrease these pressures. Few data are available, however, to evaluate the effects of these operations [12].

The numerous techniques that have been used in the past in attempts to characterize these forces are briefly reviewed next.

Indirect Measurements

The joint load is calculated by measurements of related parameters under load or by loading a model of the joint.

Miller (R.K. Miller, 1993, *A New Technique for In Vitro Measurement of the Patellofemoral Force*, unpublished work) has measured the patellofemoral force using cadaveric knees under simulated weight-bearing conditions. With this technique, the magnitude, line of action, and point of application of the patellofemoral force were calculated and found to be higher than had been suggested by other cadaver studies.

The advantage of indirect methods is that the normal joint mechanics are not disturbed while measurements are made; however, every indirect method must be validated with a direct method to be useful. Mathematical vector models represent idealized simplifications of the anatomy and force distributions, and are unable to predict exact three-dimensional force distribution and contact areas [9].

Direct Measurements

Direct measurements employ transducers adjacent to the joint surfaces that are under load. Contact pressures are difficult to measure directly in synovial joints, and there is a lack of direct experimental measurements of patellofemoral and tibiofemoral contact pressures. Few data are available about pressure values in "normal" or diseased joints, and this information is essential to quantify the effects of different reconstructive procedures on joint contact pressures.

Many different methods have been used:

1. *Dye techniques*: Goodfellow et al. [7] injected a series of dyes into the knee joint, staining the noncontact areas of the patellofemoral joint surfaces and leaving the contact areas unstained. This method defined the contact areas of the joint under load but gave no information about the magnitude of the forces involved.
2. *Pressure-sensitive film*: A number of researchers (e.g., Huberti and Hayes [6], Hehne [9], Wagner et al. [10], and Collier et al. [2]) have used Fuji film (C. Itoh, New York, USA) to measure joint contact pressures. This method consists of two layers of film, one containing microcapsules of color-forming material and another with a color-developing material. On application of pressure, the microcapsules are broken and release the color-forming material, which reacts with the color-developing material to generate a color density that can be read with a densitometer. A range of film is available to measure different pressure ranges. A limitation of this method is that abnormal loading commonly saturates the film, making exact measurements of contact pressure impossible. Mean contact areas and pressures can be calculated. Ahmed [13] postulated that the liquid-like color-forming material may equalize the pressure

hydrostatically and has questioned the effect this may have on results. Wagner et al. [10] stated that the color density produced by the pressure-sensitive film is highly dependent on the time from zero to peak pressure for loading times of less than 5 s.

3. *Indentation transducers*: Ahmed et al. [5] measured in vitro static pressure distributions at the articular surfaces of synovial joints using a transducer that interprets pressure distribution from the microindentation pattern left on a thin plastic material. The amount of deformation produced in the plastic material by the indenter is proportional to the magnitude of the pressure and the duration of its application. Shear forces produce distortion of the indentation profile, however, and thus preclude the use of this technique in joints where appreciable shear forces are produced. Once it has been indented by a high pressure the transducer is unable to record a decrease in pressure. Another major limitation is that it cannot represent steady-state pressure distributions.

4. *Piezo-resistive elastomers*: Ferguson et al. [4] embedded arrays of miniature stress sensors in the retropatellar cartilage and measured actual contact stresses in varying degrees of flexion. Because the load cells were affixed at six discrete locations, highly variable readings were obtained. Degenerative changes in the cartilage and high surface curvature near the patellar margins precluded affixing the load cells in these areas.

5. *Ultrathin load cells*: Hedman [14] developed a method to measure time-dependent loads in facet joints. The ultrathin load cell consisted of pressure-sensitive film overlying a force-sensing resistor; this cell exhibited nonlinear decreasing resistance with increasing force applied perpendicular to the device but was not able to register shear forces.

6. *Electrical resistance technique*: Szivek et al. [3] used this technique to assess the contact stresses between the femoral component and tibial insert of total knee replacements. The tibial inserts were coated with a 400-Å-thick gold-palladium layer, which allowed the electrical resistance between the two components to be measured under loading. The authors were able to show that contact stresses were higher than those suggested by pressure-sensitive film.

None of the foregoing methods can be used intraoperatively, and most measure static contact stress distributions in vitro. These techniques have produced estimates of the patellofemoral forces that differ by as much as a factor of two, with results being substantially lower than those predicted by theoretical analysis of the extensor mechanism [15]. In fact, a fully satisfactory method for measuring dynamic contact pressure distributions has not yet been developed. A new measuring device is only useful if it is able to provide more reliable, more complete, or more accurate information than existing techniques.

Our sensor for dynamic measurement of intraarticular pressure distribution will be manufactured from medical-grade silicone, allowing instantaneous feedback on joint pressures in vivo or in the research laboratory. To minimize alterations in pressure readings, the sensor will have a compliance in bending greater than that of cartilage, enabling it to conform to gross deformations of the articular surface without disturbing contact pressure distribution.

Electrical Implementation

Transducers to measure force are widely used in a variety of applications. Two of the most common techniques use a piezoresistive or capacitive structure. Piezoresitive sensors have certain advantages (including simpler electronic interfacing), but they can be very difficult to implement in the form of an array. Capacitive pressure sensors are about an order of magnitude more pressure sensitive for a given sensor device size and more than an order of magnitude less sensitive to temperature. For these and other reasons, it has been decided to utilize techniques that detect the change in capacitance. This requires each transducer to have two conductive plates separated by a dielectric layer.

The Physical Sensor

Figure 1 shows an isometric view of the sensor pad, which is under development. It has five conceptual layers composed of two different materials: silicone and a conductive medium.

The use of silicone sheets provides certain advantages, one of which is biocompatibility, at least for the short times of implantation envisaged for the sensor. Another advantage of using silicone is the insulating properties of the outer layers (layers 1 and 5). These provide electrical isolation, from the external world for the conductive layers. The encapsulation also assists in protecting the fragile conductive layers from mechanical damage (e.g., cracking or flaking).

The conductive layers (layers 2 and 4) are two separate layers that are not electrically connected. The thickness of these layers is determined by the techniques and technology used to create them. In general, however, the thicker (or more metallic) the medium used to construct the conductive layers, the more rigid and inflexible the sensor pad will be overall. This will hinder the use of the sensor pad as a general research tool that meets all situations.

While it is most likely that most of the sensor pad's thickness will derive from the silicone sheets (layers 1, 3, and 5), it will be layer 3 that largely governs the overall thickness of the sensor pad. It is envisaged that the conductive layers 2 and 4 will have negligible effect on the pad's thickness.

The central layer (layer 3) is important to the sensor pad because its actual thickness must be a compromise between several factors: (1) The thinner the layer, the thinner the whole sensor pad, and the pad must be as thin as possible so as to reduce its influence on the forces to be measured. (2) The thinner the sheet, the lower the voltages that can be used. (3) The thicker layer allows

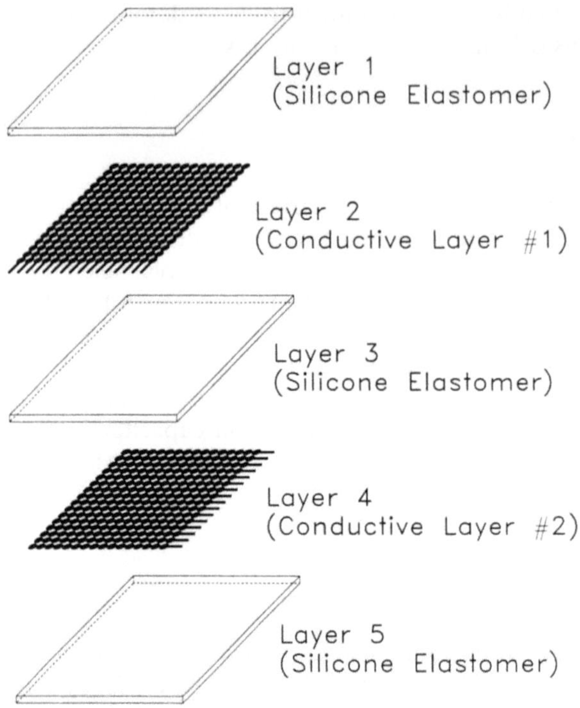

Fig. 1. Exploded view of sensor pad construction showing the five layers used

easier recording of the changes in its thickness. (4) This layer also acts as the dielectric for the capacitive transducers, and therefore its thickness influences the range of force that can be measured.

Results

A prototype of the sensor with standard commercially available electronic components (Fig. 2) was developed (by Dr. A. Cimmino) with a 16 × 16 array of sensing elements of which each element is 2.5 mm × 2.5 mm. The output of the sensing pad was connected to an AMLab (Associative Measurements, Sydney, Australia) data acquisition and processing module, which allowed real-time display of the pressure distributions.

The prototype sensor has been loaded in a tensometer with a single plastic (polymethylmethacrylate) bar and two plastic bars. The physical setup is demonstrated for the two-bar experiment in Fig. 3; the two bars (approximately 20 mm long × 3 mm wide) are arranged in a V shape with 3 mm between them at the narrowest point. On loading the bars on sensor, a real-time display of the change in the applied force was obtained. A sample of the results has been used to create three-dimensional surface graphs of the one-bar and two-bar experiments shown in Figs. 4 and 5, respectively. The effect of centrally

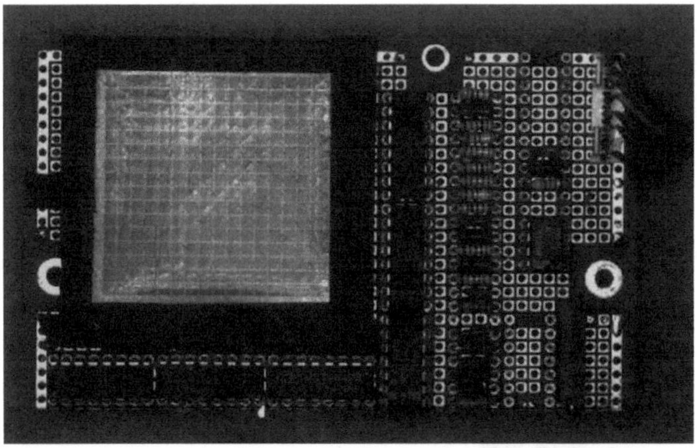

Fig. 2. Photograph of first prototype sensor showing sensing pad and associated electronics

Fig. 3. Photograph of first prototype sensor being tested with two bars compressed onto the surface by a tensometer

loading the bars can be seen in the bending of the bars and consequent reduction of pressure at the ends (this is especially visible in Fig. 4). For the two-bar experiment, the pressure distribution for each bar is clearly distinguishable (Fig. 5), indicating that there is little cross talk between sensing elements.

A second prototype sensor is being developed that is large enough to cover the retropatellar surface but thin and compliant enough to be inserted arthroscopically into the knee joint, thus minimizing the disturbance in

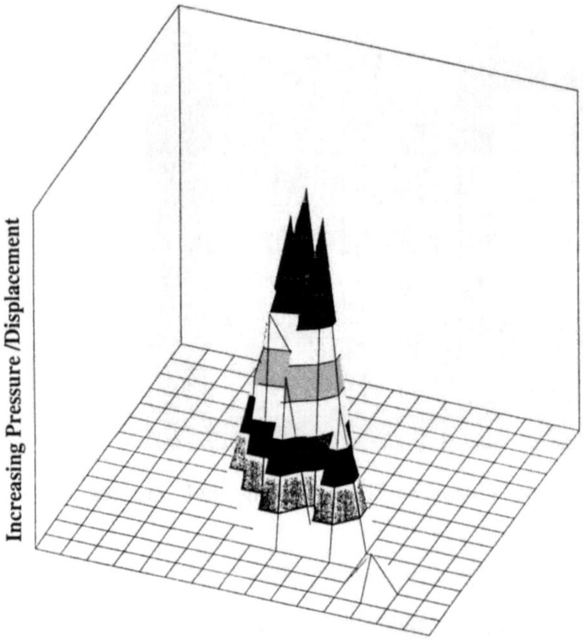

Fig. 4. Graph of the pressure distribution obtained with a single bar compressed onto the surface

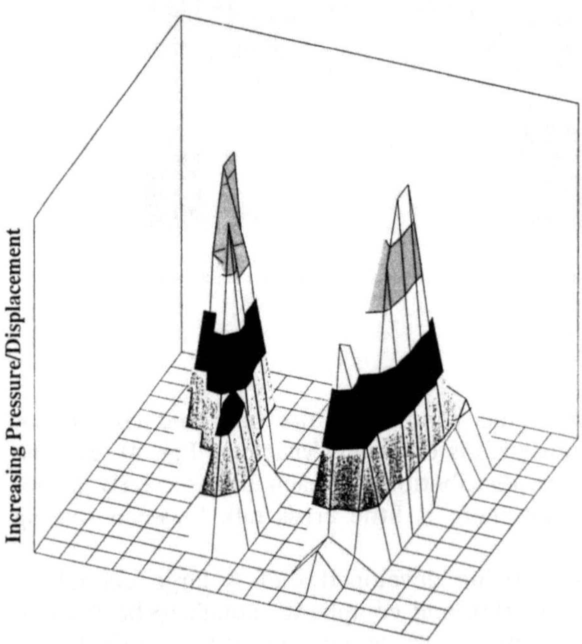

Fig. 5. Graph of the pressure distribution obtained with two bars compressed onto the surface

joint mechanics. This sensor will undergo in vitro testing and calibration with spherical as well as flat surfaces with calculable pressure distributions.

Conclusions

A pressure sensor is currently under development. The feasibility of the technique has been demonstrated on the bench using a prototype sensing pad and conventional electronic components. The sensor can be built with a sensing element less than 3 mm × 3 mm, which should provide adequate resolution for most large joints. A second prototype sensor pad suitable for arthroscopic insertion is being constructed.

The sensor will have application in joint reconstructive surgery and patellar realignment, and it will also provide an invaluable research tool to study all the larger joints.

References

1. Maquet PGJ (1976) Biomechanics of the knee. Springer, Berlin, pp 7–70
2. Collier JP, Mayor MB, McNamara JL, Surprenant VA, Jensen RE (1991) Analysis of the failure of 122 Polyethylene Inserts from Uncemented Tibial knee components. Clin Orthop Relat Res 273:232–242
3. Szivek JA, Cutignola L, Volz RG (1993) Tibio-femoral contact stress distribution evaluation of artificial total knees. In: Proceedings of the 39th annual meeting Orthopaedic Research Society, San Franciso, California, 15–18 February 1993
4. Ferguson AB, Brown TD, Fu FH, Rutkowski R (1979) Relief of patellofemoral contact stress by anterior displacement of the tibial tubercle. J Bone Joint Surg 61A:159–166
5. Ahmed AM, Burke DL, Yu A (1983) In-vitro measurement of static pressure distribution in synovial joints. Part II: Retropatellar surface. J Biomech Eng 105:226–236
6. Huberti HH, Hayes WC (1984) Patellofemoral contact pressures. J Bone Joint Surg 66A:715–724
7. Goodfellow J, Hungerford DS, Zindel M (1976) Patellofemoral joint mechanics and pathology. J Bone Joint Surg 58B:287–290
8. Perry J, Antonelli D, Ford W (1975) Analysis of knee joint forces during flexed knee stance. J Bone Joint Surg 57A:961–967
9. Hehne HJ (1990) Biomechanics of the patellofemoral joint and its clinical relevance. Clin Orthop Relat Res 258:73–85
10. Wagner UA, Sangeorzan BJ, Harrington RM, Tencer AF (1992) Contact characteristics of the subtalar joint: load distribution between the anterior and posterior facets. J Orthop Res 10:535–543
11. Ficat RP, Hungerford DS (1979) Disorders of the patellofemoral joint. Williams and Wilkins, Baltimore

12. Kurasowa H, Fukubayashi T, Nakajima H (1980) Load-bearing model of the knee joint: physical behaviour of the knee joint with or without menisci. Clin Orthop Relat Res 149:283–290
13. Ahmed AM (1983) A pressure distribution for in-vitro static measurements in synovial joints. J Biomech Eng 105:309–314
14. Hedman TP (1992) A new transducer for facet force measurement in the lumbar spine: benchmark and in-vitro test results. J Biomechan 25:69–80
15. Ahmed AM, Burke DL, Hyder A (1987) A force analysis at the patellar mechanism. J Orthop Res 5:69–85

Part 3. Artificial Joint

Stress Analysis of the Proximal Tibia After Unicompartmental Knee Arthroplasty with Finite-Element Method

Nobuyuki Yoshino[1], Shinro Takai[1], Nozomu Inoue[1], and Tsutao Katayama[2]

Summary. The finite-element method (FEM) has recently been employed to optimize the design of prostheses and to examine the mechanical problems of total knee arthroplasty. Stress analyses of unicompartmental knee arthroplasty (UKA) using FEM have not been previously reported, although UKA is a more popular procedure. The objective of this study is to determine the factors affecting the stress distribution of the tibia after UKA. A two-dimensional FEM in consideration of the heterogeneous and anisotropic properties of cancellous bone was applied for stress analysis at the proximal tibia after UKA. Two types of implants, an ultrahigh molecular weight polyethylene (UHMWPE) type and a metal-backed type, were compared in three different alignments (neutral, varus, and valgus). Two different widths of implants and four different levels of the osteotomy (6–20 mm distal to the joint line) were then compared in neutral alignment. Higher stress concentrations were observed in the cancellous bone beneath the UHMWPE implants than beneath the metal-backed implants in all alignments. Stress concentrations in cancellous bone beneath implants in varus were higher than beneath bone in neutral alignment. High stress concentration at the intercondylar eminence was found in valgus alignment. There were no remarkable differences with respect to the level of the osteotomy.

Key words: Stress analysis—Finite-element method—Unicompartmental knee arthroplasty—Implant design—Metal backing—Alignment of the lower limb—Coverage of the implant—Level of tibial osteotomy

---

[1] Department of Orthopedic Surgery, Kyoto Prefectural University of Medicine, Kawara, Hirokoji, Kamigyo-ku, Kyoto, 602 Japan
[2] Department of Mechanical Engineering, Doshisha University, 1-3 Tataramiyakodani, Tanabe, Tsuzuki-gun, Kyoto, 610-03 Japan

Introduction

Unicompartmental knee arthroplasty (UKA) is an attractive procedure that has advantages over bicompartmental or tricompartmental knee arthroplasty (TKA), such as preservation of bone stock and the anterior and posterior cruciate ligaments as well as the patellofemoral joint and the opposite compartment to maintain normal knee kinematics with a functional range of motion. Despite these advantages, UKA has been a somewhat controversial procedure in regard to its efficacy and indication. Early in the evolution of the operation some authors reported excellent clinical results [1], but others reported deterioration of results because of implant breakage and loosening [2,3]. These failures were derived partially from misjudgment in the selection of patients, partially from postoperative malalignment, and partially from poor implant designs, especially inadequate implant thickness without metal backing. In recent years, indications for UKA have been severely limited, implant designs have been modified to include metal backing [4–7], and numerous publications have reported successful use of UKA [5,8–10]. The results of UKA are superior to those of a high tibial osteotomy, which may not last as long [8]. Thus, interest in the UKA is on the rise again.

The objective of the current study was to determine the factors affecting the stress distribution of the proximal tibia after UKA. For this objective, the finite-element method (FEM) in consideration of the heterogeneous and anisotropic material properties of cancellous bone in the proximal tibia [11] was employed to assess the following four factors: (1) two type of implants, the all-ultrahigh molecular weight polyethylene (UHMWPE) type and the metal-backed type, (2) postoperative alignment, (3) coverage of tibial bone, and (4) the level of the tibial osteotomy.

Methods

The two-dimensional FEM analysis was performed to study stress distribution in the proximal tibia under loaded conditions. Figure 1 shows FEM models. Figure 1a is assumed to be in neutral alignment, Fig. 1b is assumed to be in varus alignment due to undercorrection, Fig. 1c is assumed to be in valgus alignment due to overcorrection, and Fig. 1d is assumed to be in neutral alignment, with the tibial cortex covered by the implant. Figure 2a shows the mean trabecular orientation, and Fig. 2b shows the distribution of Young's modulus in the mean trabecular orientation angle of cancellous bone [11]. These data in Fig. 2 were calculated by previously reported numerical analysis of a soft radiograph of the sliced proximal tibia [11]. The material properties of cancellous bone, cortical bone, UHMWPE, and the metal (Co-Cr-Mo) are shown in Table 1. Figure 3 shows the loading and boundary conditions for the models of the proximal tibia after UKA. Concerning loading conditions, unicondylar loading and bicondylar loading were applied to the neutral align-

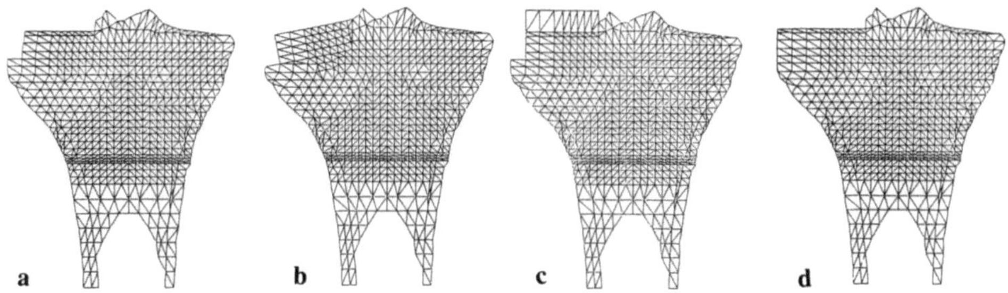

Fig. 1a–d. Finite-element mesh. **a** Neutral alignment (610 nodal points, 1114 elements). **b** Varus alignment due to under correction (614 nodal points, 1123 elements). **c** Valgus alignment due to overcorrection (619 nodal points, 1130 elements). **d** Neutral alignment with cortical coverage (615 nodal points, 1124 elements)

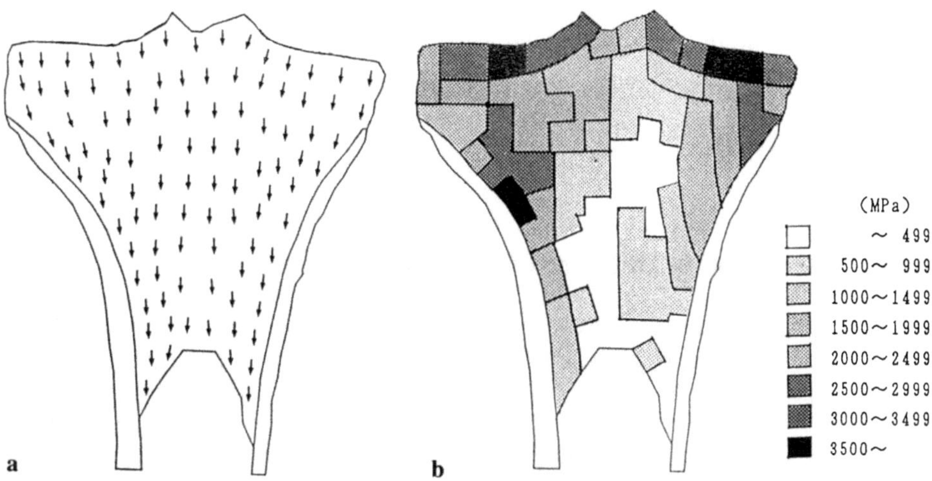

Fig. 2a,b. Material properties of cancellous bone. **a** Mean trabecular orientation. **b** Distribution of Young's modulus of cancellous bone

Table 1. Material properties

Young's moduli	(MPa)
Cancellous bone	131–4059
Cortical bone	12000
UHMWPE	500
Co-Cr-Mo	200000
Poisson's ratio	0.3

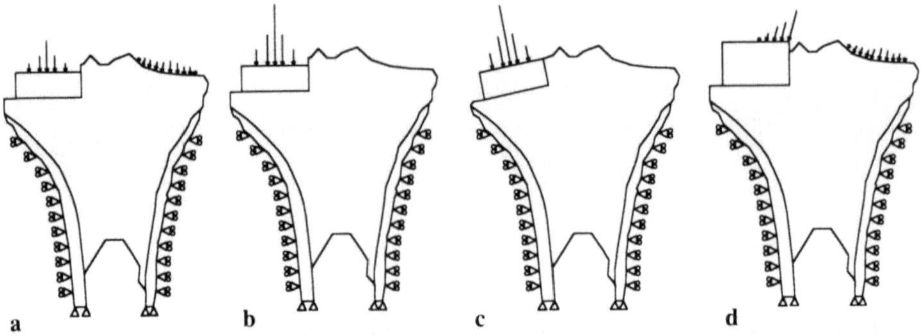

Fig. 3a–d. Loading and boundary conditions. **a** Bicondylar loading for neutral alignment. **b** Unicondylar loading for neutral alignment. **c** Unicondylar loading for varus alignment. **d** Bicondylar loading for valgus alignment

ment, unicondylar vertical loading with respect to the implant was applied to the varus alignment, and bicondylar shear loading was applied to the valgus alignment.

Results

Stress Distribution in Different Postoperative Alignments

Figure 4b,c,f,g shows von Mises' equivalent stress distribution per given load and per unit bone density considering the volume fraction of the trabeculae in neutral alignment. The stress concentration was found to be in the middle portion of the replaced condyle in all UHMWPE types (Fig. 4b,c). However, stress concentration was not found in the metal-backed type under bicondylar loading (Fig. 4g). Stress was distributed to the stem and both edges in the metal-backed type under unicondylar loading (Fig. 4f). In varus alignment, stress concentration beneath the implant was found to be shifted to the medial side of tibia and to be more remarkable than that in neutral alignment (Fig. 4a,e). In valgus alignment, stress concentration was found at the eminence of the tibia (Fig. 4d,h).

Stress Distribution in Different Coverages of the Tibia

Figure 5a–d shows von Mises' equivalent stress distribution of narrow coverage of the tibia, and Fig. 5e–h shows von Mises' equivalent stress distribution of wide coverage of the tibia. The wider the implant coverage, the less stress concentration was found in the cancellous bone beneath the implant.

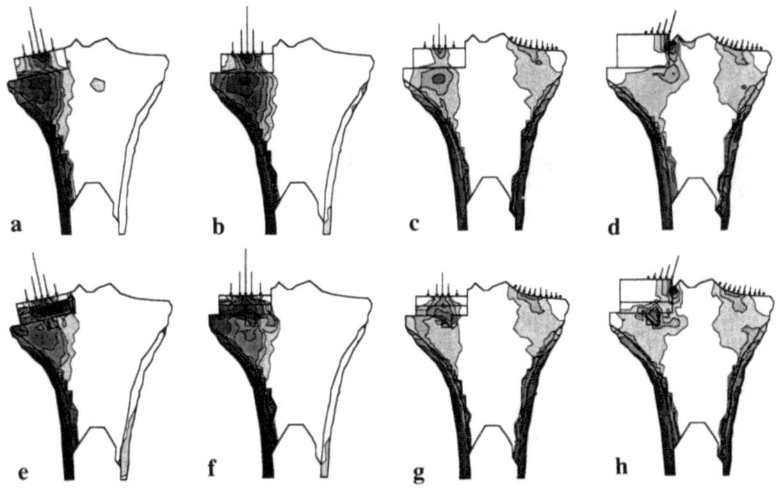

Fig. 4a–h. Von Mises' equivalent stress distribution per given load and per volume fraction of trabeculae (difference of the alignment). **a** Unicondylar loading for UHMWPE-type UKA in varus alignment. **b** Unicondylar loading for UHMWPE-type UKA in neutral alignment. **c** Bicondylar loading for UHMWPE-type UKA in neutral alignment. **d** Bicondylar loading for UHMWPE-type UKA in valgus alignment. **e** Unicondylar loading for metal-backed-type UKA in varus alignment. **f** Unicondylar loading for metal-backed-type UKA in neutral alignment. **g** Bicondylar loading for metal-backed-type UKA in neutral alignment. **h** Bicondylar loading for metal-backed-type UKA in valgus alignment

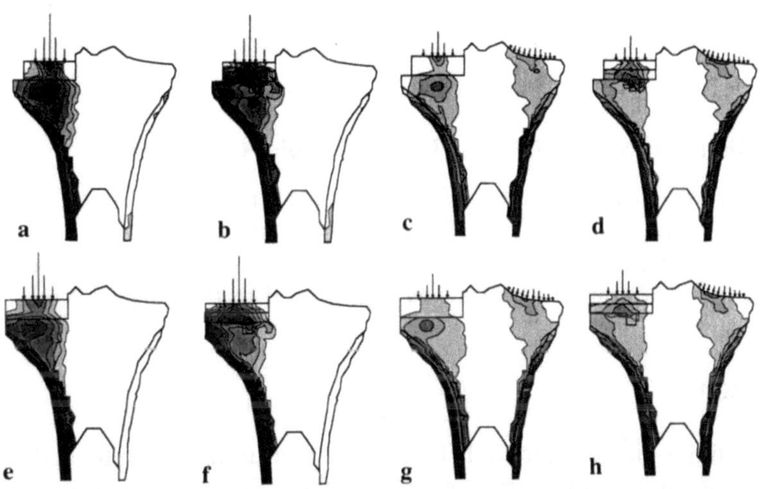

Fig. 5a–h. Von Mises' equivalent stress distribution per given load and per volume fraction of trabeculae (difference of the coverage of the component in neutral position). **a** Unicondylar loading for UHMWPE-type UKA. **b** Unicondylar loading for metal-backed-type UKA. **c** Bicondylar loading for UHMWPE-type UKA. **d** Bicondylar loading for metal-backed-type UKA. **e** Unicondylar loading for wide UHMWPE-type UKA. **f** Unicondylar loading for wide metal-backed-type UKA. **g** Bicondylar loading for wide UHMWPW-type UKA. **h** Bicondylar loading for wide metal-backed-type UKA

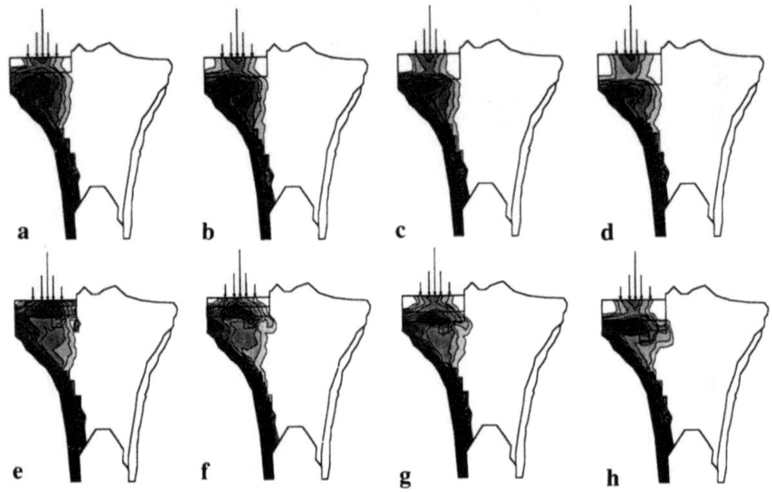

Fig. 6a–h. Von Mises' equivalent stress distribution per given load and per volume fraction of trabeculae (difference of the level of the osteotomy line). **a** Unicondylar loading for 6-mm UHMWPE-type UKA. **b** Unicondylar loading for 10-mm UHMWPE-type UKA. **c** Unicondylar loading for 15-mm UHMWPE-type UKA. **d** Unicondylar loading for 20-mm UHMWPE-type UKA. **e** Unicondylar loading for 6-mm metal-backed-type UKA. **f** Unicondylar loading for 10-mm metal-backed-type UKA. **g** Unicondylar loading for 15-mm metal-backed-type UKA. **h** Unicondylar loading for 20-mm metal-backed-type UKA

Stress Distribution at Different Level of Tibial Osteotomy

Figure 6a–d shows von Mises' equivalent stress distribution of an all-UHMWPE type implant at different heights of osteotomy, and Fig. 6e–h shows that of the metal-backed type. The lower the level of the osteotomy line became, the more directly stress was transmitted to the cortical bone.

Discussion

Since the finite-element method (FEM) was first utilized for the analysis of the structure of aircraft by Turner et al. in 1954, FEM has been employed for many mechanical analyses of materials and structures. In the orthopedic field, FEM has recently been employed to evaluate and optimize the design of prostheses and to examine the causes of prosthetic mechanical problems such as loosening and sinking of the tibial component of total knee arthroplasty (TKA). In reports comparing various designs, there is general agreement that all-plastic tibial components result in higher bone and cement stresses when compared with metal-backed types [12,13]. Many authors demonstrated that increasing the length of a central stem decreases the proximal trabecular bone stresses [13,14]. Similar results were found for components with a larger surface area [15]. These results from finite-element analysis in TKA are supported by clinical results. Thus, FEM is useful in the prediction of the mechanical problems of the implant and its fixation mechanism and, furthermore, for an estimation of the acceptable range of the error.

The stress distribution in finite-element analysis depends not only on the loading conditions but also on the geometry of the implant, the properties of the material, and the boundary and interface conditions. The material properties of the cancellous bone for the FEM in many previous studies were based on the data from compression tests. Therefore, the elastic moduli of cancellous bone could be obtained in only limited portions, and trabecular orientation was not assessed. Askew and Lewis [14] reported on stress analysis including heterogeneity and anisotropy, but they divided the proximal tibia into four regions. Their report concluded that if the heterogeneity is taken into account correctly, the effect of anisotropy is only minor. In our previous study, comparison of two modes of finite-element analysis, with and without taking heterogeneity and anisotropy of the cancellous bone into account, showed the importance of the detailed material properties of cancellous bone [11]. These properties are of greater importance in stress analysis of the tibia after UKA than after TKA for the following reasons: (1) the implant is smaller than in TKA and the opposite compartment is preserved, and (2) the osteotomy line is more proximal than that in TKA, and the material property of the cancellous bone varies widely in the proximal portion.

The causes of the mechanical failure of the UKA were reported to be derived from many factors, as follow: (1) technical failure, such as varus positioning of the implant and overcorrection of the postoperative leg alignment; (2) design and material of the implant; and (3) errors in selection of patients, such as those with severe varus deformity, involvement of the patellofemoral joint, panarthritis such as rheumatoid arthritis and chondrocalcinosis, ligament insufficiency, severe obesity, high activity, osteoporosis, and other problems. In the current study, the design and material of the implant, and the positioning of the implant in postoperative alignment can be simulated using FEM.

In regard to the metal backing and thickness of the implant, Marmor [1,10] and Shurley et al. [16] reported that the 6-mm tibial component without metal backing is subject to deformation and loosening. Knutson et al. [17] reported that plastic deformation of a thin tibial component will not only deform the proximal surface of the component but also eventually induce a deeper deformation ("cold creep"). Using roentgen stereophotogrammetric analysis, Ryd et al. reported that cold flow was found in all 9- to 12-mm-thick UHMWPE implants and not in metal-backed types [7]. In the current study, metal backing has proven to be beneficial for stress distribution, similar to the results of the analysis in TKA.

In a study of alignment of the limb in patients who had a UKA, good results were obtained when the mechanical axis was on center or slightly medial to the center [9]. Many reports warned against overcorrection, which leads to excess load in the lateral compartment and its early degeneration [5,18,19]. The stress to the implant from inadequate postoperative alignment acts as shear stress and leads to loosening in both varus and valgus knees. During a UKA procedure, the alignment is corrected not only by the removal of osteophytes and balancing of the soft tissues but also by means of the thickness of the implant. As a result, abnormally high pressures may be transmitted through the opposite compartment, and subluxation may occur if the implant is too tight because of overcorrection. In the present study, undercorrection revealed higher stress concentrations in the underlying cancellous bone than in neutral alignment, and overcorrection revealed stress concentrations at the eminence of the tibia where the bone density is low. These findings are reflected in clinical results.

The width of the implant is predicted to have little effect on the stress distribution, and this prediction is similar to the results of finite-element analysis of TKA [13]. Clinically the coverage of the implant rarely affects the sinking and loosening in TKA.

Because a severe varus deformity is a contraindication for UKA, the osteotomy line of the tibia is usually within the range of 20 mm distal to the joint line. Our study suggested the acceptable range of the osteotomy, but these results should be proven by clinical outcome.

The results of the current study were substantially equivalent to the clinical results mentioned previously; however, there may be many factors that remain to be defined with finite-element analysis. In the present analysis, material connections such as prosthesis–bone interfaces and UHMWPE–metal tray junctions were assumed to be rigidly bonded. Loading conditions were also defined to be static. These conditions had to be simplified, and are not equivalent to actual conditions. Therefore, boundary and loading conditions should be established in future studies. The highly complicated geometry of knee joint should be also simplified to provide a basis of better mechanical understanding.

Conclusions

The two-dimensional finite-element model in consideration of heterogeneous and anisotropic properties of cancellous bone was applied to analysis of stress distribution in the proximal tibia after UKA. Metal-backed implants were shown to provide stress distribution superior to that of all-UHMWPE implants in this study. The proper postoperative alignment must be achieved in UKA. The osteotomy line is assumed to be acceptable in a range of as much as 20 mm from the articular surface of the tibia.

References

1. Marmor L (1979) Marmor modular knee in unicompartmental disease. J Bone Joint Surg 61A:347–353

2. Insall J, Aglietti P (1980) A five- to seven-year follow-up unicondylar arthroplasty. J Bone Joint Surg 62A:1329–1337
3. Laskin RS (1978) Unicompartmental tibiofemoral resurfacing arthroplasty. J Bone Joint Surg 60A:182–185
4. Bernasek TL, Rand JA, Bryan RS (1988) Unicompartmental porous coated anatomic total knee arthroplasty. Clin Orthop 236:52–59
5. Kozinn SC, Marx C, Scott RD (1989) Unicompartmental knee arthroplasty. A 4.5–6 year follow-up study with a metal-backed tibial component. J Arthroplasty 4:S1–10
6. Magnussen PA, Bartlett RJ (1990) Cementless PCA unicompartmental Joint arthroplasty for osteoarthritis of the knee. J Arthroplasty 5:151–158
7. Ryd L, Lindstrand A, Selvik G (1990) Cold flow reduced by metal backing. Acta Orthop Scand 61:21–25
8. Broughton NS, Newman JH, Baily RAJ (1986) Unicompartmental arthroplasty and high tibial osteotomy for osteoarthritis of the knee. A comparative study after 5–10 year follow up. J Bone Joint Surg 68A:447–452
9. Kennedy WR, White RP (1987) Unicompartmental arthroplasty of the knee: preoperative alignment and its influence on overall results. Clin. Orthop 221: 278–285
10. Marmor L (1988) Unicompartmental knee arthroplasty: ten to 13-year follow-up study. Clin Orthop 226:14–20
11. Yoshino N, Inoue N, Watanabe Y, Yamashita F, Hirasawa Y, Hirai T, Katayama T (1992) Stress analysis of the proximal tibia after total knee arthroplasty with a finite element method. In: Niwa S, Perren SM, Hattori T (eds) Biomechanics in orthopedics. Springer, Berlin Heidelberg New York Tokyo, pp 253–262
12. Lewis JL, Askew MJ, Jaycox DP (1982) A comparative evaluation of tibial component designs of total knee prostheses. J Bone Joint Surg 64A:129–135
13. Murase K, Crownishield RD, Pedersen DR, Chang T-S (1982) An analysis of tibial component design in total knee arthroplasty. J Biomech 16:13–22
14. Askew MJ, Lewis JL (1981) Analysis of model variables and fixation post length effects on stresses around a prosthesis in proximal tibia. J Biomech Eng 103:239–245
15. Garg A, Walker PS (1986) The effect of the interface on the bone stresses beneath tibial components. J Biomech 19:957–967
16. Shurley TM, O'Donogue DH, Smith WD, Payne RE, Grana WA (1982) Unicompartmental arthroplasty of the knee. Clin Orthop 164:236–246
17. Knutson K, Jonsson G, Anderson JL, Larusdottir H, Lidgren L (1981) Deformation and loosening of the tibial component in knee arthroplasty with unicompartmental endoprostheses. Acta Orthop Scand 52:667–673
18. Emerson RH Jr, Head WC, Peter PC Jr (1992) Soft tissue balance in medial unicompartmental knee arthroplasty. J Bone Joint Surg 74B:807–810
19. Thornhill TS (1986) Unicompartmental knee arthroplasty. Clin Orthop 205:121–131

Bone Remodeling Around Implanted Materials

Masanori Oka[1], Yongshun Chang[1], Takashi Nakamura[1], Zhonglian Li[1], Toshiaki Kitsugi[1], Sadami Tsutsumi[1], and Haruki Takagi[2]

Summary. Bone formation and remodeling around implanted materials are influenced by the kinds of materials, surface properties of the materials, anatomical sites of implantation, and the load-bearing condition. In this study, three kinds of materials were implanted into the femoral condyles of dogs, and bone formation and remodeling were observed for 6 months. Uniform thickening of lamellar bones was observed around bead-coated alumina while thick fibrous tissues surrounded by corticalized bones formed around smooth alumina. This distinct difference in remodeling pattern was subjected to mechanical simulation by using finite-element analysis (FEA). Although stress distribution around bead-coated alumina simulates the bone density, the remodeling pattern around an uncoated implant could not be determined by the stress. To describe the latter remodeling pattern, it is necessary to analyze the strain surrounding the implants. Abundant bone ingrowth into titanium fiber under a similar loading condition was obtained 8 weeks after surgery. This ingrowth allowed firm attachment to the bones. As the tibial joint surface remained intact against polyvinyl alcohol (PVA) hydrogel, this artificial osteochondral composite material seems to be a very promising joint prosthetic material.

Key words: Bone remodeling—Wolff's law—FEA—Artificial articular cartilage—Bone ingrowth—Bone formation under load-bearing condition

Introduction

Bone formation and remodeling around implanted materials are influenced by the kind of material, the material surface properties, and the anatomical sites

[1] Research Center for Biomedical Engineering, Kyoto University, 53 Kawahara, Shogoin, Sakyo-ku, Kyoto, 606 Japan
[2] Department of Orthopedic Surgery, Fukui Red Cross Hospital, 2-4-1 Tsukimi, Fukui, 910 Japan

of implantation. In addition, bone remodeling depends on the loading condition. In most studies on bone remodeling, materials were implanted not in the position that would be under loading condition but in marrow spaces or transcortically [1]. According to Wolff's law, mechanical stress influences the remodeling process of bone and subsequently its structure and strength [2]. By Wolff's law, the materials should be implanted into the bones under loading condition insofar as the bone remodeling process is concerned.

In a previous paper [3], the differences of bone remodeling around various alumina surfaces were studied. Alumina was implanted into the load-bearing portions of the medial and lateral femoral condyles of adult mongrel dogs. As a result, uniform thickening of trabecular bones in a lamellar pattern was observed around bead-coated implants, while thick fibrous tissues surrounded by corticalized bones were observed around smooth implants [3]. These prominent differences in bone remodeling patterns aroused our interest because the latter remodeling pattern was also observed in the fracture ends of nonunions, as well as in subchondral bone plate [4]. We attempted to analyze the stress distribution on the surrounding bones around the materials using finite-element analysis (FEA) and to simulate the bone remodeling process numerically.

The second objective of this study is to examine the speed and volume of new bone ingrowth into the pores of artificial osteochondral composite material. The reason why we have developed an artificial articular cartilage has been described in detail in our previous publications [5–8]. In short, it is necessary to preserve, as much as possible, the cancellous bones, which play an important role in the shock-absorbing effect of the joints. The most important aim in developing an artificial articular cartilage is to fix the material quickly and firmly to the bones. For that purpose, we manufactured an artificial osteochondral composite material, which is composed of porous titanium fiber with polyvinyl alcohol (PVA) infiltrated into the pores. We implanted this composite material under the same loading condition.

Finally, to make clear some pathological changes of articular cartilage when it is placed against hard materials, as observed clinically in the acetabulum in the case of Moore's endoprosthesis, the differences of tibial joint surface whether sliding against alumina or sliding against an artificial articular cartilage were histologically examined and compared.

Materials and Methods

Implant Preparation and Characterization

Three kinds of rectangular parallelepiped test pieces were used. The first material was alumina with bead-coated sides and a mirror-finished joint surface. The mean pore size of the coated layer was 260 μm; its porosity was 38%. The second type of alumina test pieces were of the same size and shape with smooth side surfaces (Fig. 1). The third type of test pieces of the same shape

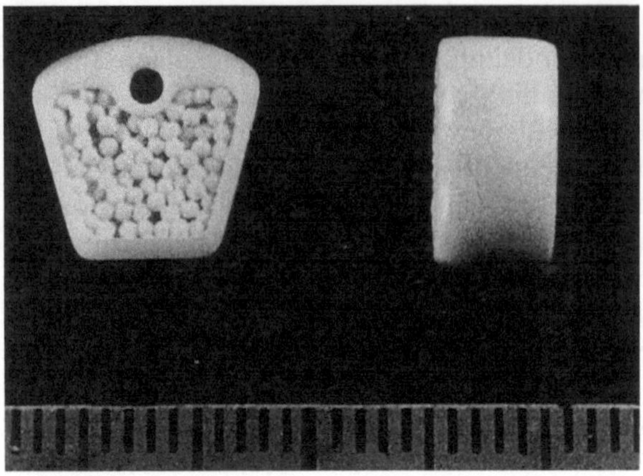

Fig. 1. A small hole was made in implants to provide an anchor to the jig for the pull-out test

were manufactured of pure titanium fiber of porosity of 70% and mean pore size of 170 μm; the joint surfaces were covered by an artificial articular cartilage, polyvinyl alcohol hydrogel (PVA-H). To make the composite material, PVA (molecular weight, 220 000) solution was first infiltrated into the pores of the titanium fiber (distal half) and then bound to the fiber by gelling the PVA. The shear strength between the PVA-H and the titanium fiber was about 2.2 MPa. The test pieces were sterilized in ethylene oxide gas and the alumina pieces in an autoclave.

Surgery

The three kinds of test pieces were implanted into the femoral condyles of adult mongrel dogs weighing between 10 and 15 kg. Under standard aseptic conditions, the femoral condyles were exposed via the medial parapatellar approach, and holes for the implants were made using a broach with cutting surfaces and a dental burr. After an undersized trial, the test pieces were inserted by tapping in the state of press-fit to achieve initial implant stability into the weight-bearing portion of the medial and lateral femoral condyles. In each animal, both hind legs were operated on during the same operation so that the animal would be obliged to stand on the operated legs. As shown in Fig. 2, these test pieces, including defects left open, were implanted in various combinations. No postoperative external fixation was performed. Dogs were killed 4, 8, and 24 weeks after surgery. Both femurs were extracted and examined by radiography. Maintenance of these animals and the animal experiments were carried out according to the guidelines for animal experiments of Kyoto University in Japan.

Histological Examination

All the specimens retrieved from the animals were examined using radiography, and all segments of the femur containing implants were then excised

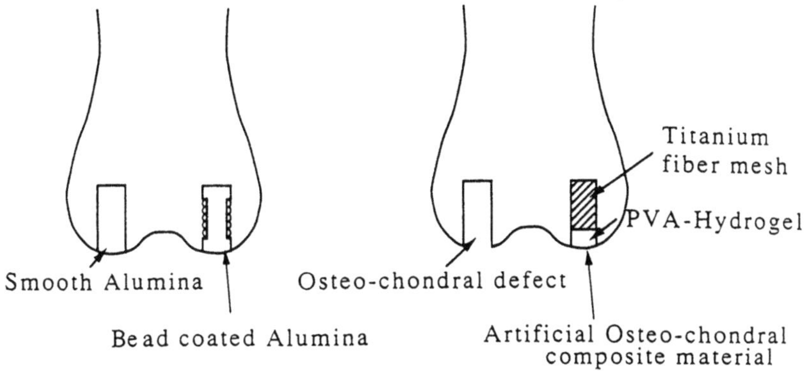

Fig. 2. Three kinds of materials were implanted into the load-bearing portion of the femoral condyles of dogs in various combinations. *PVA*, polyvinyl alcohol

and fixed in 10% phosphate-buffered formalin solution for histologic processing. Samples were dehydrated in serial concentrations of ethanol and embedded in polyester resin. Undecalcified sections of tibia including the implants were cut and then ground into thicknesses ranging from 100 to 150 µm. Giemsa surface staining and contact microradiographic examination of the same cut section were performed to distinguish the new bone formation around the implant surface and the fibrous tissue interface. Histological changes of tibial joint surfaces, including menisci, were examined using thionin staining [9]. By the thionin staining method, proteoglycan could be stained specifically also in undecalcified sections.

Finite-Element Analysis

The COSMOS/M from SRAC (Structural Research and Analysis, Santa Monica, CA, USA) was used for the quasi-three-dimensional FEA. Hardware utilized was the PC-9801RA from NEC (Tokyo, Japan) [10]. Considering the symmetry of the implanted portion, three-dimensional FEA models of only one-fourth of the femoral condyle, including materials, were created (Fig. 3). In this model, a thin layer of fibrous tissue (0.1 mm) was assumed to exist in the interface between the smooth-surfaced alumina and the cancellous bones, while the bead-coated surface was covered with new bone growth (Fig. 4). The properties of the materials used in the analysis are listed in Table 1.

The loading conditions were determined using the free body diagram shown in Fig. 5. According to this technique, the resultant force on the knees of the dogs was calculated to be 150 N, which could represent twice the body weight. Based on these models and parameters, induced stress and strain distribution in the cancellous bones adjacent to the implants were calculated.

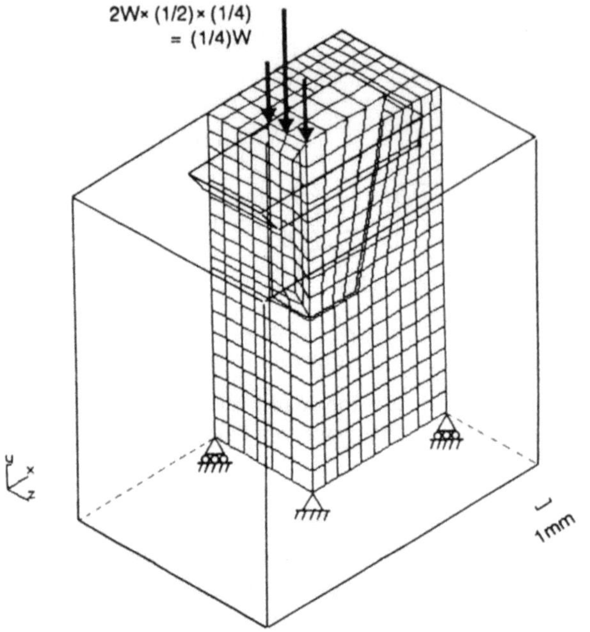

Fig. 3. Finite-element mesh of the implants and cancellous bones (1276 elements, 1779 nodal points)

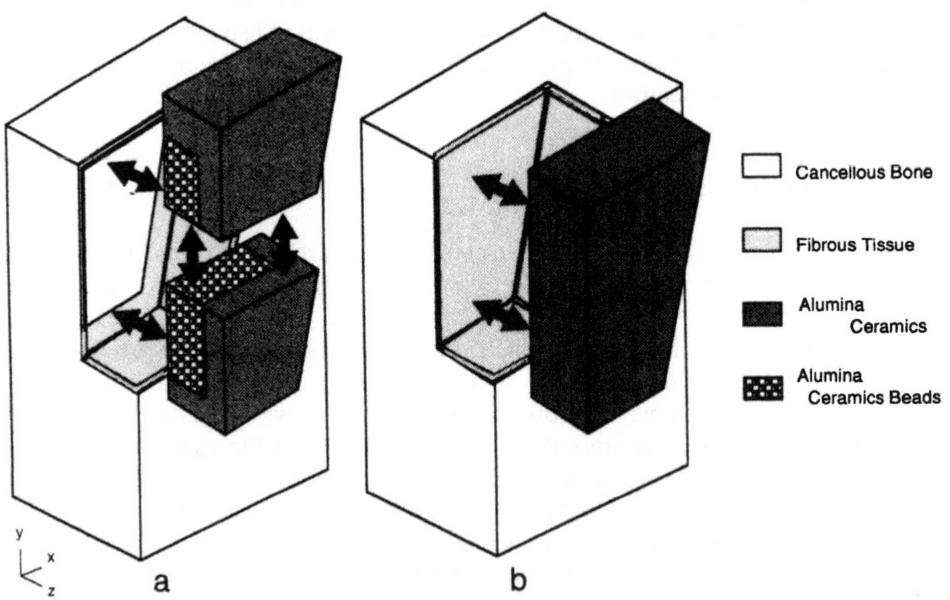

Fig. 4a,b. Models of bead-coated and uncoated implants show regions with different material properties. **a** Bead-coated implant model. **b** Uncoated implant model

Table 1. Parameter values of material properties

Materials	Young's modulus	Poisson's ratio
Cancellous bone	300 MPa	0.30
Fibrous tissue	5 MPa	0.49
Alumina ceramics	350 GPA	0.19
Alumina ceramic beads	200 GPa	0.19
Truss element	0.01 MPa	0.30

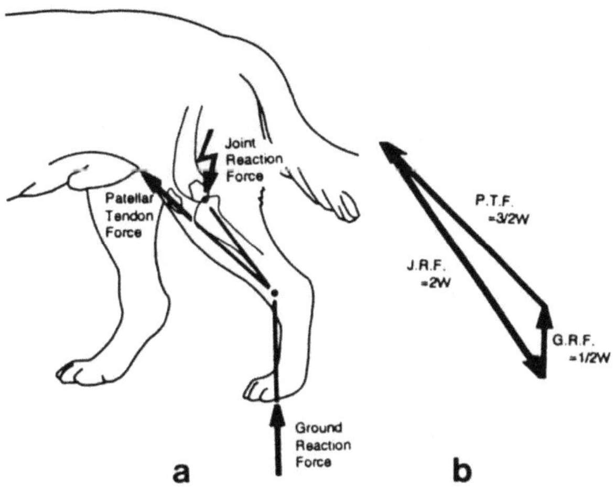

Fig. 5a,b. Schematic of loads on the implants. **a** Free body diagram. **b** Triangle of vector forces

Results

Clinical Observations

All the animals were able to bear their body weight upon standing within 1 week, and were allowed to walk outside their cages 2 weeks after the operation. No infection was observed when the animals were killed. The range of motion of the knee joints was nearly normal.

Radiographic Findings

Four and 8 weeks after implantation, no marked change was observed in the radiographs of bead-coated implants or those of uncoated implants. After 24 weeks, radiographs of the bead-coated implants showed good fixation. However, those of uncoated implants showed a radiolucent zone around the implant and radiopaque changes in the surrounding bone (Fig. 6). No findings such as sinking or loose beads were noted in any animal.

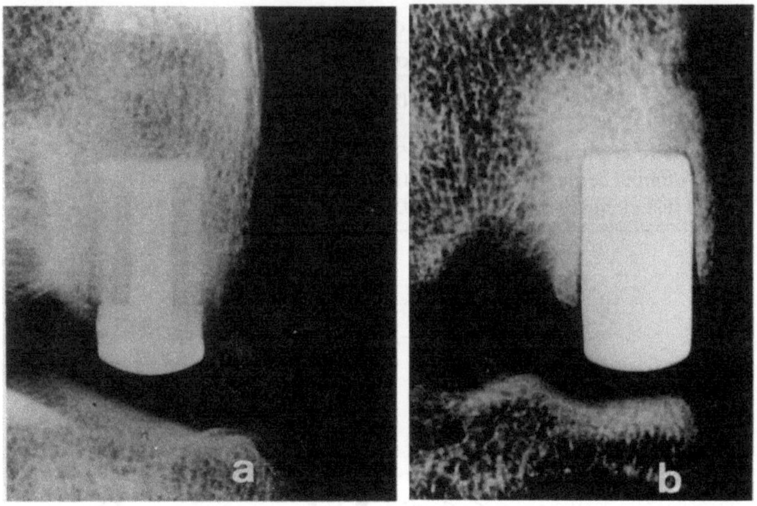

Fig. 6a,b. Postmortem radiographs of a knee joint 24 weeks after implantation. **a** Bead-coated implant. **b** Uncoated implant. Note radiolucent zone and sclerotic bone formation surrounding the uncoated implant

Histological Findings

In the sections of bead-coated implants, irregular ingrowth of immature bone was observed in pores 4 weeks after implantation; some of the ingrowth reached the lowest layer. After 8 weeks, mature bone that had partially remodeled along the surface of the beads was noted, and bone ingrowth was completed as far as the bottom of the beads (Fig. 7). After 24 weeks, remodeling of trabecular bone was further advanced. At the bone–implant interface, thickening of trabecular bone in a lamellar pattern and direct contact in parts where fibrous tissue was not interposed were observed. On the other hand, the uncoated implants were surrounded by immature bone 4 and 8 weeks after implantation (Fig. 8). However, after 24 weeks, the radiolucent zone and surrounding radiopaque changes observed on radiographs became more apparent, and the thickness of the fibrous layer at the bone–implant interface increased with time.

Newly formed bone surrounding titanium fibers was found 8 weeks after the implantation (Fig. 9). With higher magnification, mature bone ingrowth into deep pore spaces could be seen (Fig. 10). Although no direct bonding between material and artificial articular cartilage (PVA hydrogel) was found, composite material was firmly attached to neighboring bones because of the bone ingrowth.

Changes of the Tibial Joint Surface

By visual observation, notable ulceration was observed in the tibial articular surface of dogs in which alumina with or without bead coating was implanted

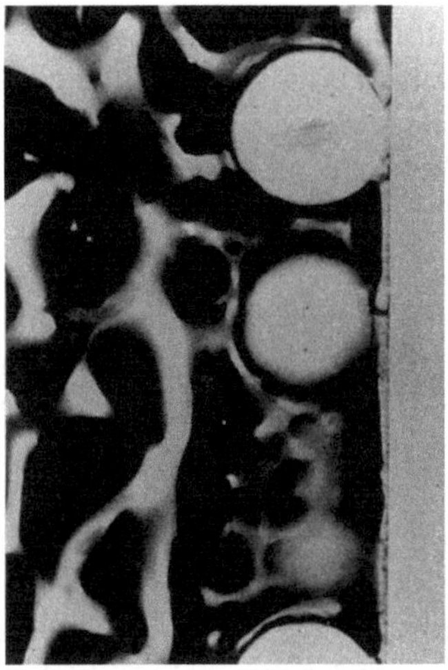

7

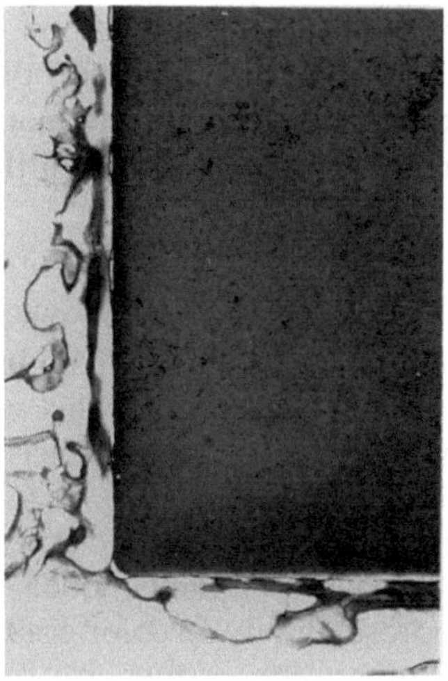

8

Fig. 7. Histological appearance of the bead-coated implant 8 weeks after implantation. Thickening of mature bone in lamellar pattern surrounds beads. Contact microradiograph; original magnification ×40

Fig. 8. Histological appearance of the uncoated implant 8 weeks after implantation. Interface between bone and implant is clearly demarcated by immature fibrous tissue. Giemsa surface staining, original magnification ×20

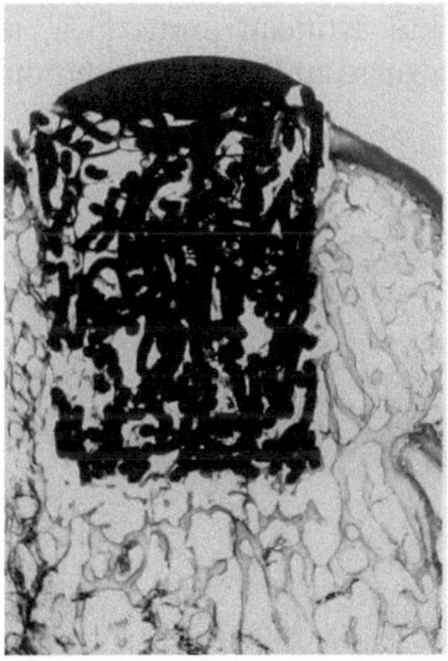

9

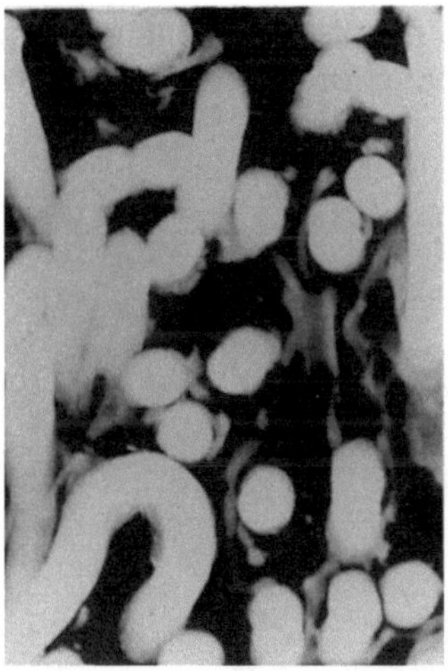

10

Fig. 9. Histological appearance of artificial osteochondral composite material 8 weeks after implantation. Dark blunt portion on top of the titanium fiber is the artificial cartilage, PVA hydrogel, which shows firm attachment to the titanium fiber. Giemsa surface staining, original magnification ×2

Fig. 10. Mature bone that has grown into the pores of the titanium fiber 8 weeks after implantation. Contact microradiography, original magnification ×40

(Fig. 11). Histological examinations by thionin staining revealed the loss of cartilage. On the other hand, the tibial articular surface placed against the artificial osteochondral composite materials showed no pathological changes at all. This phenomenon was shown in the histological specimen (Fig. 12).

Stress and Strain Distribution

By using FEA, eight mechanical parameters such as minimal principal stress, von Mises equivalent stress, maximum shear strain, and strain energy density (SED), etc., were calculated in this study. Only the minimal principal stress is shown, however, because there was not much difference among the other parameters.

The minimal principal stress distribution in the bones adjacent to the bead-coated and uncoated implant models are shown in Fig. 13. Stress values were originally represented by 16 color gradations. However, because of the difficulty of printing color, these values are represented here in Fig. 13 by black-and-white gradations. In the bead-coated model, an average stress of (−)0.48 MPa was demonstrated around cancellous bones, while (−)0.24 MPa was seen around the uncoated model. This stress value is about one-half of the bead-coated model, although the stress at the bottom of the uncoated implant was about 1.5 times larger than that of the bead-coated model.

Discussion

Because of the development of cementless artificial joints [11], much more interest in bone formation and remodeling surrounding implanted materials has

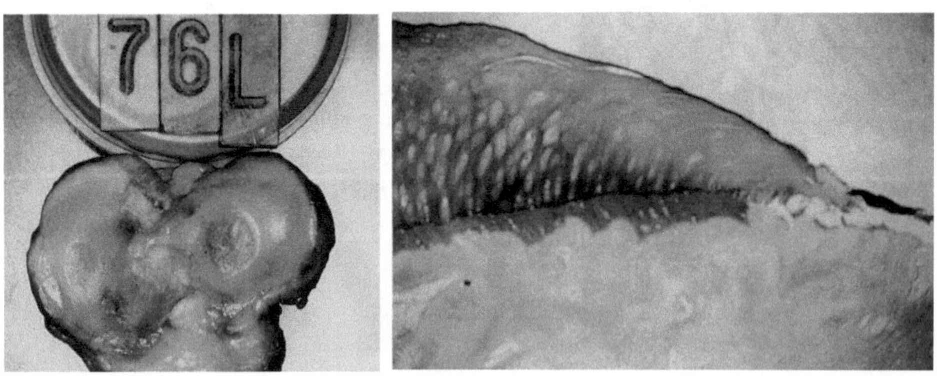

a b

Fig. 11a,b. Tibial joint surface against alumina implant 8 weeks after surgery. **a** Macroscopic finding shows remarkable ulceration of cartilage. **b** Histological appearance, thionin staining; original magnification ×100. Right half of articular cartilage was lost, and stainability of proteoglycan by thionin decreased in the surrounding articular cartilage

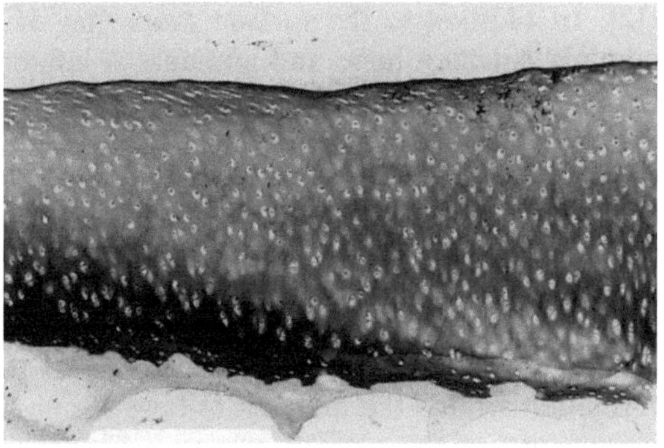

Fig. 12. Tibial articular surface against artificial osteochondral composite material 8 weeks after implantation. Cartilage structure is almost normal and stains well with thionin. Thionin, ×100

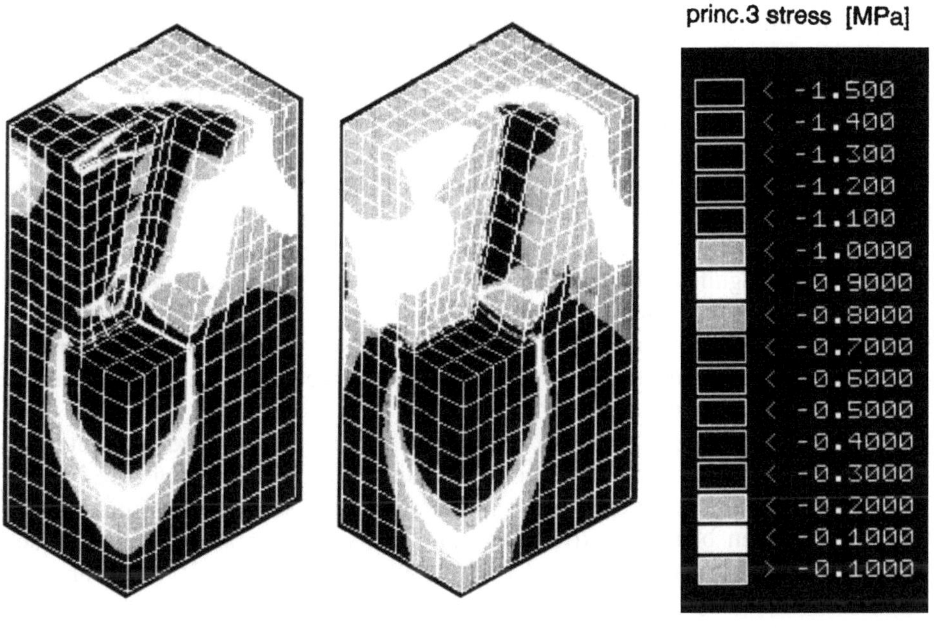

Fig. 13. Minimal principal stress distribution in the bones adjacent to implant model

arisen. In cementless artificial joints, quick, firm attachment of the implants to the bones is desirable. For that purpose, various methods, such as porous treatment of the material surface including coating with bioactive ceramics, have been proposed [12].

It is now generally accepted that the optimal pore size is approximately 150–250 μm and that coating with bioactive ceramics promotes bone ingrowth

[12]. In addition to the surface properties of the materials, we think that the bonding between bone and implants is influenced by the osteoconductivity of the implants and the osteogenic potency of the tissues around implants.

In another study (Chang et al., to be published), we investigated the difference of bone formation depending on the kinds of materials and on the anatomical site of implantation. In short, when implanting various kinds of ceramics transcortically into rabbit tibia, the greatest bone formation was found adjacent to the periosteum. The second and third most common bone formation was seen adjacent to the endosteum and in bone marrow, respectively. At an early stage, which is up to 6 weeks after surgery, the bone formation around the material follows the biological process of fracture healing irrespective of the type of material. However, the newly formed bone thereafter showed various remodeling patterns, depending on the mechanical stimuli according to Wolff's law. For example, the bone formed in the bone marrow showed a tendency to be resorbed as time passed. The experiments just described were, however, performed under nonload-bearing conditions. To elucidate bone remodeling under mechanical stimuli, we think the materials should be implanted under load-bearing condition.

Takagi et al. [3] implanted alumina test pieces with or without bead-coating into the load-bearing portions of medial and lateral femoral condyles of dogs. They observed that the uncoated implants showed signs of loosening similar to those observed clinically. In contrast, the bonding strength of the bead-coated implants was significantly greater than that of the uncoated implants, and increased with time after implantation. Results of histological examination suggested that increase in bonding stress was caused by thickening of ingrown trabeculae, which could be observed as early as 4 weeks after implantation, at the bone–implant interface.

According to Takagi's method, three kind of test pieces were implanted in our study. From the aspect of bone remodeling, the X-ray finding (see Fig. 6) 24 weeks after implantation is very interesting because of the prominent differences between bead-coated and smooth implants. The particularly characteristic remodeling pattern around the smooth implants, that is, a radiolucent zone surrounded by radiopaque changes, is very interesting, because similar findings are often clinically observed in the case of fracture ends of nonunion and also in the subchondral bone plate.

FEA has recently also been applied to the bone remodeling process. An example is bone loss around a femoral prosthesis that was described by stress shielding [13,14]. It is generally believed that the bone resorption results from stress shielding and adaptive bone remodeling according to Wolff's law. Expecting that the remodeling process described in the previous paragraph could be analyzed mechanically, we attempted to simulate these conditions numerically using a new FEA program.

As shown in Fig. 13, stress distribution around the bead-coated implant was similar to the bone density found in the X-ray (see Fig. 6). In this case the stress meant minimal principal stress (compressive), but four other parameters of the stress showed little difference. This result seems to support Pauwels' remark [4] that bone is formed and remodeled in adaptation to applied stress intensity, irrespective of the kind of stress such as compression, tension, and shear.

Although stress distribution around bead-coated alumina could simulate the bone density, the remodeling pattern around the uncoated implant could not be described by the stress. Minimal principal stresses just adjacent to the uncoated implant were one-half of those around the bead-coated material. This could explain the radiolucent zone, that is, fibrous tissues, to some extent. However, the surrounding radiopaque zone could not be described by the results of stress distribution. Therefore, we paid more attention to strain value around the implant. As shown in Fig. 14, fibrous tissues that formed around the uncoated implant showed a much higher value of mean principal three strain and mean shear strain energy density (SED). The mean shear SED, which was to be the most influential parameter on bone formation, was proposed by Huiskes et al. [13,15]. They also proposed that the interposed fibrous tissues caused excessive stress concentration in surrounding bones [15]. It is of course necessary to investigate further the strain distribution surrounding implants, which may give us a possible explanation for the corticalized bone formation around fibrous tissues. We now believe that there may be some

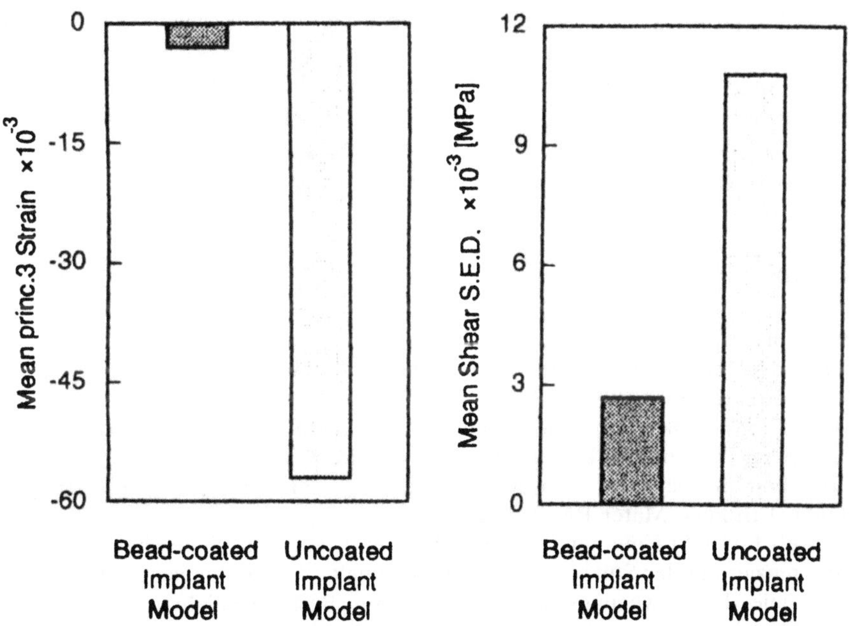

Fig. 14. Comparison of strain values between bead-coated model (bone) and uncoated model (fibrous tissue). Mean principal three strain, (−) value; mean shear SED, (+) value

mechanical sensor in the fibrous tissues, and that this sensor may promote bone formation if the strain exceeds a certain limit (threshold).

Subchondral cancellous bones remodel themselves constantly according to Wolff's law, and transmit the load most effectively to the distal bone structures. In current artificial joints, however, all these important structures including healthy portions, are resected. This technique causes loss of the shock-absorbing effect as well as of bone stock.

For the purpose of preserving the cancellous bones as much as possible, we have attempted to develop an artificial articular cartilage, PVA-H, which is a rubber-like gel [5,6]. With respect to some of the mechanical properties of this material for an artificial articular cartilage, such as lubricating and shock-absorbing functions, we have obtained very encouraging results, which are described in detail in other papers [5–7]. The important problem that still remains is how to attach the material to the bone firmly. This problem is solved by making an artificial osteochondral composite material, as has been investigated in this study. As shown in Figs. 9 and 10, sufficient bone ingrowth into the pores of titanium fiber mesh was observed 8 weeks after the implantation. Neither a demarcated nor a radiolucent zone around the material was observed. Because the wear-resistant properties of PVA hydrogel have been improved strikingly through γ-radiation, this composite material seems to be a very promising joint material.

Finally, turning the discussion to the changes in the tibia joint surface, remarkable pathological changes were observed in the case of the alumina implant. On the other hand, the tibial joint surface remained intact when placed against the artificial articular cartilage. These results showed good biocompatibility of PVA hydrogel as an artificial articular cartilage that could be used for surface replacement arthroplasty. Thus, artificial osteochondral composite material has now become a very promising material.

References

1. Nakamura T, Yamamuro T, Higashi S, Kokubo T, Ltoo S (1985) A new glass-ceramic for bone replacement: evaluation of its bonding to bone tissue. J Biomed Mater Res 19:685–698
2. Wolff J (1892) Das Gesetz der Transformation des Knochen. Hirsehwald, Berlin
3. Takagi H, Yamamuro T, Hyakuna K, Nakamura T, Kotoura Y, Oka M (1989) Bone bonding behavior of bead-coated alumina ceramic under load-bearing condition. J Biomed Mater Res 23:161–181
4. Pauwels F (1965) Eine neue Theorie über den Einfluss mechanischer Reize auf die Differenzierung der Stützgewebe. In: Gesammelte Abhandlungen zur funktionellen Anatomie des Bewegungsapparates. Springer, Berlin Heidelberg New York, pp 481–518
5. Oka M, Noguchi T, Kumar P, Lkeuchi K, Yamamuro T, Hyon SH, Ikada Y (1990) Development of an artificial articular cartilage. Clin Mater 6:361–381

6. Noguchi T, Yamamuro T, Oka M, Kotoura Y, Hyon SH, Ikada Y (1991) Poly(vinyl alcohol) hydrogel as an artificial articular cartilage: evaluation of biocompatibility. J Appl Biomater 2:101–107
7. Oka M, Ikeuchi K, Tsutsumi S, Yamamuro T, Noguchi T, Nakamura T (1992) Biomechanics of natural and artificial joints. In: Niwa S, Perren SM, Hattori T (eds) Biomechanics in orthopedics. Springer, Berlin Heidelberg New York Tokyo, pp 282–298
8. Ikeuchi K, Oka M (1988) A study of cartilage deformation with squeeze film effect. In: Tsuruta T (ed) Transactions of 3rd world biomaterials congress. Business Center for Academic Societies, Tokyo, p 270
9. Bulstra SK, Drukker J, Kuijer R, Buurman WA, van der Linden AJ (1993) Thionin staining of paraffin and plastic embedded sections of cartilage. Biotech & Histol 68:20–28
10. Oka M (1993) Load bearing mechanisms of natural and artificial joints. In: Imura S (ed) Hip biomechanics. Springer, Berlin Heidelberg New York Tokyo, pp 255–263
11. Engh CA, Griffin WL, Marx CL (1990) Cementless acetabular components. J Bone Joint Surg B-72:53–59
12. Tompkins G, Rosenberg A (1993) Porous ingrowth and implant fixation. Semin Arthroplasty 4:223–237
13. Huiskes R, Hollister S (1993) From structure to process, from organ to cell: recent developments of FE analysis in orthopedic biomechanics. J Biomed Eng 115:520–527
14. Weinans H, Huiskes R, Grootenboer H (1992) Effects of material properties of femoral hip components on bone remodeling. J Orthop Res 10:845–853
15. Weinans H, Huiskes R, Grootenboer J (1990) Trends of mechanical consequences and modeling of a fibrous membrane around femoral hip prostheses. J Biomech 23:991–1000

The Energy-Absorbing Function of Cancellous Bone and Its Influence on the Loosening of Artificial Joints

TAKASHI HAYAMI[1], MASANORI OKA[2], KEN IKEUCHI[2], TAKASHI NAKAMURA[2], HIROYUKI HOJO[3], and KAZUHIKO SAKAGUCHI[3]

Summary. Deformation behavior of trabecular architecture (cancellous structure) was visualized using an ultrahigh-speed video recording camera during both static and impact compression tests of subchondral cancellous bone at nominal strain rates from 0.008 to $25\,s^{-1}$. The mechanical energy-absorbing mechanisms of cancellous bone could be roughly classified into two types: viscoelastic hysteresis and release of elastic energy from microfractures of the trabeculae. The energy-absorbing ability of degenerated cancellous bone was approximately 50% that of normal bone. The osteoarthritic process results in a change in the cancellous structures such that a pronounced fabric direction is not evident. In the case of normal cancellous bone, the structural stiffness of the bone was maintained by each trabecula functioning as a long column in the cancellous structure. It was confirmed visually that, for the in vitro case, aseptic loosening of the prosthesis was caused by a decrease in the cancellous bone stock or reduction in the energy-absorbing ability of the joints. Thus, the energy-absorbing properties of subchondral cancellous bone were dependent on the morphology and deformability of the cancellous structure.

Key words: Cancellous bone—Energy-absorbing mechanism—Trabecula—Degeneration—Impact deformation—Loosening

Introduction

Among the bony structures in a joint to have a load-bearing function, subchondral cancellous bone has an important responsibility in dissipation and

[1] Department of Mechanical Systems Engineering, Kinki University, 1 Takayaumenobe, Higashihiroshima, Hiroshima, 729-17 Japan
[2] Research Center for Biomedical Engineering, Kyoto University, 53 Kawahara, Syogoin, Sakyo-ku, Kyoto, 606 Japan
[3] Department of Mechanical Engineering, Doshisha University, 1-3 Tataramiyakodani, Tanabe, Tsuzuki-gun, Kyoto, 610-03 Japan

distribution of mechanical energy [1,2]. Radin et al. [3,4] studied the relationship between impact loading on a joint and the degeneration of the subchondral cancellous bone. Serink et al. [5] reported on the relationship between the deformation characteristics of trabeculae and osteoarthritis. Although their conclusions differed on the development and progression of joint degeneration, they agree on the shock-absorbing mechanisms being closely related to the deformation behavior of the trabeculae. Miyanaga [6] compared normal and degenerated cancellous bone and found a significant difference in both the dynamic elastic modulus and internal damping. Such investigations of the biomechanical properties of cancellous bone not only clarify the development and progression of joint degeneration, but also suggest etiologies for aseptic loosening that may be the most serious and ongoing complication in current artificial joints. However, there has been very limited discussion of the influence of structural morphology, intertrabecular fluid, and the impact deformation behavior of cancellous bone [7,8]. In addition, there are few studies on the relationship between the apparent elastic modulus of cancellous bone and the intraosseous pressure of the intertrabecular fluid [6, 9].

To elucidate the energy dissipation mechanisms of cancellous bone, static and impact compression tests were performed on normal and degenerated cancellous bone and the implant model. During the impact tests, the deformation behavior of cancellous bone was recorded using an ultra high-speed video recording camera.

Materials and Methods

Specimens

Human subchondral cancellous bone specimens were obtained from femoral heads that were surgically excised for transcervical fracture. The femoral heads were divided into several portions in the frontal plane as shown in Fig. 1. The trabecular trajectory of the principal compression group (the principal compressive direction) was determined from radiographs. The cubic specimens, 6.5 mm on a side, that were removed had the vertical axis aligned with the principal compressive direction [10,11]. Similarly, cubic specimens were obtained from femoral heads that displayed advanced osteoarthritis, as determined by the presence of bone cysts and osteophyte formation on radiographic examination. Impact test specimens of normal cancellous bone were also obtained from a human proximal tibia that had been amputated after a motor vehicle accident. To simulate aseptic loosening of a prosthesis, ultrahigh molecular weight polyethylene (UHMWPE) was bonded into a sample of normal femoral head cancellous bone with bone cement (polymethyl methacrylate, PMMA) (Fig. 2). The outer surface of the UHMWPE had V-shaped grooves approximately 0.25 mm deep with a pitch of approximately 1 mm to improve adhesion to the PMMA.

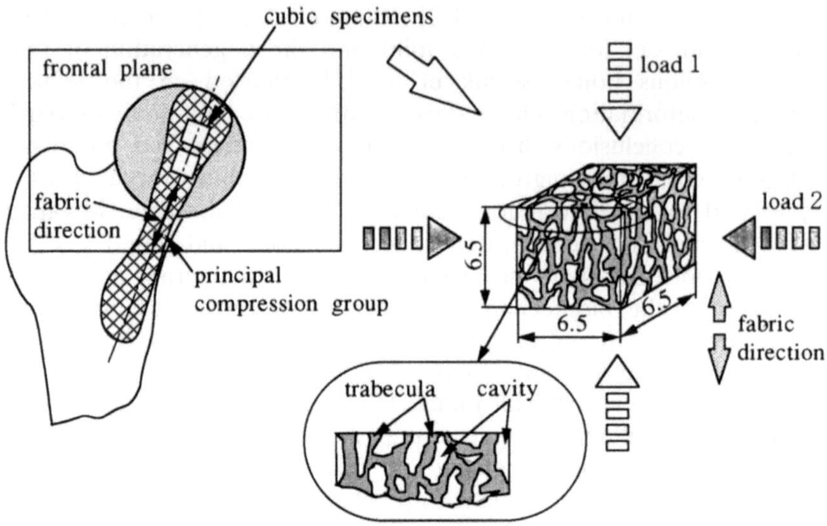

Fig. 1. Geometry of the cancellous bone specimens. More than 17 normal specimens and 13 degenerated specimens were excised from human femoral heads

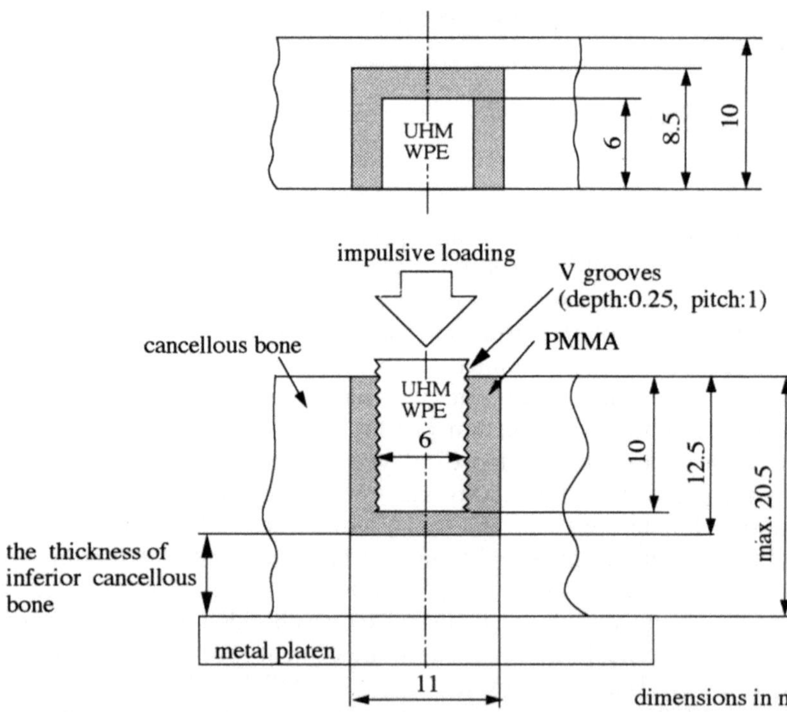

Fig. 2. Specimen configuration to model loosening of a prosthesis. Impact energy was 0.32 J

Static Compression Test

Compression tests were performed using a servodriven-type testing system [11] that compensates for the machine compliance. Each specimen was compression tested both in the principal compressive direction of the cancellous structure (load 1) and in its normal direction (load 2) (as shown in Fig. 1). A thin oil film was coated on the metal platen surface and a static preload of 8 N applied to the specimens. All specimens were loaded to a strain of 0.025 at one of five nominal strain rates: 0.008, 0.01, 0.015, 0.02, and $0.025\,s^{-1}$.

The stress–strain curves were plotted for three loading-unloading cycles. The first cycle was used to precondition the specimen, and the dissipated mechanical energy (energy loss) was determined from the areas enclosed by stress–strain curves for the second and third cycles. The deformation behavior of the cancellous structure was also recorded using a charge coupled device (CCD) camera.

Impact Compression Test

For the impact compression tests, a drop weight impact tester [12] was used that permits an 850-g mass to drop from a height of several centimeters. A nominal impact strain of 0.04 was applied at nominal strain rates from 18 to $25\,s^{-1}$.

An ultrahigh-speed video camera (FASTCAM, Photron, Tokyo) was used to record the impact deformation behavior of the cancellous structure. The images were converted, using a computer processing system, into binary images for analysis of the instantaneous deformation properties. In other words, the maximum widths were determined by projecting the specimen and cavity profiles on the vertical plane (Feret's diameter on the vertical plane). The ratio of Feret's diameter (the deformation ratio) was calculated as the ratio of the predeformation to postdeformation projected width.

For the impact test on the implant model, the thickness of the cancellous bone on the lower surface of the specimens was removed in fixed increments of 1 mm. A constant impact energy of 0.32 J was applied to the upper surface of the UHMWPE only. The average crack opening displacement (COD), the crack width, was determined for cracks or separations at the material boundaries.

Results

Static Compression Behavior

A typical set of stress–strain curves from the compression test for a normal cancellous bone in the principal compressive direction is shown in Fig. 3. Stress–strain curves for all specimens were nonlinear and convex on the lower side.

The relationship between energy loss and strain rate, calculated as shown in Fig. 3, is presented in Fig. 4. The energy loss increased linearly with strain

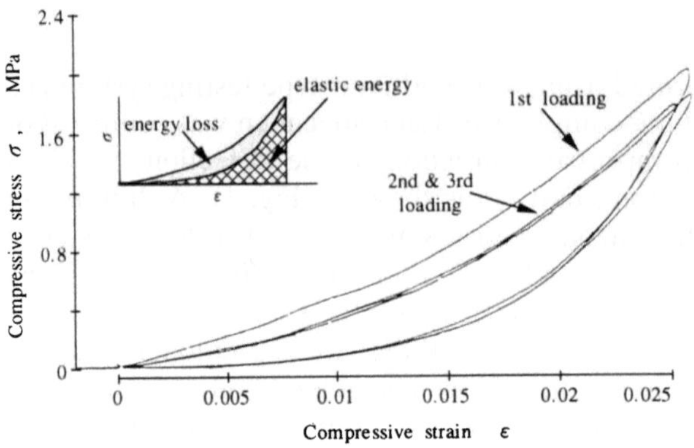

Fig. 3. Example set of stress–strain curves for a normal cancellous bone specimen at nominal strain rate of $0.01\,s^{-1}$. A planimeter was used to measure the area of the loops

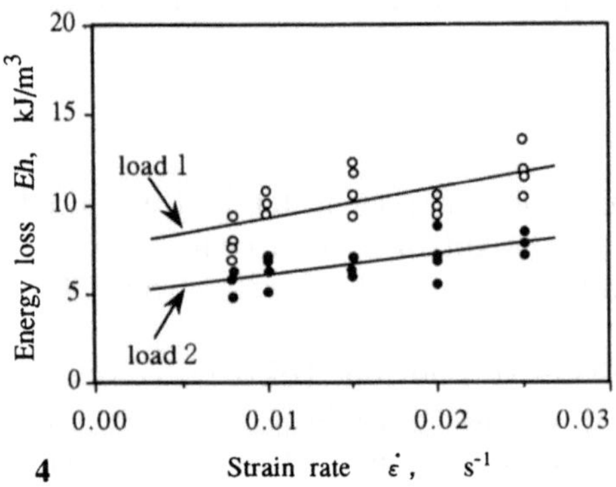

4

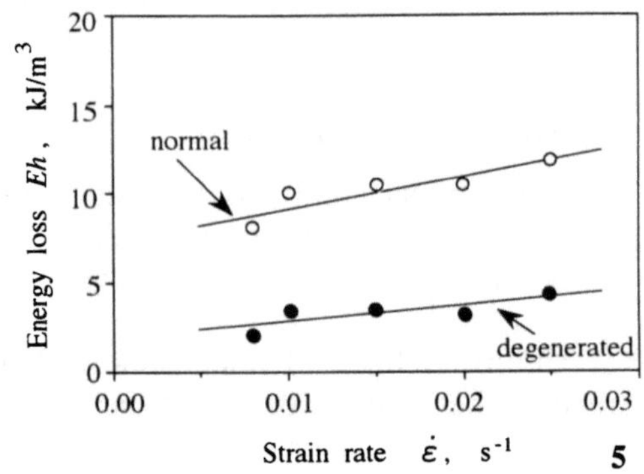

5

Fig. 4. Dissipated energy for normal cancellous bone

Fig. 5. Dissipated energy for normal and degenerated cancellous bone. Plotted data were the average of more than five compressive tests on different specimens at each strain rate

rate. No significant influence of fabric direction in the increase of energy loss was found. Energy loss in the principal compressive direction was about 1.5 fold that in the direction normal to the principal compressive direction.

Figure 5 shows the energy losses for compression tests in the principal compressive direction for both normal and degenerated cancellous bone. Energy loss for degenerated cancellous bone was less than 50% that of normal cancellous bone.

The compressive deformation behavior of the normal cancellous structure in the principal compressive direction is shown in Fig. 6. The compressive strain was maintained until one of the trabeculae shown in the video image fractured. Figure 7 shows the stress–strain curve that was recorded during the test shown in Fig. 6. The trabecula started to deflect to the left when the nominal strain was approximately 0.13 (Fig. 6b). The trabecula fractured and separated at the

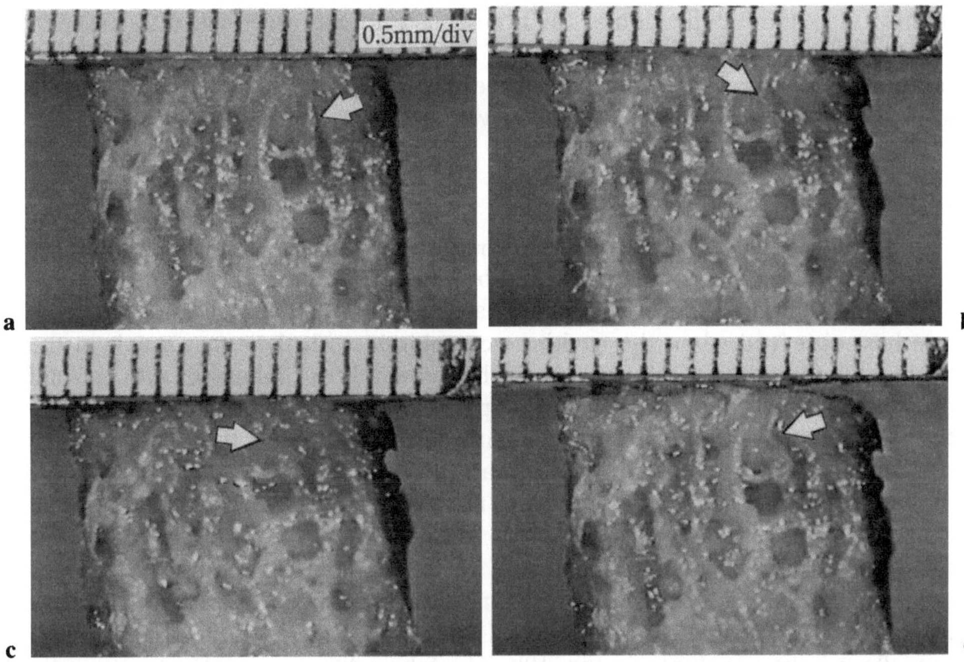

Fig. 6a–d. Deformation of normal cancellous bone recorded using a CCD video camera. Specimen was loaded in the vertical direction. **a** Before deformation. **b** Nominal strain of 0.08. **c** Nominal strain of 0.12. **d** After removal of the load

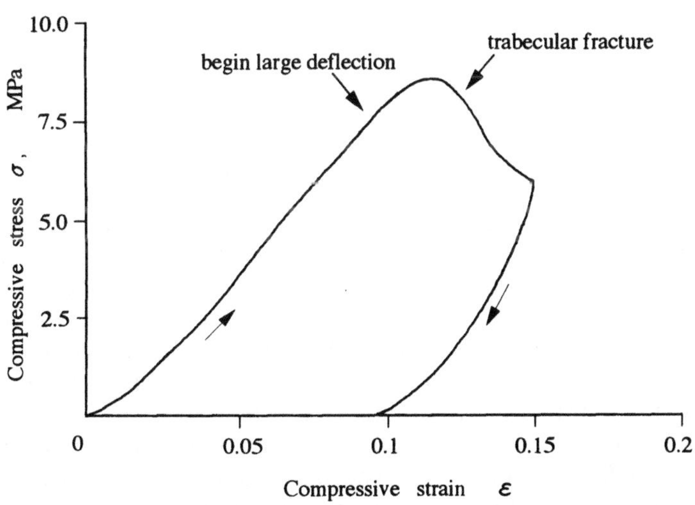

Fig. 7. Stress–strain curve showing fracture of a trabecula at nominal strain rate of $0.01\,s^{-1}$

point of maximum deflection as the nominal strain was reaching 0.15 (Fig. 6c). As shown in Fig. 7, after fracture of the trabecula the compressive stress rapidly decreased. Fracture of the trabecula results in a large increase in the area enclosed by loading-unloading curves.

Impact Compression Behavior

The instantaneous compressive deformation of the cancellous structure is shown in Fig. 8. The cavities enclosed by trabeculae were observed to deform even during impact (high speed; strain rate of $20\,s^{-1}$) loading. The deformation was best observed when the cavities were oblong with their major axis in the horizontal direction as shown. A further example is indicated with a solid arrow in Fig. 9. In this figure the while particles, indicated by open arrows, are the intertrabecular fluid within the cavities. Intertrabecular fluid exuded from the cavities during compressive deformation.

The ratios of Feret's diameter for the two loading directions in normal cancellous bone are shown in Fig. 10a. In the principal compressive direction, the time required for elastic recovery after the maximum deformation was approximately 0.7 ms, whereas approximately 2.7 ms was required for recovery in the direction normal to the principal direction. The results for degenerated cancellous bone are shown in Fig. 10b. The fabric direction of the structure did not affect the time required for elastic recovery.

The deformation of a single trabecula under impact loading is shown in Fig. 11. The trabecula has deformed by various means, the principal of which is buckling. A continuous increase in trabecular thickness can be observed. The complex deformation of the trabecula is principally caused by buckling, although the influence of bending is also evident.

After repeated impact loading (19 times), the trabecula shown in Fig. 11 was observed to crack (Fig. 12). With removal of the load, the deformation almost completely elastically recovered such that the crack is no longer visible.

Impact Deformation Behavior Around a Prosthesis

Figure 13 shows the instantaneous deformation of the implant model under the maximum strain when the thickness of the cancellous bone had been reduced by 2.5 mm. A large crack can be observed at the boundary between the cancellous bone and the PMMA (arrow A). Separation can be also seen at the boundary of the UHMWPE from the PMMA (arrow B). Cracking and separation at the boundaries were not observed if the thickness of the cancellous bone below the UHMWPE was greater than 8 mm.

Figure 14 shows the change of COD versus elapsed time at both the C1 and C2 cracks (as defined in Fig. 13b). The maximum COD of crack C1 reached more than 1 mm, caused by its continuation to the bottom of the model (Figs. 14a and 13a). Crack C2, however, had only a small maximum COD, and its deformation velocity was approximately twice that of crack C1 (Fig. 14b).

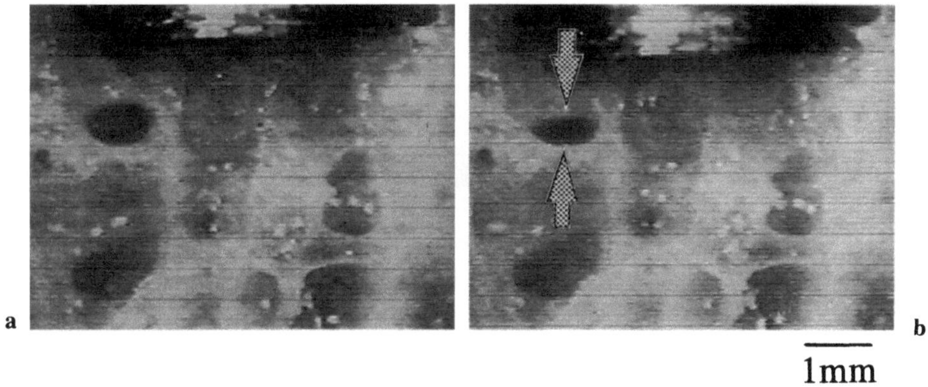

Fig. 8a,b. Instantaneous deformation of normal cancellous bone specimen from a human proximal tibia. **a** Before deformation. **b** At maximum deformation. Specimen was loaded in the vertical direction at nominal strain rate of $20\,s^{-1}$. Frame rate of video recording was 4500 frames per second (fps)

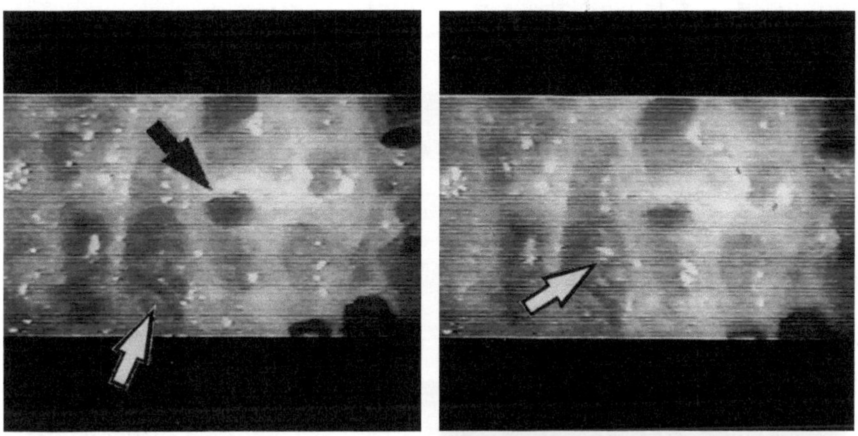

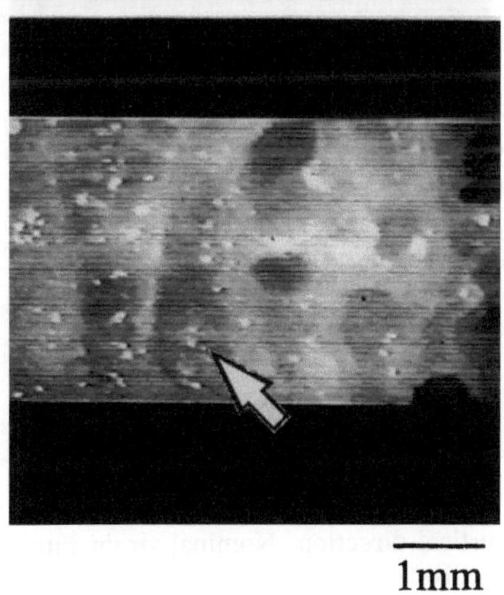

Fig. 9. Movement of intertrabecular fluid caused by impact deformation of the cavity. The *solid (black) arrow* marks a cavity, whereas the *open (white) arrows* mark the intertrabecular fluid within cavities. Nominal strain rate of $25\,s^{-1}$. Frame rate was 9000 fps

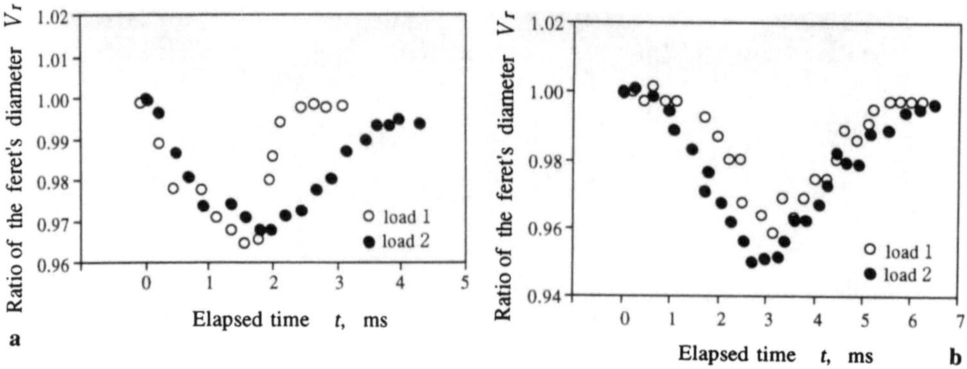

Fig. 10a,b. Variation with time of the compressive deformation ratios for normal (**a**) and degenerated (**b**) cancellous bone. Nominal strain rate of $20\,s^{-1}$

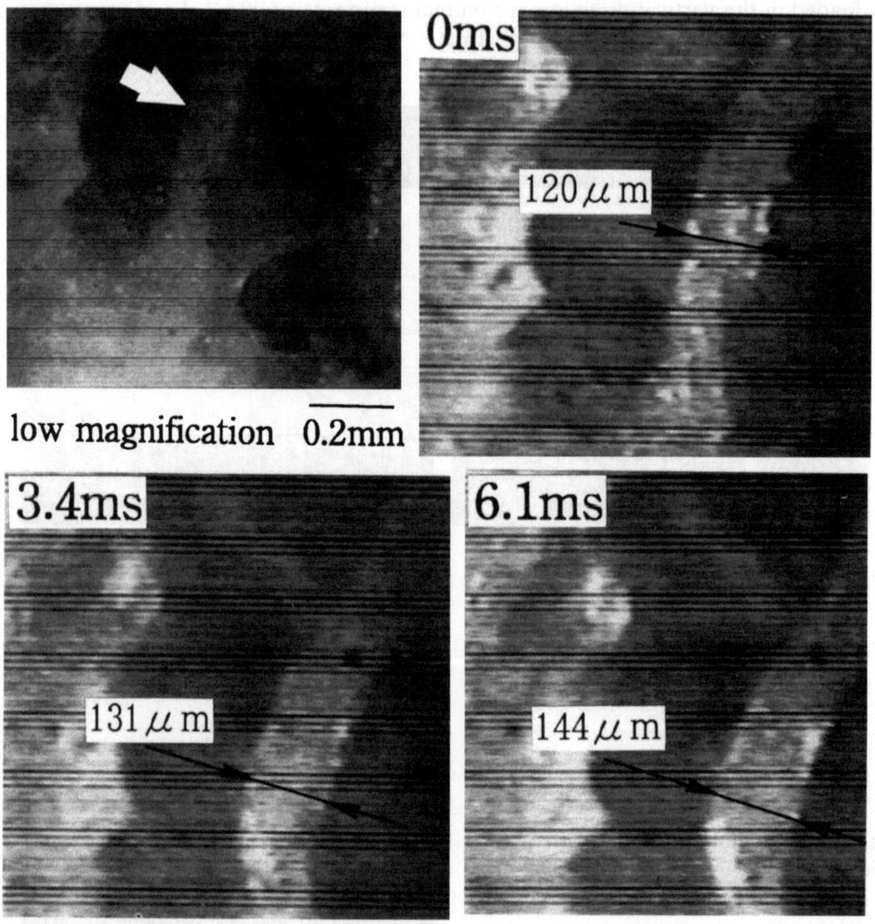

Fig. 11. Deformation of a trabecula caused by impact loading. Specimen was loaded in the vertical direction. Nominal strain rate of $18\,s^{-1}$. Resolution of image analyzer was 1.5 μm/pixel

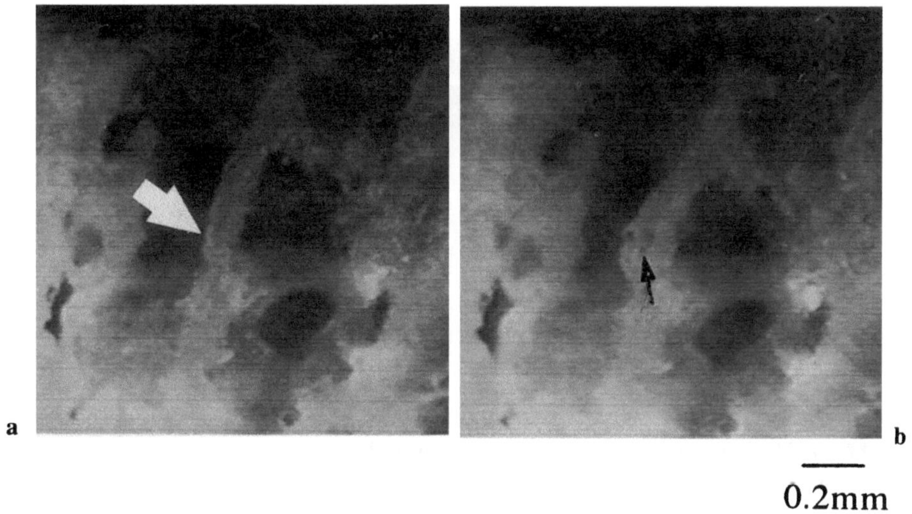

Fig. 12a,b. A fatigue crack in a trabecula after repeated impact loading (19 times). **a** Before deformation, the *white arrow* marks the largest deformation that occurred. **b** At maximum deformation, the *black arrow* marks a fatigue crack. Frame rate was 13 500 fps

Fig. 13a,b. Visualization of loosening in the implant model. **a** Original image. **b** Partial binary image. Resolution of image analyzer was 33 μm/pixel. Frame rate was 4500 fps

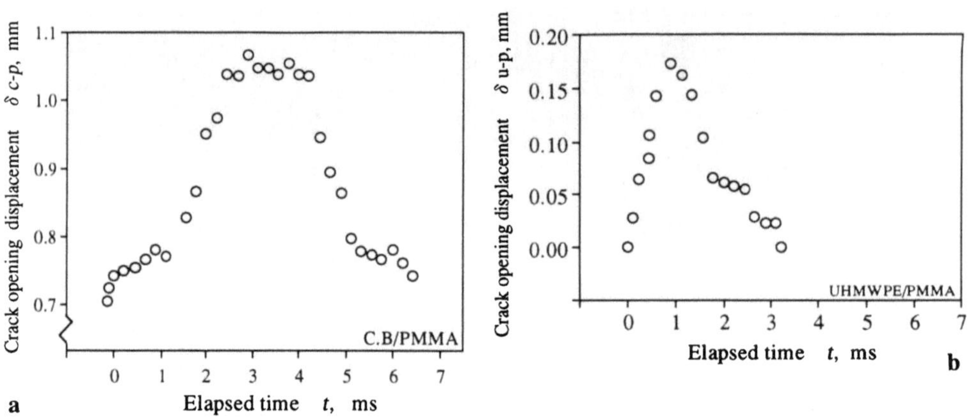

Fig. 14a,b. Variation with time of crack opening displacement (COD) of the crack at the boundary between the cancellous bone and the PMMA (**a**) and between the UHMWPE and the PMMA (**b**)

These CODs are relative displacement between the cancellous bone and the PMMA or between the UHMWPE and the PMMA. Translation displacement of the PMMA was observed; however, neither deformation nor fractures were evident. The cracks resulted from differences in the deformation of the materials.

Discussion

The Effect of Cancellous Structure on the Dissipation of Mechanical Energy

Viscoelastic hysteresis is one of the major mechanical energy dissipation mechanisms of cancellous bone [13] in that the force and displacement to be loaded cause the strain and strain rate within to not fracture the trabeculae. This energy dissipation mechanism results in the crescent-shaped hysteresis loop of the stress–strain curves (see Fig. 3).

Figure 15 shows the distribution ratio of energy loss and elastic energy in the total mechanical energy applied to the specimens for compression in the principal compressive direction. Degeneration-induced trabecular sclerosis may result in a decrease in the energy dissipation ability of the affected cancellous structure. It is highly probable that degeneration makes trabecula more brittle rather than elastic. This corresponds with the clinical findings that the energy-absorbing ability of degenerative joints is significantly lower than that of normal joints, and that embrittlement and bone sclerosis of cancellous bones become greater with the progression of osteoarthritis [14]. It can be expected that the main reason for this is change in the morphology of the cancellous

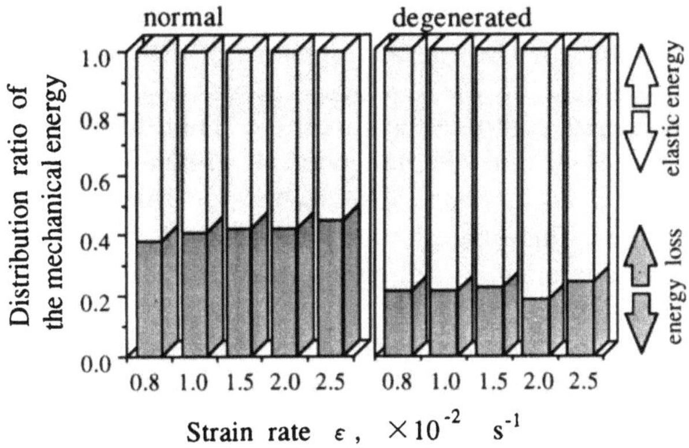

Fig. 15. Influence of cancellous bone quality on the mechanical energy distribution

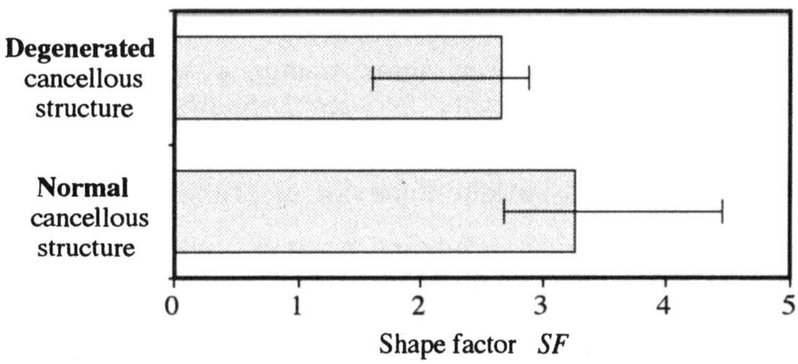

Fig. 16. Geometric differences in the cancellous structures between normal and degenerated cancellous bone. The shape factors (*SF*) for 94 cavities in normal bone and for 47 cavities in degenerated bone were calculated

structure. To confirm this hypothesis, the differences between the cancellous structure of normal and degenerated cancellous bone were quantitatively compared. The degree of roundness of the longitudinal cross section (along the principal compressive direction) of the cavities in the cancellous structure was measured. The shape factor *SF* is used to express the degree of roundness of cavities and is defined by the following equation:

$$SF = \frac{(Wmax)^2 \cdot \pi}{4 \cdot S}$$

where *Wmax* is the maximum width along the major axis and *S* is the sectional area of cavities in cancellous structure in the frontal plain. Thus, a shape factor of 1 implies completely round, whereas ellipsoids are described by a *SF* greater than 1. The cavities in the degenerative cancellous structure were more round than those in the normal cancellous structure (Fig. 16). As the increase in roundness of the cavities directly relates to the progression of osteoarthritis, this phenomenon therefore cannot be explained only from a mechanical standpoint. Degenerative change of the cancellous structure (an increase in the roundness of the cavities) results in a decreased structural stiffness in the principal compressive direction. Therefore the deformability of the structure increases, and the mechanical stimulation to the cancellous bone becomes larger. As a result, the trabeculae thickens and the cancellous structure is varied. This suggests that one reason for the loss of energy dissipation ability is the decrease in toughness that results from accelerating sclerosis of the trabeculae.

Figure 7 shows that the failure of a single trabecula significantly influences the mechanical property of a specimen and thus proves that individual trabeculae play an important mechanical role. Fracture of a trabecula is caused not only by compression but also by buckling (see Fig. 6). In the case of normal cancellous bone, because the trabeculae are oriented along the principal compressive direction they behave as a long plate in supporting the external loading. The large loop shown in Fig. 7 indicates that the energy dissipated by the release of stored elastic energy during fracture is very large. This dissipation mechanism is different from that of viscoelastic hysteresis.

Impact Deformation and Fatigue Behavior of Trabeculae

Several reports on the impact properties of natural joints or joint tissues have been published, but a few of them have used unrealistic physiological test conditions such as a strain rate that was too high.

Even at high but physiologically normal strain rates, the cancellous structure showed a degree of deformability (see Fig. 8). Such deformation of cancellous structure permits intertrabecular fluid flow (see Fig. 9). The intertrabecular fluid flows within the complicated structure of cancellous bone such that the pressure is reduced and the mechanical energy is dissipated [7]. This form of energy dissipation is different from the mechanisms described earlier here. In addition, it is expected that the mechanical energy dissipated by this mechanism is less than the energy absorbed by viscoelastic hysteresis or fracture of the trabeculae, because the permeability of cancellous bone is very large.

For normal cancellous bone the time required for elastic recovery varies depending on the fabric directions of the cancellous structure (see Fig. 10a). The reason for these differences may be in the different capacities of the specimens to store elastic energy depending on the fabric direction of the cancellous structure. As described, this results from the increased buckling resistance of the trabeculae in the principal compressive direction such that the stored elastic energy is greater than for deformation in other directions. For compression in the direction normal to the principal compressive direction, it can be assumed that the trabeculae are beams that are subjected to bending stress. When compressed to the same elastic strain, the trabeculae that are behaving as a long plate store more elastic energy, and elastic recovery occurs rapidly. On the other hand, if this logic is applied to the deformation of degenerated cancellous structure, where both the cavities are round and the fabric direction is unclear (see Fig. 16), the loading direction should have little influence on the velocity of elastic recovery. In practice, the influence of fabric direction was not confirmed experimentally (Fig. 10b).

Figure 11 shows the instantaneous elastic deformation of a trabecula that is behaving as a long column. Both buckling and bending deformation occurred for several milliseconds, and the trabecula thickened by as much as 20%. The crack shown in Fig. 12 appears as though the fascicles that constitute a trabecula are split by the repeated changes in trabecular thickness. This shows the crack is caused by fatigue failure. During the fatigue process of such trabeculae, many microfractures occur [4,8], and it would appear that the trabecular failure in the fatigue process causes the energy dissipation of cancellous bone.

The Influence of Cancellous Bone Removal in Total Hip Replacement Surgery on Aseptic Loosening of a Prosthesis

The primary reason for aseptic loosening that frequently occurs in current artificial joints is considered to be the reduction in shock-absorbing ability [1,15]. In other words, the loss of a large volume of cancellous bone that includes, healthy and normal bone and its subsequent replacement by artificial materials with limited energy-absorbing ability have a significant effect on the success of total hip replacement (THR) surgery.

In this study, loosening at the boundaries between the implant materials have been successfully visualized during impact tests on an implant model (see Fig. 13). This phenomenon occurred because of the reduction in the thickness (apparent volume) of inferior cancellous bone and clarifies that the mechanical cause of aseptic loosening is the reduction of the energy dissipation ability caused by the decrease in cancellous bone stock. From the results for the COD shown in Fig. 14, it appears the crack results from deformation of the cancellous bone and UHMWPE but not of the PMMA. Therefore, the differences between the COD and the deformation speed for cracks at the two boundaries resulted because the apparent elastic modulus of cancellous bone and the elastic modulus of UHMWPE were significantly different from that of PMMA.

By preserving as much cancellous bone as possible during THR surgery, it is expected that not only will the energy dissipation ability be maintained, but also the imbalance in the degree of deformation between different materials can be compensated for by the deformability of the cancellous structure. Thus, the development of a surface replacement arthroplasty that would preserve the maximum amount of healthy cancellous bone [15] may significantly reduce the incidence of aseptic loosening and result in a decrease in the need for joint revisions.

References

1. Oka M, Ikeuchi K, Tsutsumi S, Yamamuro T, Noguchi T, Nakamura T (1992) Biomechanics of natural and artificial joints. In: Niwa S, Perren SM, Hattori T (eds) Biomechanics in orthopedics. Springer, Berlin Heidelberg New York Tokyo, pp 282–298
2. Brandt KD, Radin E (1987) The physiology of articular stress: osteoarthritis. Hosp Pract 15:103–126
3. Radin EL, Paul IL, Tolkoff MJ (1970) Subchondral bone changes in patients with early degenerative joint disease. Arthritis Rheum 13:400–404
4. Radin EL, Parker HG, Pugh J, Steinberg R, Paul I, Rose R (1973) Response of joints to impact loading: III. J Biomech 6:51–57
5. Serink MT, Nachemson A, Hansson G (1977) The effect of impact loading on rabbit knee joints. Acta Orthop Scand 48:250–262
6. Miyanaga Y (1979) Mechanical behaviour of the subchondral bone in the experimentally induced osteoarthritis. J Jpn Orthop Ass 53:681–695
7. Hayami T, Oka M, Ikeuchi K (1993) Shock absorbing ability of joint materials: II. J Jpn Soc Biomater 11:14–19
8. Michel MC, Guo XDE, Gibson J, McMahon TA, Hayes WC (1993) Compressive fatigue behavior of bovine trabecular bone. J Biomech 26:453–463
9. Ochoa JA, Sanders AP, Heck DA, Hillberry BM (1991) Stiffening of the femoral head due to intertrabecular fluid and intraosseous pressure. J Biomech Eng 113: 256–262
10. Linde F, Hvid I, Madsen F (1992) The effect of specimen geometry on the mechanical behaviour of trabecular bone specimens. J Biomech 25:359–368
11. Hayami T, Oka M, Ikeuchi K, Nakamura T, Hojo H, Sakaguchi K (1994) The mechanical behaviour and functions of subchondral cancellous bone in human hip joint. J Jpn Clin Biomech 15 (in press)
12. Hayami T, Oka M, Ikeuchi K, Nakamura T, Hojo H, Sakaguchi K (1994) Visualization of shock absorbing mechanisms in joint tissues using an ultra high speed camera: II. J Jpn Clin Biomech 15 (in press)
13. Linde F, Hvid I, Pongsoipetch B (1989) Energy absorptive properties of human trabecular bone specimens during axial compression. J Orthop Res 7:432–439
14. Peyron JG, Altman D (1992) The epidemiology of osteoarthritis. In: Moskowitz RW (ed) Osteoarthritis, 2nd edn. Saunders, Philadelphia, pp 15–37
15. Oka M, Noguchi T, Kumar P, Ikeuchi K, Yamamuro T, Hyon SH, Ikada Y (1990) Development of an artificial articular cartilage. Clin Mater 6:361–381

Part 4. Epiphyseal Plate

The Effect of Maturation on Dynamic Viscoelastic Properties of Epiphyseal Plate in the Rabbit

Munehiko Yoshida, Wookcheol Kim, Nozomu Inoue, Yoshiyuki Arai, and Yoshinobu Watanabe

Summary. Dynamic viscoelastic tests were performed on the epiphyseal plates of rabbits under compression and tension loads to study the effect of maturation on the mechanical properties of the epiphyseal plates. The test materials were dissected from immature Japanese white rabbits of various body weights. The dynamic absolute modulus, the storage modulus, the loss modulus, and the tangent phase-shift angle δ were measured in the frequency range 0.1–40 Hz. The values of the dynamic compression modulus of the epiphyseal plates varied from 0.6 to 2.6 MPa (strain 5%). The tangent δ indicated the smallest at the frequency range 1–10 Hz, which corresponds to normal physical activity. The tangent δ also indicated negative correlation with body weight ($r = -0.72$, -0.66, -0.64, -0.67, -0.65, -0.65, -0.62, -0.6, -0.57, and -0.49 at the frequency rate of 0.1, 0.2, 0.4, 0.8, 1, 2, 4, 8, 10, and 20 Hz, respectively, $P < 0.01$). In conclusion, the epiphyseal plate in the rabbit is more elastic with increase in body weight. Furthermore, the epiphyseal plate has greater elasticity at speeds associated with normal physical activity.

Key words: Epiphyseal plate—Viscoelasticity—Mechanical property—Maturation—Epiphyseal separation

Introduction

The epiphyseal plate plays an important role in the longitudinal growth of immature bones. The biomechanical strength of the epiphyseal plate is weaker than that of bone, and the clinical separation of the epiphysis is encountered. It is well known that epiphyseal separation and slipped femoral capital epiphysis

Department of Orthopedic Surgery, Kyoto Prefectural University of Medicine, Kawara, Hirokoji, Kamigyo-ku, Kyoto, 602 Japan occur more often at adolescence. It is therefore worthwhile to investigate the effect of maturation on the mechanical properties of the epiphyseal plate to elicit the mechanism of epiphyseal separation or slipped femoral capital epiphysis.

Materials and Methods

Twenty-eight Japanese while rabbits weighing 470–1930 g were used. The test materials were dissected from the distal ends of radius and ulna. There were 43 specimens composed of a bone–epiphyseal plate–bone complex of about 7 mm in length, and the bilateral ends were flattened by the motor file. Dynamic viscoelastic tests were performed with a Viscoelastic spectrometer (Iwamoto Manufacturing, Japan). This machine system made it possible to perform dynamic viscoelastic tests under compression/tension and shearing loads.

The dynamic absolute modulus, the storage modulus, the loss modulus, and the tangent phase-shift angle δ were measured for each specimen at frequency rates of 0.1, 0.2, 0.4, 0.8, 1, 2, 4, 8, 10, 20, and 40 Hz under compression/tension loads. The preload was 100 g under compression and the maximum displacement was 50 µm (strain was almost 0.05 of the epiphyseal plate).

Results

The values of the dynamic absolute modulus, the storage modulus and the loss modulus of the specimens ranged from 5.9 to 33 MPa, from 5.8 to 33 MPa, and from 0.5 to 5.9 MPa, respectively. In the literature, the values of the dynamic modulus of compact bone were reported as being from 7 to 20 GPa [1]. These values are about 1000-fold the magnitude of the data obtained from the current tests. Thus, the data obtained in this study may need revision to take into account the length of the specimens and the widths of the epiphyseal plates. The values of the dynamic absolute modulus, the storage modulus, and the loss modulus of the epiphyseal plates were found to range from 0.6 to 2.6 MPa, from 0.6 to 2.6 MPa and from 0.1 to 0.5 MPa, respectively. The tangent phase-shift angle δ was found at 0.06–0.31.

Variations in the dynamic absolute modulus, the storage modulus, and the loss modulus with frequency rate in two representative specimens are shown in Figs. 1 and 2. Variations of the tangent phase-shift angle δ with frequency rate are shown in Fig. 3. The tangent phase-shift angle δ indicated lower values in the frequency range from 1 to 10 Hz.

Variations in the tangent phase-shift angle δ with body weight at fixed frequency rates (0.1 and 10 Hz) are shown in Figs. 4 and 5. The tangent phase shift angle δ indicated negative correlation with body weight in the frequency range 0.1–20 Hz (0.1 Hz, $r = -0.72$; 0.2 Hz, $r = -0.66$; 0.4 Hz, $r = -0.64$; 0.8 Hz, $r = -0.67$; 1 Hz, $r = -0.65$; 2 Hz, $r = -0.65$; 4 Hz, $r = -0.62$; 8 Hz, $r = -0.6$; 10 Hz, $r = -0.57$; 20 Hz, $r = -0.49$, $P < 0.001$).

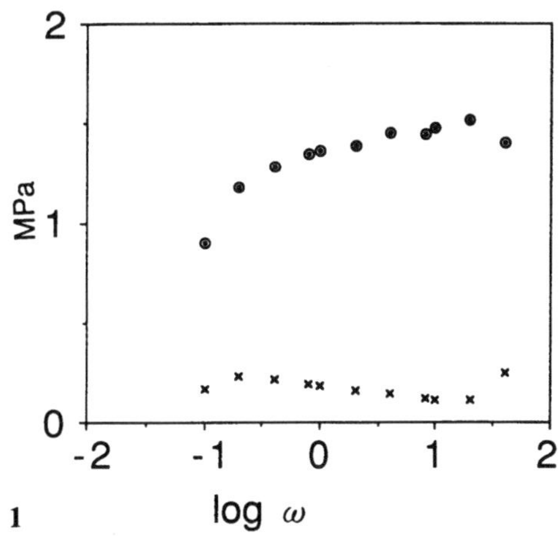

1

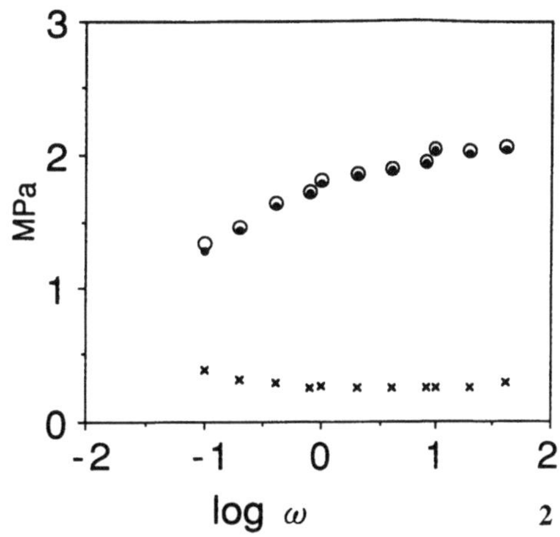

2

Fig. 1. Variations in the dynamic compression modulus, the storage modulus, and the loss modulus with frequency rate in rabbit weighing 470 g. Dynamic absolute modulus, *open circles*; storage modulus, *small black circles*; loss modulus, *x*; frequency, ω

Fig. 2. Variations of the dynamic compression modulus, the storage modulus, and the loss modulus with frequency rate in rabbit weighing 1930 g. Dynamic absolute modulus, *open circles*; storage modulus, *small black circles*; loss modulus, *x*; frequency, ω

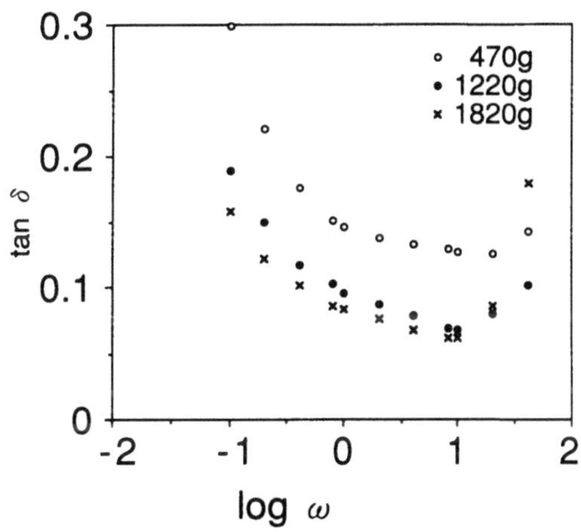

Fig. 3. Tangent δ (δ, phase-shift angle) indicates lower values in the frequency range 1–10 Hz (a material is purely elastic when δ = 0° and purely viscous when δ = 90°). Frequency, ω

Discussion

Bone and cartilage are thought to be viscoelastic materials; their behavior can be described as being time-dependent. Several parameters may be used to describe the viscoelastic behavior of these tissues. In most previous studies, indentation methods were commonly used. Such methods are effective in

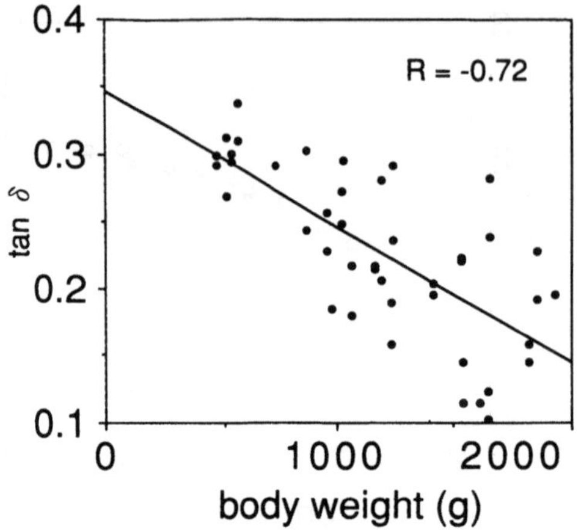

Fig. 4. The relationship between body weight and tangent δ was a negative linear relation ($r = -0.72$, $P < 0.001$) at the frequency rate of 0.1 Hz

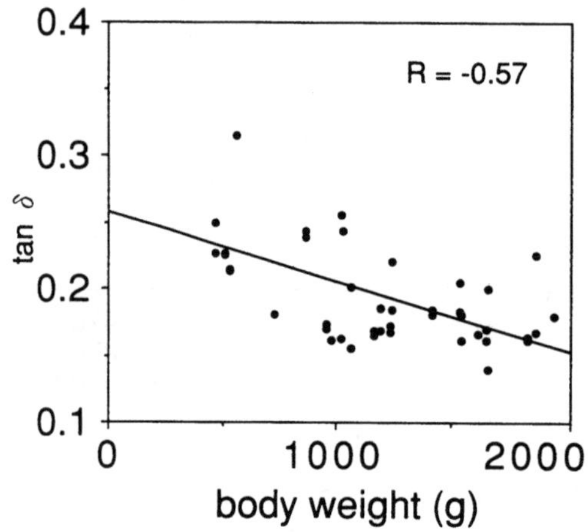

Fig. 5. The relationship between body weight and tangent δ was a negative linear relation ($r = -0.57$, $P < 0.001$) at the frequency rate of 20 Hz

making measurements during slower activities, but they are not as effective in measuring the viscoelastic behavior associated with activities such as walking or running. The dynamic viscoelastic test has proven to be a more appropriate method to investigate these responses [2].

In dynamic viscoelastic tests, the material is subjected to sinusoidally varying strain and the resulting sinusoidal reactions are measured. A geometric relationship between the dynamic modulus, the storage modulus, the loss modulus, and the tangent δ is shown in Fig. 6. The real part of the dynamic modulus is referred to as the storage modulus and reflects the maximum energy stored in each cycle of deformation. The imaginary part is called the loss modulus and corresponds to the energy dissipated in the material per cycle of deformation. Kobayashi [3] reported the changes in the dynamic viscoelastic properties of

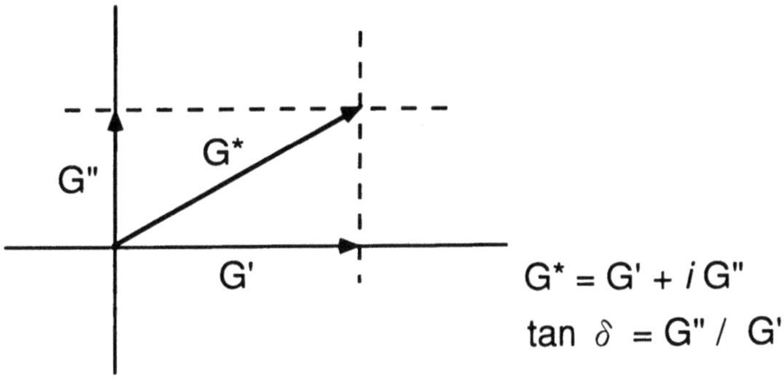

Fig. 6. Geometric relationship between the dynamic modulus (G*), the storage modulus (G′), the loss modulus (G″), and the tangent δ

the callus during its maturation. He concluded that this test was useful to evaluate the mechanical consolidation of the callus.

Torzilli and Mow [4,5] modeled articular cartilage as a combination of the mechanical interactions of a solid matrix and an interstitial fluid [4,5]. Hayes and Bodine [6] reported flow-independent viscoelastic properties of articular cartilage matrix. In our current study, no reference can be made to the relationship between the mechanical properties and the composed materials. However, it is intended to elicit the actual viscoelastic behavior of the epiphyseal plate during ordinary daily activities of motion.

Mow et al. [7] reported the dynamic viscoelastic properties of articular cartilage under shearing load in their textbook. According to Mow's study, the values of dynamic modulus varied from 0.2 to 2.5 MPa for bovine articular cartilage under infinitesimal shearing strain. In our study, the values of the dynamic absolute compression modulus ranged, approximately, from 0.6 to 2.6 MPa. These results imply that the epiphyseal plate also has a role as a shock absorber.

The phase-shift angle δ characterizes the energy dissipation relative to the energy storage within the tissue. In our study, the tangent phase-shift angle δ indicated negative correlation with body weight at fixed frequency rates ranging from 0.1 to 20 Hz ($P < 0.001$). These results imply that greater elasticity emerges with increase in body weight or maturity of the epiphyseal plate. Furthermore, the phase-shift angle δ indicated lower values at the frequency range from 1 to 10 Hz in this study. These results mean that the epiphyseal plate acts more elastically at these ordinary activity speeds. In these dynamic viscoelastic tests, it was not necessary to destroy the specimens and so they could be measured repeatedly. Almost the same results were obtained with all specimens for the tangent δ-frequency curves when the tests were repeated.

References

1. Tateishi T, Shirasaki Y, Kimura T, Ogawa K (1980) Viscoelasticity and strength of compact bones (in Japanese). Nihon Kikai Gakkai A:97–106
2. Miyanaga Y (1979) Measurement of strength and viscoelasticity of subchondral bone (in Japanese). Seikeigeka 30(10):1199–1205
3. Kobayshi A (1987) Study of the biomechanical characteristics of the fracture callus (in Japanese). J Jpn Orthop Assoc 61(11):1273–1284
4. Torzilli PA, Mow VC (1976) On the fundamental fluid transport mechanism through normal and pathological articular cartilage during function: I. The formulation. J Biomech 9:541–552
5. Torzilli PA, Mow VC (1976) On the fundamental fluid transport mechanism through normal and pathological articular cartilage during function: II. The analysis, solution and conclusion. J Biomech 9:587–606
6. Hayes WC, Bodine AJ (1978) Flow-independent viscoelastic properties of articular cartilage matrix. J Biomech 11:407–419
7. Mow VC, Zhu W, Ratcliffe A (1991) Structure and function of articular cartilage and meniscus. In: Mow VC, Hayes WC (eds) Basic orthopaedic biomechanics. Raven, New York, pp 143–196

The Relationship Between Tensile Properties and the Microstructure of the Epiphyseal Plate in an Experimental Model

Yasusuke Hirasawa, Shinro Takai, Toshiyuki Fujii, and Yoshiyuki Arai

Summary. The relationship between the tensile properties and microstructure for the resting zone, proliferating zone, and hypertrophic zone in the distal radial and ulnar epiphyseal plates of rabbits was investigated. Each epiphysis–epiphyseal plate–metaphysis complex (EPMC) was obtained from rabbits weighing from 940 to 1080 g. After the periosteum and perichondrial ring were excised, four dye markers were placed on the surface of the epiphyseal plate itself. The EPMC was distracted to failure using a tensile testing machine, and the strain behavior was recorded using a microscope, CCD camera, and video dimension analyzer system, simultaneously. The location of each dye marker was then identified by hematoxylin-eosin staining after mechanical testing. Epiphyseal plate tissue for analyses of the collagenous fiber architecture of each zone was examined using a scanning electron microscope (SEM). In tension, the majority of the specimens failed at the level of the hypertrophic zone adjacent to the proliferating zone. The mean values of the tangent modulus of the resting zone, proliferating zone, and hypertrophic zone were found to be 27.0 MPa, 14.9 MPa, and 15.1 MPa, respectively. The tangent modulus of the resting zone was significantly higher than those of the other two zones. Both the strain at failure and the energy-absorbing capability until failure of the hypertrophic zone were significantly lower than those of the other two zones. This difference may correlate with differences in the matrix volume and the collagenous fiber architecture of each zone. The noncontact micromotion measurement method developed in this study demonstrated that the resting zone, proliferating zone, and hypertrophic zone in the epiphyseal plate vary dramatically in microstructural tensile properties.

Key words: Epiphyseal plate—Tensile properties—Microstructure

Department of Orthopaedic Surgery, Kyoto Prefectural University of Medicine, Kawara, Hirokoji, Kamigyo-ku, Kyoto, 602 Japan

Introduction

The epiphyseal plate is a unique organ in the immature long bone. The peculiar microstructure and mechanical properties of the epiphyseal plate cause epiphyseal separation. However, the mechanism of separation is not clear. According to a study in an experimental model, the ultimate stress and tangent moduli of the bovine distal femoral epiphyseal plate have been shown to vary with the anatomical site [1]. The epiphyseal plate consists of three zones: the resting zone, the proliferating zone, and the hypertrophic zone. The microstructural properties of each of the three zones are not well defined. Therefore, the purpose of this study was to clarify the microstructural tensile properties and morphological architecture of each zone in the epiphyseal plate of the rabbit.

Materials and Methods

Ten immature rabbits weighing 940–1080 g were used. Each epiphysis–epiphyseal plate–metaphysis complex (EPMC) was obtained from the ulna and radius. After the periosteum and perichondrial ring were excised, four dye markers were placed on the surface of the approximately 1-mm thick epiphyseal plate under a microscope (Fig. 1). Both ends of the EPMC were positioned within metal cylinders and embedded in polymethylmethacrylate cement. After all specimens were preloaded to 1.5 N and five cyclic loadings to 3.0 N were performed, the EPMC was distracted to failure at a displacement rate of 0.5 mm/min using a tensile testing machine (Autograph, model AGS-500B, Shimadzu Co., Kyoto, Japan). The strain behavior between the four dye markers were simultaneously recorded using a microscope (IMZ-EA, Kyoto Kogaku Co., Kyoto, Japan), CCD camera (CS-3440, Tokyo Electronic Industry

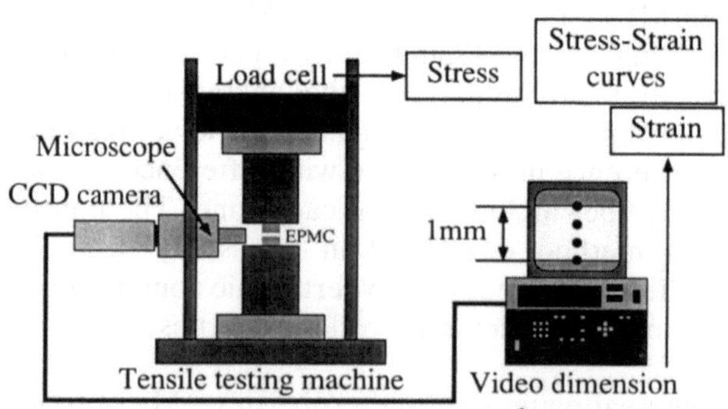

Fig. 1. Schematic diagram of the experimental apparatus. The strain behavior of four dye markers was recorded with a noncontact micromotion measurement system. *EPMC*, epiphysis–epiphyseal plate–metaphysis complex

Co., Ltd., Tokyo, Japan) and a video dimension analyzer system (X-Y tracker C-3162, Hamamatsu Photonics K.K., Hamamatsu, Japan). The cross-sectional areas of the EPMC were estimated as the elliptic area obtained from the minor axis and the major axis. The microstructural tensile properties for various zones in epiphyseal plates were then calculated from the stress-strain curves (Fig. 2). The specimens were tested on the same day they were obtained and were moistened with physiological saline at a temperature of about 24°C before and during testing. The location of each dye marker was then identified by hematoxylin-eosin (HE) staining after mechanical testing. Epiphyseal plate tissue for analyses of the collagenous fiber architecture of each zone was chemically macerated, then air dried, gold spatter-coated, and examined using a scanning electron microscope (SEM).

Results

Accuracy of Measurement

The number of dots in the working area of the video monitor was 1932 in the longitudinal direction and 2556 in the transverse direction. The centers of gravity in dye markers were homed by X-Y tracker and were represented by X and Y coordinates. The average number of dots between the centers of gravity was 612 ± 42 in the transverse direction. The dispersion of homing by X-Y tracker was within about 2%.

Sites of Epiphyseal Separation

All of the failure patterns were Salter and Harris type I injuries [2]. The majority of the specimens failed by traction at the level of the hypertrophic

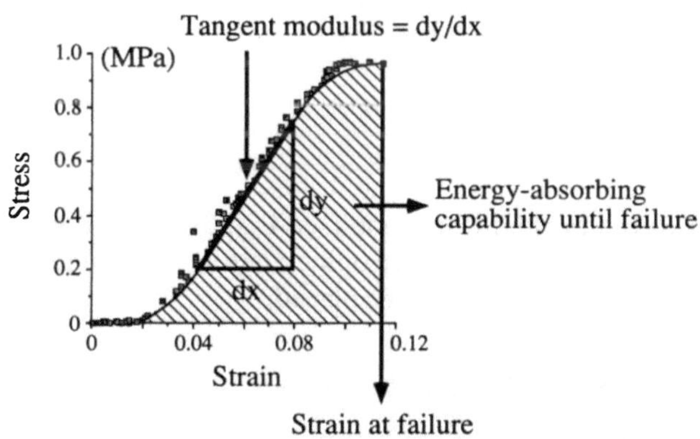

Fig. 2. Typical tensile stress-strain curve of the hypertrophic zone. Tangent modulus, strain at failure, and energy-absorbing capability until failure were calculated from each of the stress-strain curves

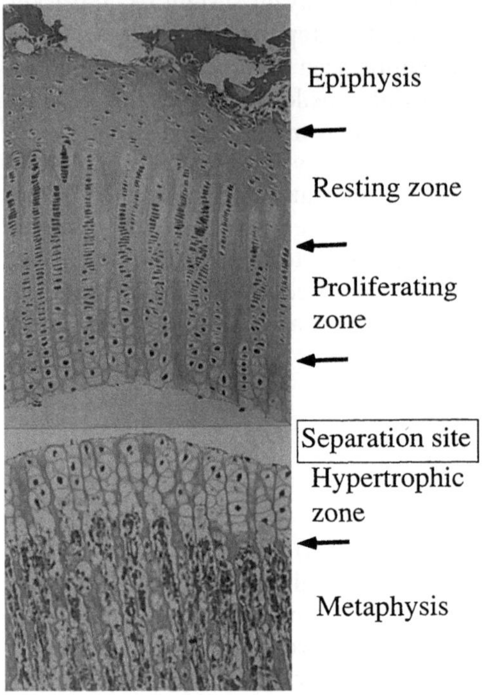

Fig. 3. Site of epiphyseal separation. The majority of the specimens failed by traction at the level of the hypertrophic zone adjacent to the proliferating zone. Each *arrow* shows where the dye markers were placed. H&E, ×100

Table 1. Cross-sectional area and ultimate stress of the distal radial and ulnar epiphyseal plate in rabbits

Region	Cross-sectional area (mm^2)		Ultimate stress (MPa)	
Radius ($n = 10$)	29.7 ± 1.1	N.S.	1.05 ± 0.12	N.S.
Ulna ($n = 10$)	27.9 ± 1.9		1.03 ± 0.09	

Data are expressed as mean ± standard error of mean.
N.S., not significant ($P > 0.1$).

zone adjacent to the proliferating zone (see Fig. 3). Each border of the resting zone, proliferating zone, and hypertrophic zone corresponded approximate to the sites where the dye markers were placed.

Cross-Sectional Area and Ultimate Stress

As shown in Table 1, the mean cross-sectional area of the epiphyseal plate was 29.7 mm^2 in the radius and 27.9 mm^2 in the ulna. The ultimate stress was 1.05 MPa in the radius and 1.03 MPa in the ulna. For biomechanical parameters, there were no significant differences between the radius and ulna.

Tangent Modulus, Strain at Failure, and Energy-Absorbing Capability Until Failure

As shown in Table 2, the tangent moduli of the resting zone, proliferating zone, and hypertrophic zone were found to be 27.0 MPa, 14.9 MPa, and 15.1 MPa, respectively. The tangent modulus of the resting zone was significantly higher than that of the other two zones. The strain at failure and the energy-absorbing capability until failure of the resting zone, proliferating zone, and hypertrophic zone were found to be 0.043, 0.081, and 0.096, and 32.9 kJ/m^3, 66.2 kJ/m^3, and 88.5 kJ/m^3, respectively. Both the strain at failure and the energy-absorbing capability until failure of the resting zone were significantly lower than those of the other two zones.

Matrix Volume and Collagenous Fiber Architecture

In the resting zone, the abundant cartilage matrix separated the chondrocytes. In the proliferating zone, cartilage cell columns were formed in parallel with the long bone axis of the shaft (see Fig. 4). The cartilage matrix separated each cell column, and the matrix volume diminished rapidly in the hypertrophic zone. The resting zone had a network structure of collagen fibers, arranged in a random pattern. The collagenous architectures of the other two zones were laid along the tensile direction.

Discussion

Harris [3] tested the proximal epiphyseal plate of immature rat tibiae to failure in bending and reported the maximal load in bending tests in 1950. Morscher et al. [4] likewise reported the maximal load in tensile tests using the proximal epiphyseal plate of immature rat tibiae in 1965. Bright et al. [5] analyzed the ultimate bending stress and energy-absorbing until failure by the improved bending-testing machine of Harris. It was noteworthy that these early studies showed pilot results in biomechanical studies of the epiphyseal plate. But the accuracy of these studies are not always considered satisfactory today when the microcomputer and mechanical testing machine are available.

Table 2. Tensile properties of various zones of epiphyseal plate in rabbits

Region	Tangent modulus (MPa)		Strain at failure		Energy-absorbing capability	
Epiphyseal plate (n = 10)	17.7 ± 6.1		0.091 ± 0.008		84.5 ± 8.1	
Resting zone (n = 10)	27.0 ± 1.3	⎤	0.043 ± 0.006	⎤	32.9 ± 5.7	⎤
Proliferating zone (n = 10)	14.9 ± 1.4	⎥ *	0.081 ± 0.009	⎥ *	66.2 ± 4.9	⎥ *
Hypertrophic zone (n = 10)	15.1 ± 1.6	⎦	0.096 ± 0.010	⎦	88.5 ± 7.8	⎦

Data are expressed as mean ± standard error of mean.
$^*P < 0.01$.

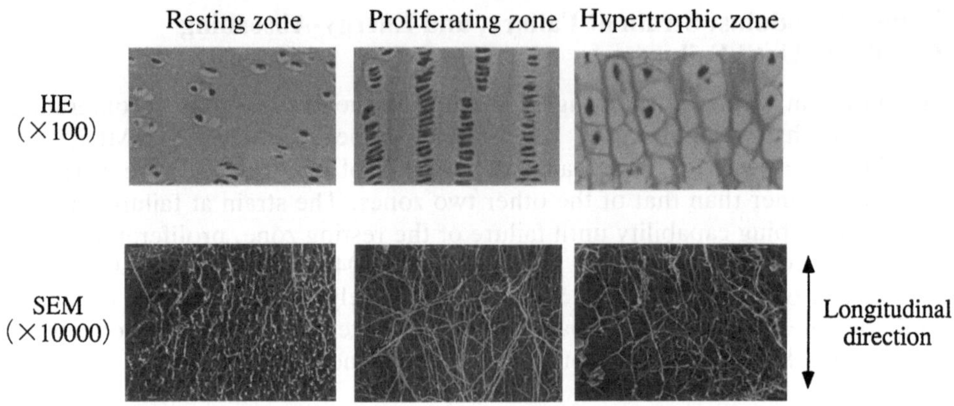

Fig. 4. Matrix volume and collagenous fiber architecture of each zone. In the resting zone, the abundant cartilage matrix separated the chondrocytes. The matrix volume diminished rapidly in the hypertrophic zone. The resting zone had a network structure of the collagen fibers. The collagenous architectures of the other two zones were laid along the tensile direction. *SEM*, scanning electron microscope. H&E, ×100, SEM, ×10 000

Cohen et al. [1] recorded the strain of the epiphyseal plate using a noncontact video dimensional analyzer system. This system is an excellent method to eliminate the mixture of the strain in other parts except the target area. We have always used this method whenever the strain was determined. This system has another advantage in that the strain in a minute area can be analyzed by means of a microscope. In the present study, the analysis of strain behavior in various zones of the epiphyseal plate was successful using the noncontact micromotion measurement system.

The major findings of this study were that the tangent modulus of the resting zone was a significant 75% stiffer than that of the other two zones, and that the hypertrophic zone showed the highest values of strain at failure and energy-absorbing capability until failure. This difference may correlate with differences in the matrix volume and the collagenous fiber architecture of each zone. The network structure of the collagen fiber in the matrix makes the resting zone stiffest. The stiffness was assumed to be important to protect the germinal cells in the resting zone. In the hypertrophic zone, the collagen fiber was aligned parallel with the tensile direction. This orientation of collagen fiber essentially reinforces the matrix when the tensile stress is applied in the longitudinal direction. However, the lowest matrix volume leads to the highest strain at failure.

The following mechanism for epiphyseal separation in the present model is proposed: Rapid diminution of the cartilage matrix leads to stress concentration at the level of the hypertrophic zone adjacent to the proliferating zone. Cracking at the boundary grows. Finally, breakage occurs (see Fig. 5). Moen

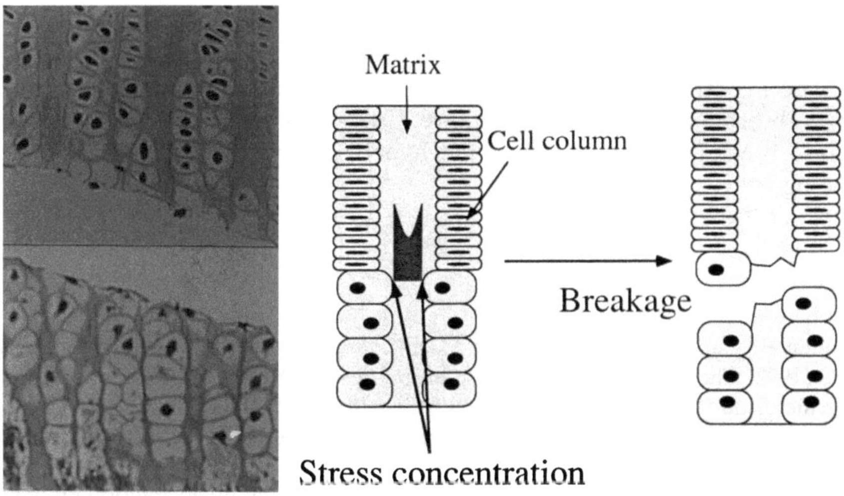

Fig. 5. Epiphyseal separation mechanism. The rapid diminution of the cartilage matrix leads to stress concentration at the level of the hypertrophic zone adjacent to the proliferating zone

and Pelker [6] showed that the epiphyseal plate of bovine femora and tibiae separated at the level of the proliferating zone under tensile load. However, their testing speed (0.5 mm/s) was higher than that used in our present study (0.5 mm/min). The change of testing speed may cause the change in manner of separation.

Using the noncontact micromotion measurement system it was demonstrated that the resting zone, proliferating zone, and hypertrophic zone in the epiphyseal plate varied dramatically in microstructural tensile properties. However, there are many subjects for future studies, such as quantification of collagen content, collagen fiber orientation and proteoglycan content, compressive and torsional testing, impact loading testing, viscoelastic testing, age-related changes, and the effect of growth hormone.

Conclusion

1. The noncontact micromotion measurement method developed in this study was beneficial to the assessment of the mechanical properties in various zones of the epiphyseal plate for small animals.

2. It was demonstrated that the resting zone, proliferating zone, and hypertrophic zone in the epiphyseal plate varied dramatically in microstructural tensile properties. These differences may correlate with differences in the matrix volume and the collagenous fiber architecture of each zone.

3. The resting zone showed the highest value for tangent modulus. The stiffness was assumed to be important to protect the germinal cells in this zone.

4. The hypertrophic zone showed the highest value in strain at failure and energy-absorbing capability until failure. The stress concentration was believed to cause breakage at the level of the hypertrophic zone adjacent to the proliferating zone.

References

1. Cohen B, Chorney GS, Phillips DP, Dick HM, Buckwalter JA, Ratcliffe A, Mow VC (1992) The microstructural tensile properties and biochemical composition of the bovine distal femoral growth plate. J Orthop Res 10:263–275
2. Salter RB, Harris WR (1963) Injuries involving the epiphyseal plate. J Bone Joint Surg 64-A:587–622
3. Harris WR (1950) The endocrine basis for slipping of the upper femoral epiphysis. J Bone Joint Surg 32-B:5–11
4. Morscher E, Desaulles PA, Schenk R (1965) Experimental studies on tensile strength and morphology of the epiphyseal cartilage at puberty. Ann Pediat 205:112–130
5. Bright RW, Burstein AH, Elmore SM (1974) Epiphyseal-plate cartilage. A biomechanical and histological analysis of failure modes. J Bone Joint Surg 56-A: 668–703
6. Moen CT, Pelker RR (1984) Biomechanical and histological correlations in growth plate failure. J Pediat Orthop 4:180–184

Part 5. Cruciate Ligament

Biomechanics of the ACL and ACL Reconstruction: New Concepts and Applications

Savio L.-Y. Woo, Glen A. Livesay, John W. Xerogeanes, Ted Rudy, Yoshitsugu Takeda, and Yasuyuki Ishibashi

Summary. The anterior cruciate ligament (ACL) possesses a complex geometry that encompasses portions of differing lengths which display differing functions depending on knee orientation and external loading. It is well documented as a major stabilizer of knee motion, and is, not surprisingly, also the most commonly injured knee ligament. Due to the inability of the ACL to heal spontaneously when torn at its mid-substance, it has been extensively studied during the last few decades. In this chapter, important biomechanical concepts and studies relating to the ACL will be reviewed and the clinical application of this knowledge will also be discussed. Particular attention will be placed on the measurement of structural properties of the femur-anterior cruciate ligament-tibia complex (FATC) as well as on the evaluation of the role of the ACL in overall knee kinematics. Assessment of ACL contribution through measurements of length, strain, and force changes in the ACL will be discussed, as well as the application of recently introduced robotic technology. Through review of these areas, a more detailed picture of the complete contribution of the ACL to total knee function will be provided.

Key words: Anterior cruciate ligament—Robotics—Knee—Kinematics—In situ force—strain—Biomechanics—insertion site

Introduction

Ligaments are strong bands of connective tissue that traverse the diarthrodial joints of the body and bind the skeleton, providing joint stability and assisting in the maintenance of normal movement. In the human knee, there are four

Musculoskeletal Research Center, Department of Orthopedic Surgery, 3471 5th Avenue, University of Pittsburgh, Pittsburgh, PA 15213, USA major ligaments, the most studied and arguably the most important being the anterior cruciate ligament (ACL). The ACL runs from the posterior medial aspect of the lateral femoral condyle to the anterior lateral portion of the medial tibial plateau. It has a complex geometry that encompasses portions of differing lengths, and its primary function is to resist anterior translation of the tibia [1,2]. the ACL also plays a significant role in resisting internal and varus-valgus rotations of the knee [3,4].

The ACL has been studied extensively because it is the knee ligament that is most commonly injured, and when torn at its midsubstance, has limited ability to heal spontaneously [5]. In the United States alone, approximately one-half million clinically significant ACL injuries occur annually [5]. The sequelae of these injuries can be severe; the patient with an ACL-deficient knee is at risk for joint instability, functional impairment, meniscal tearing, and degenerative joint disease [5–10]. ACL reconstruction is thought to interrupt the cascade because evidence suggests that secondary injury to the meniscus can be prevented by ACL reconstruction [5,11]. In a recent clinical study, Daniels and coworkers prospectively followed 236 patients with measured knee instability. The patients were grouped as operative or nonoperative on the basis of their treatment choice. Follow-up evaluation after 5 years documented that ACL reconstruction decreased measured joint instability but that patient-perceived symptoms of joint instability were not different between the groups. Surprisingly, the incidence of degenerative joint disease was actually higher in the patients with reconstructed ACLs than in the nonoperative group. The authors believed that this was because the patients who underwent reconstruction had a higher preoperative incidence of meniscal injury [5].

The most throughly studied ACL replacement graft has been that of the patellar tendon. Many authors have characterized the biomechanical properties of the human patellar tendon graft [12–17]. It should be recognized that when selecting a replacement graft for the ACL, it is important to consider the entire load-elongation curve, rather than concentrating on only the ultimate load a graft can sustain. Figure 1 compares the load-elongation curves for two hypothetical grafts together with that for a human femur–ACL–tibial complex (FATC). The ultimate load of graft B is equal to that of the FATC, but one-half that of graft A. However, the load-elongation curves clearly reveal that graft A has a much greater stiffness than that of the human FATC, and that graft B is considerably less stiff. Therefore, under physiological loading neither will perform in the same manner as the native ACL. As a result, neither graft would serve well as an ACL replacement. During physiological loading, a graft has to undergo a large number of cyclic elongations, and an implanted graft that is too stiff or too compliant will not guide the motion of the knee joint properly. Therefore, it is important to match the entire load-elongation curve for the ACL graft to that for the human FATC.

In this chapter, important biomechanical concepts and studies relating to the ACL are reviewed. The clinical application of this knowledge is also discussed. Emphasis is placed on measuring the structural properties of the FATC as well

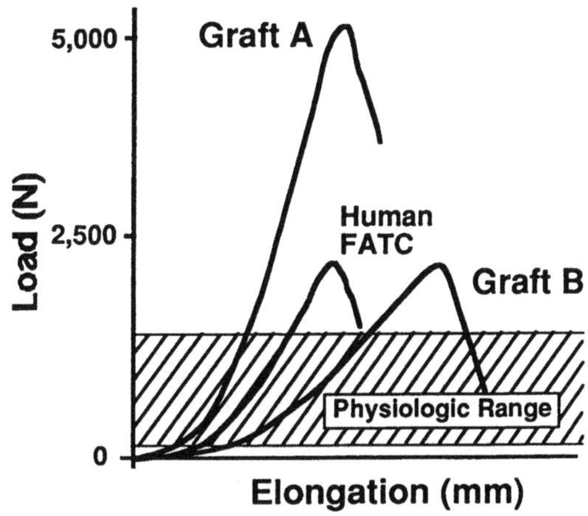

Fig. 1. Schematic diagram of the structural response of the human (femur–anterior cruciate ligament (ACL)–tibie comples (*FATC*) shown with two idealized replacement grafts. Although graft A would allow the same elongation to failure, and graft B would allow the same load to failure, both grafts are poorly suited for replacements because they have drastically different stiffness in the physiologic range

as knee kinematics with thc newly developed robotic technology. This chapter is divided into five sections: structural properties of the FATC, contribution of the ACL to knee kinematics, use of robotics to study 6-DOF joint motion, effects of muscular stabilization, and future directions and conclusions.

Structural Properties FATC

Knowledge about the tensile properties of the ACL is essential not only for the choice of biological grafts and the design of synthetic grafts but also for understanding thc kinematics of thc knee [18]. The structural properties of the entire FATC have been frequently represented by load-elongation curves [1,19–21]. Common parameters obtained from the curve include stiffness, ultimate load, ultimate elongation, and energy absorbed at failure [21]. These parameters are affected by factors such as environmental temperature, specimen orientation, skeletal maturity, and aging of the specimens. Among these factors, specimen orientation and age have recently been identified as being important [21].

Effects of Specimen Orientation

As many studies have noted [22–24], the ACL has a complex structure. These studies show that the ACL is composed of multiple fiber bundles that have different lengths in situ and attachments to bone that are broad. Thus, these anatomical complexities make the uniform loading of the entire ACL in simple uniaxial tensile testing very difficult. When the tensile test is conducted, as much of the ACL should be loaded as possible to avoid underestimation of its structural stiffness and strength. Effects of specimen orientation on tensile properties has been well illustrated by testing the FATC of rabbits in different orientations [25].

In a follow-up study, our laboratory studied 27 pairs of human cadaver knees. The FATC pairs were tested at 30° flexion, one in anatomical orientation and the other in tibial orientation [21]. In the anatomical orientation, the tensile load is applied along the axis of the ACL, preserving the normal angles of the insertion with respect to the bone. In the tibial orientation, the tensile load is applied in line with the tibial and femoral insertion site, distorting the original angles of the ACL insertion. The linear stiffness, ultimate load, and energy absorbed at failure for the FATC were found to be significantly higher in specimens with the tensile load applied along the anatomical orientation than in those with the tensile load applied along the tibial orientation. Therefore, by maintaining the anatomical angles of the femoral and tibial insertions, a greater portion of the ACL fibers were simultaneously loaded.

In our study, the ultimate load of the specimens (22–35 years) was significantly greater in those tested in the anatomical orientation (2160 ± 157 N) than those tested in the tibial orientation (1602 ± 167 N). The latter value was similar to that reported by Noyes and Grood [1]. However, it is believed, for the structural properties of the FATC, that the higher values obtained (more than 30% greater) obtained by testing in the anatomical orientation are most representative and should be considered as the basis for selecting an ACL replacement graft.

The testing orientation also affected failure mode. The FATCs tested in the tibial orientation had a higher incidence of insertion failure by the sequential "peeling off" of individual fiber bundles (48%) than those tested in the anatomical orientation (4%). This is further evidence for using the anatomical orientation during structural testing.

Effects of Specimen Age

The age of the specimen also has a profound effect on the tensile properties of the ACL. Trent et al. [20] tested the femur–ACL–tibia complex (FATC) in knees of humans between 25 and 55 years of age and reported an average stiffness of 141 N/mm and an ultimate load of 633 N. In the same year, Noyes and Grood [1] showed that the linear stiffness and ultimate load was significantly higher in 6 knees from young donors between 16 and 26 years of age (182 ± 56 N/mm and 1730 ± 660 N, respectively) than those from 20 knees from persons ranging from 48 to 86 years in age (129 ± 39 N/mm and 734 ± 266 N, respectively).

In our laboratory, we have also investigated the age effect on the structural properties of FATC [21]. We stratified 27 pairs of cadaver knees by age into three groups: younger (22–35 years), middle-aged (40–50 years), and older (60–97 years). The structural properties of the FATC were tested in anatomical orientation and found to decrease significantly as specimen age increased. In the younger specimens, linear stiffness (242 ± 28 N/mm) and ultimate load (2160 ± 167 N) values were significantly higher than those of the older specimens (180 ± 25 N/mm and 658 ± 129 N, respectively).

There were also age-related differences in the failure modes of the ligaments. The older specimens had a higher incidence of substance failures (67%) than the younger (28%) or middle-aged (44%) specimens. However, 50% of the younger specimens tested failed by tibial avulsion. These differences in failure modes show that ligaments deteriorate at a faster rate than bone.

Contribution of the ACL to Knee Kinematics

The recent heightened interest in knee kinematics results, in part, from issues that have surfaced as a result of advanced treatment options now available for injuries and ailments of this joint. Perhaps the best example of this involves the refinement of surgical reconstructive techniques for the ACL and the choice of optimal graft placement and tension, as well as postoperative rehabilitation. A complete understanding of the response of the normal knee to mechanical loads applied externally or induced by musculature is required so that the degree of normal knee function restored by ACL reconstruction may be evaluated. The information obtained detailing the contribution of the ACL to normal knee function should enhance the evaluation of ligament reconstruction and replacement graft selection.

Length of the ACL

Several approaches have been used to quantify changes in the length of the ACL during normal knee motion and various external loading conditions. Wang et al. [26] used a graphical approach, calculating the distance between pin tips placed to approximate the femoral and tibial insertions of the ACL, and found that the ACL lengthened 7% between 0° and 90° passive flexion. Trent et al. [20] extended this method to use multiple pins, and noted preferential loading on the anterior portion of the ACL with flexion. Butler et al. [27] digitized multiple points along the midsubstance of the ACL and determined that the medial and lateral portions of the ACL lengthened with increasing knee flexion, while the anteromedial and posterolateral portions shortened. Lew and Lewis [28] estimated distance changes between the insertions of the ACL in vivo by relating ligament insertion site information obtained from cadaveric specimens to live human subjects who underwent laboratory testing, and found that knee flexion slightly lengthened the anteromedial fibers and shortened the posterolateral fibers on the order of 1 mm.

In our laboratory, changes in the length of the anteromedial (AM), intermediate (IM), and posterolateral (PL) portions of the human ACL as a function of knee flexion angle have been examined in vitro with a 6-degree-of-freedom (6-DOF) kinematic linkage system [29]. We have found that during passive flexion (0°–90°) the length of the AM portion increase, the PL portion decreases, and the IM portion remains relatively constant in length (Fig. 2). Valgus moments applied to the knee lengthened the AM and PL portions of the ACL significantly more than the IM portion. In contrast, under 100 N of anterior load, the AM and PL portions lengthened less than the IM portion. The lengthening of these three portions during valgus and anterior loading was greater at higher flexion angles. We have also examined the change in length of the anterior and posterior portions of the ACL during unconstrained (5-DOF) and constrained (1-DOF) AP drawer up to 100 N at 0°, 30°, 45° and 90° of flexion [18,30]. In eight intact cadaver knees, the anterior portion of the ACL lengthened an average of 3.3 mm while the posterior portion shortened an average of 1.5 mm between 0° and 90° of flexion. Limiting knee motion to 1 DOF (AP translation) and again applying the same 100 N anterior tibial drawer led to increased lengths in both portions of the ACL at all flexion angles when compared to the unconstrained 5-DOF case.

Strain Within the ACL

Investigators have also measured strains in the ACL with knee motion and under external loading conditions. Two devices that have been used to measure ligament strain are liquid mercury strain gauges (LMSGs) and Hall-effect strain transducers (HESTs). Flexion-induced changes in strain of the ACL have been examined with liquid mercury gauges. Kennedy et al. [31] found that the strain on the AM surface was least at 35° of knee flexion, and later Meglan et al. [32] observed increased ligament strain toward extension. Other groups have employed a device called HESTs to measure ligament strain. Using this device, Arms and coworkers also determined the strain of the anteromedial bundle of the ACL to be lowest at 35° flexion [33]. Additionally, they demonstrated that when internal tibial rotation as well as varus and valgus moments were applied to the knee, strains in the ACL increased.

In vivo ligament strains have been recently examined using HESTs [34–38] and differential variable reluctance transducers (DVRTs) [39]. Fleming et al. [37] implanted this type of transducer in the AM portion of the ACL and evaluated the effects of different types of rehabilitation exercises on AM strain magnitudes. Isolated hamstring activation did not strain the ACL at any flexion angle, while isolated quadriceps activation induced strain in the AMB at both 15° and 30° flexion. Co-contraction of the quadriceps and hamstrings was not found to induce any AM strain between 50° and 90° knee flexion.

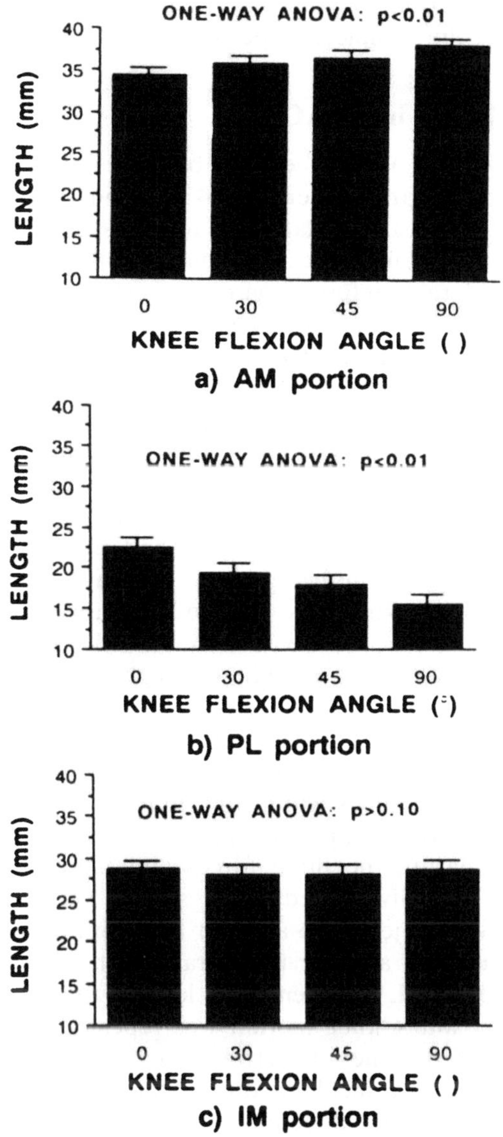

Fig. 2a–c. Lengths of the (**a**) anteromedial (*AM*), (**b**) posterolateral (*PL*), and (**c**) intermediate (*IM*) portions of the ACL during knee flexion-extension (from [29])

While providing information of the changes in strain on the surface of the ACL, these measurements are noted to report information for a single location during testing, and have been primarily limited to the AM portion of the ligament. Because the insertions of the ligament are strained significantly more than the midsubstance during loading, and also as the force distribution within the ACL has been shown to be nonuniform under external loading of the knee, the strain of the surface of the ACL is highly nonuniform. For this reason, it is important to note that the local measurement of strain is not a true indication of the complete response of the ACL.

Forces in the ACL

Several approaches have been used to measure forces within the ACL, including buckle transducers [40–43], strain gauges affixed near ligament insertion sites [44], instrumentation of the tibial insertion with a load cell [45–47], and also implantable force transducers [48,49]. These approaches have provided considerable information concerning the in situ forces developed within the ACL, but they are also invasive in nature and require the attachment of mechanical devices on or near the ligament that may alter force values.

Other investigators have chosen to use noncontact methods, such as radiographic or kinematic approaches, to measure the in situ forces within the human ACL [18,50]. Vahey and Draganich [50] placed silastic tubes filled with mercury along the ACL and took radiographs during testing to estimate the direction of the ACL. This information was then combined with the forces applied to the knee before and after removal of the ACL to determine the in situ forces in the ligament. We have employed a conceptually similar approach using a kinematic linkage to digitize the femoral and tibial insertions of the ACL that are then combined with force information to allow calculation of in situ ACL forces [18]. Although avoiding alteration of the ACL during testing, these methods either report only magnitudes of force or require a geometric assumption for the direction of force.

Our laboratory recently introduced a direct, noncontact method determining in situ forces within the human ACL using a universal force-moment sensor (UFS) [51,52]. This approach involves rigidly attaching a UFS to the tibia to measure the three forces and three moments at the knee that result from externally applied loadings (e.g., anterior tibial drawer). First, the intact knee was subjected to anterior tibial drawer up to 100 N at a specified flexion angle and the anterior tibial translation was recorded. Then, all soft tissues except the ACL were removed, leaving an "ACL-only" specimen, and the exact path of intact knee motion was repeated. Because the ACL is the only structure resisting motion between the tibia and femur, the forces and moments measured by the UFS during this repeated path can be used to determine the magnitude, direction, and point of application of the in situ forces in the ligament [51,52]. For a 1-DOF (AP only) anterior drawer of the intact knee at 30° flexion, the in situ forces in the ACL were found to be 46.4 ± 10.1 N and 114.8 ± 7.4 N under 50 N and 100 N of applied drawer, respectively. The ratio between the in situ force in the ACL and the applied anterior drawer increased with the applied force; the in situ force in the ACL was smaller than the applied anterior loads less than 65 N, whereas it was greater than the applied anterior loads greater than 85 N.

To describe the direction of the in situ force, the line of action of the ligament force was transformed to two angles, termed elevation and deviation.

The elevation, α, represents the angle between the tibial plateau and the projection of the line of force on a sagittal plane. Physically, this is the angle that the ACL appears to make with the tibial plateau from a medial or lateral view. The deviation, β, represents the angle between the line of force and the sagittal plane. The direction of the in situ force in the ACL was determined to be 41.7° ± 4.9° and 12.5 ± 4.9° under 100 N of anterior tibial drawer, elevation (α) and deviation (β), respectively. This is shown schematically in Fig. 3, and was noted to correspond approximately to the insertion direction of the AM portion of the ACL [52]. During anterior loading, the point of application of the in situ force in the ACL was located in the posterior section of the anteromedial portion of the tibial insertion site of the ACL (Fig. 4).

Force Distribution Within the ACL

A few investigators have chosen to examine the force distribution within the ACL under external loading [18,53], but have previously reported only force magnitudes or required anatomical assumptions for the direction of in situ force. Our UFS approach can be further used to examine the contributions of the various portions of the ACL under external loading of the knee [54]. By further dissecting the "ACL only" mentioned above to an "AM portion of the ACL only" specimen, a similar UFS methodology can be applied to determine the in situ forces in the AM and PL portions of the ACL.

Under 100 N of a 1-DOF anterior drawer, the AM portion was found to support approximately 95% of the in situ force in the ACL, whereas the PL portion carried almost no force. Further dissection revealed that the posterior section of the AM portion (P-AM) portion carried approximately three-fourths

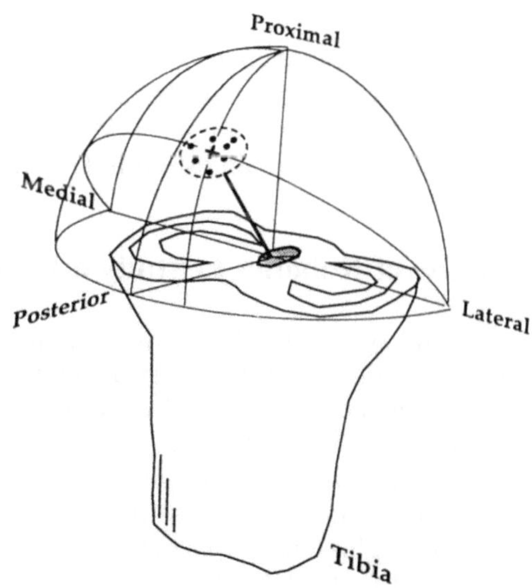

Fig. 3. Schematic diagram showing the mean direction of the in situ force in the ACL under 100 N of anterior tibial drawer at 30° flexion, from a posterior oblique view on a right knee (from [52]). The *bold line* indicates the mean direction, while the *smaller filled circles* represent the directions from each of the seven specimens

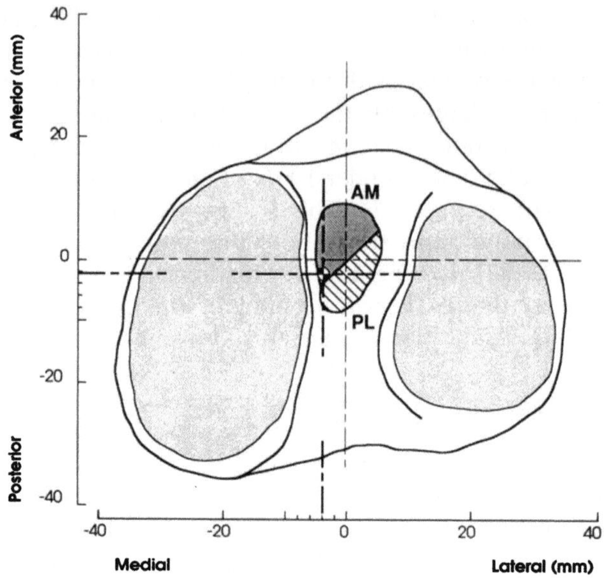

Fig. 4. Superior view of the tibial plateau showing the point of application of the in situ force within the tibial insertion of the ACL for a representative right knee under 100 N applied anterior load. the centroid of ligament loading is shown, and AM and PL represent the anteromedial and posterolateral portions of the tibial insertion site of the ACL, respectively. Note that the axes are referenced to the centroid of the tibial insertion site (from [52])

of the total ACL force, with the anterior section of the AM portion (AAM) supported the remaining one fourth [54]. Further, the directions of the in situ forces in the A-AM and P-AM portions of the ACL under 100 N of anterior tibial load were found to be different. For the P-AM portion, the force followed the direction of the AM portion, while that for the A-AM portion displayed a direction almost entirely within the sagittal plane.

The Use of Robotics to Study 6-DOF Joint Motion

To extend our methodology to include 6-DOF joint motion, we have developed a testing system that uses a robotic manipulator [55]. The use of robotics in biomechanics was originally introduced by Fujie et al. [56] and the primary advantage of this method of joint manipulation is that more complex and less constrained motions of the intact knee joint can be studied. Further, the 6-DOF path of intact joint motion can also be recorded and later reproduced by the robot. This allows testing to be performed in a manner similar to that described for the 1-DOF case, except that 6-DOF knee motions can now be considered. The intact knee is subjected to an external loading condition (e.g., anterior tibial loading) and the path of joint motion recorded; then, all soft tissues except the ACL can be removed and the intact path of knee motion reproduced in the "ACL only" specimen. When combined with the previously described UFS methodology, the in situ forces in the ACL can then be determined using the UFS data [51,52]. Thus, a UFS approach, used in conjunction with robotic manipulation, becomes a powerful tool both to record force and moment information and also to determine the magnitude, direction, and the point of application of the in situ force in the ACL.

To demonstrate this robotic approach, we have used porcine cadaver knee joints. Following dissection of musculature that left the joint capsule and knee ligaments intact, the tibia and femur were secured within aluminum cylinders. The femoral cylinder was then rigidly affixed to a custom clamp that allowed 6-DOF positioning and rigid fixation of the femur relative to the robot manipulator base. The UFS was mounted between the tibial cylinder and the end-effector of the manipulator to serve a role in force-moment feedback and also in determining the in situ forces in the ACL. The relationship between the UFS, the robot, and the intact joint is shown schematically in Fig. 5.

Cyclic AP loading to ±110 N at a rate of 20 mm/min was then applied by the robot through the use of force-moment control [55]. Essentially, the robot was instructed to "learn" for the path of natural, unconstrained joint motion during anterior tibial loading (e.g., characteristic internal rotation). All structures except the ACL were then dissected away, and the robot was used to reproduce the previous loading path of the intact knee. The UFS was used to record the complete force-moment condition of the "ACL only" knee as it went through the same path of joint motion. Because the manipulator was following

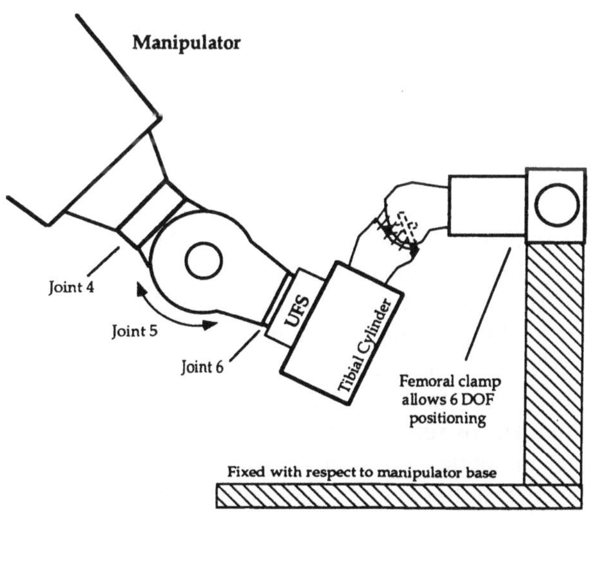

Fig. 5. Schematic of complete experimental setup demonstrating the placement of the UFS between the manipulator and the test joint to provide force-moment feedback for control of the manipulator

the same path, the in situ force in the ACL could then be determined, as described previously, by means of the principle of superposition [51,52].

The use of multi-DOF force control with the robot resulted in characteristic internal rotation under anterior tibial loading (5 DOF) and overall greater anterior translations. This indicates that natural knee kinematics are significantly different than the more restrained 1-DOF case. The latter has been used frequently in the past. Under 110 N of anterior tibial drawer, in situ force magnitudes were found to be 110 ± 14 N. The application of natural motion was also noted to alter the determined direction of in situ force in the ACL; the mean directions of the in situ forces in the ACL were found to be 15° and 6° (elevation and deviation, respectively). In the 5-DOF case, magnitudes of force are slightly lower, with directions lying more parallel with the tibial plateau. The most important conclusion to be drawn from the comparison of constrained (1-DOF) versus robotic (5-DOF) testing is that the apparent role of the ACL in the intact knee is affected by the way in which joint motion is controlled during testing.

We believe that this multi-DOF testing system will prove valuable in obtaining more realistic in situ forces in the ACL in the intact knee. Further, the capability of this approach represents a valuable step in the evolution of the cutting study. The maintenance of the exact path of joint motion makes the results of such a study independent of the cutting order, while enabling the application of complex loadings to the intact knee. Finally, it will also allow evaluation of ACL replacement grafts in the reconstructed knee by means of comparison of these data with those obtained for the native ACL in the normal knee.

Effects of Muscular Stabilization

In addition to the soft tissues that guide and help to restrain excessive knee motion, the muscles which cross the knee play a significant role in the physiological function of the knee. The activity of muscles can introduce large changes and forces experienced by the ACL or its replacement. Smidt [57] reported that the quadriceps exerts an anterior shear force on the tibia when the knee is flexed between 5° and 60° and a posterior shear force with the knee flexed between 75° and 90°. Other investigators have also documented the interaction between the quadriceps and the ACL in the initial arc of joint flexion [33,58 60]. In general, quadriceps muscle forces induce increased anterior translation of the tibia, while hamstrings muscle forces have the opposite effect. Markolf et al. [45] reported that passive extension of the knee generated forces in the ACL only during the last 10° of extension (at 5° of hyperextension, the mean force was found to be 118 N). When a 200 N force generated through the quadriceps tendon was applied to the knee, the force in the ligament increased at all angles of knee flexion, with the largest changes in the passive case occurring between 5° and 40° of flexion.

In our laboratory, we have studied the role of quadriceps and hamstring co-contraction in the "unloading" of the ACL during knee flexion-extension [61]. Through the use of our Oxford Rig assembly [62], unconstrained 5-DOF motion was allowed while simulating knee flexion-extension during vertical stance loading. In this study, muscle forces were applied to a cadaveric knee to simulate a squat exercise with isolated quadriceps activation, and then also with quadriceps/hamstrings co-contraction. By testing the knee in intact, ACL-deficient, and ACL-reconstructed states, it was demonstrated that muscle co-contraction has a significant effect on knee kinematics. The introduction of hamstring force with quadriceps force significantly reduces anterior tibial translation in the ACL-deficient knee, and also reduces forces in the reconstructed ACL during extension (Fig. 6a,b). The hamstrings have the potential of negating the increased strains in the ACL caused by quadriceps activity, and may indicate the usefulness of closed-chain (co-contraction) kinetic exercises during rehabilitation after ACL reconstruction.

From these studies, it should be clear that muscle activities substantially alter the kinematics of the knee. These changes may offset or accentuate other conditions present at the knee, such as ACL deficiency, and should therefore be considered.

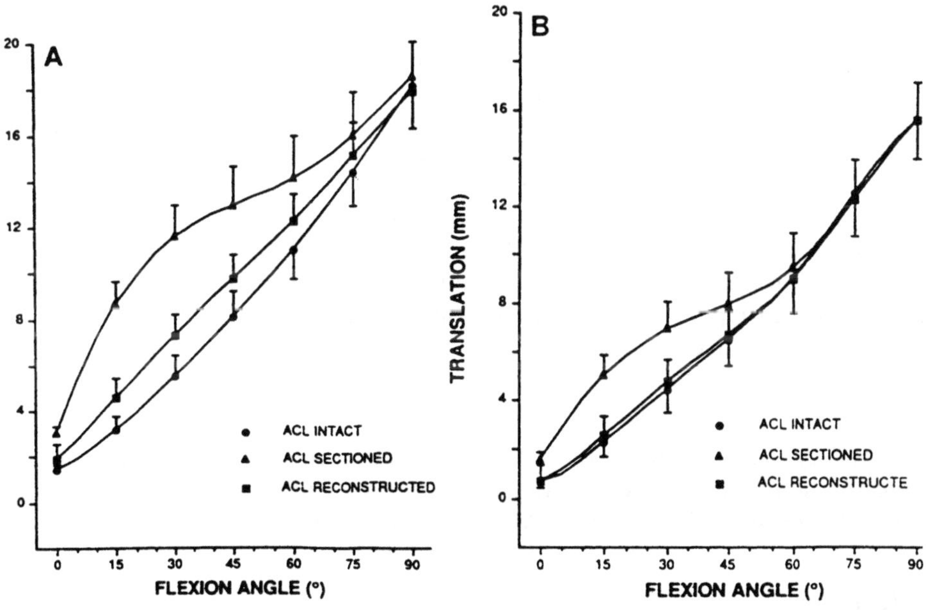

Fig. 6a,b. Anterior tibial translation versus flexion angle for the intact, ACL-deficient, and ACL-reconstructed knee under quadriceps-activated extension. **a** Results with no hamstring activation. **b** 90 N hamstring load (from [61])

Future Directions and Conclusions

Soft tissues in the knee, and the ACL in particular, play both primary and secondary roles in maintaining joint stability. Tremendous advances have been made in the surgical treatment of ACL deficiency, although the focus in evaluation of ACL reconstruction has thus far been placed on the reestablishing of its primary role in the restraint of anterior translation of the tibia. For this reason, most postoperative evaluations of ACL reconstruction involve only a single-DOF test of the knee joint (e.g., the Lachman test), which evaluates only the primary function of the ACL replacement graft. As we increase our understanding of the multiple roles played by the ACL in the maintenance of joint stability, surgical reconstruction technique and replacement graft materials can be improved to address the additional roles the ACL plays in the normal knee. Thus, clinical evaluation of knee kinematics and the eventual success of reconstructive procedures may in the future involve more complex multi-DOF testing.

Despite recent technical advances in the measurement of the normal kinematics of the human knee, as well as developments in the determination of the in situ lengths, strains, and forces within discrete knee structures, very little is known about the response of this joint to complex external loadings. A detailed understanding of the kinematics of the knee during physiological function and under external loadings requires much further work.

The advent of multi-DOF testing machines such as a six-axis articulated manipulator [55] holds promise for the advanced kinematics testing of all synovial joints. Through the use of hybrid force/position control, complex external loadings of the intact knee may be simulated and the resulting kinematics measured. As the path of the intact knee is repeated by the manipulator, information on the in situ forces occurring in ligaments about the joint may also be determined [52]. Further, the effects of muscular stabilization are also important, because they dictate the design and implementation of rehabilitation protocols. As a specific example, the variables affecting the success of ACL reconstruction may also be investigated in this manner, and such factors as the changes in graft placement or initial tension may be evaluated, One of the significant advantages of such an approach is that all data can be obtained from the same specimen, thereby reducing testing variability.

Acknowledgments. Some of the work contained in this chapter was performed while the senior author was at the University of California, San Diego. The financial support of NIH grant #AR39683 is gratefully appreciated.

References

1. Noyes FR, Grood ES (1976) The strength of the anterior cruciate ligaments in humans and rhesus monkeys: age-related and species-related changes. J Bone Joint Surg 58A:1074–1082

2. Reuben JD, Rovick JS, Schrager RJ, Walker PS, Boland AL (1989) Three-dimensional dynamic motion analysis of the anterior cruciate ligament deficient knee joint. Am J Sports Med 17:463–471
3. Markolf K, Bargar W, Shoemaker S, Amstutz H (1981) The role of joint load on knee stability. J Bone Joint Surg 63A:57–585
4. Markolf KL, Mensch JS, Amstutz HC (1976) Stiffness and laxity of the knee—the contributions of the supporting structures. J Bone Joint Surg 58A:583–593
5. Jackson D (ed) (1993) The anterior cruciate ligament: current and future concepts. 1st edn. Raven, New York, 462
6. Fetto JF, Marshall JL (1980) The natural history and diagnosis of anterior cruciate ligament insufficiency. Clin Orthop 147:29–38
7. Hirshman H, Daniel D, Miyasaka K (1991) The fate of un-operated knee ligament injuries. In: Daniel D, Akeson W, O'Connor J (eds) Knee ligaments: structure, function, injury and repair. Raven New York, pp 481–503
8. Jacobson K (1977) Osteoarthritis following insufficiency of the cruciate ligament in man: a clinical study. Acta Orthop Scand 48:520–526
9. Noyes F, Matthews D, Mooar P, Grood E (1983) The symptomatic anterior cruciate deficient knee: Part I. The long-term functional disability in athletically active individuals. J Bone Joint Surg 65A:154–162
10. Satku K, Kumar V, Ngoi S (1986) Anterior cruciate ligament injuries: to counsel or to operate? J Bone Joint Surg 68B:458–461
11. DeHaven K (1985) Meniscus repair in the athlete. Clin Orthop 198:31–35
12. Belkoff S, Haut R (1992) Microstructurally based model analysis of gamma-irradiated tendon allografts. J Orthop Res 10:461–464
13. Bulter DL, Kay MD, Stouffer DC (1986) Comparison of material properties in fascicle-bone units from human patellar tendon and knee ligaments. J Biomech 19(6):425–432
14. France E, Paulos L, Rosenberg T, Harner C (1988) The biomechanics of anterior cruciate allografts. In: Friedman M, Ferkel R (eds) Prosthetic ligament reconstruction of the knee. Saunders, Philadelphia
15. Haut R, Powlison A (1990) The effects of test environment and cyclic stretching on the failure properties of human patellar tendons. J Orthop Res 8:532–540
16. Haut R, Powlison A, Rutherford G, Kateley J (1989) Order of irradiation and lyophilization on the strength of patellar tendon allografts. Trans Orthop Res Soc 14:514
17. Johnson G, Choi N, Tramaglini D, McMahon P, Woo SL-Y (1992) Viscoelastic properties of the human patellar tendon: age related changes. N Am Congr Biomech 16:9–10
18. Takai S, Livesay GA, Woo SL-Y, Adams DJ, Fu FH (1993) Determination of the in-situ loads on the human anterior cruciate ligament. J Orthop Res 11:686–695
19. Kennedy JC, Hawkins RJ, Willis RB, Danylchuk KD (1976) Tension studies of human knee ligaments, yield point, ultimate failure, and disruption of the cruciate and tibial collateral ligaments. J Bone Joint Surg 58A:350–355
20. Trent PS, Walker PS, Wolf P (1976) Ligament length patterns, strength and rotational axes of the knee joint. Clin Orthop 117:263–270
21. Woo SL-Y, Hollis JM, Adams DJ, Lyon RM, Takai S (1991) Tensile properties of the human femur-anterior cruciate ligament-tibia complex: the effect of specimen age and orientation. Am J Sports Med 19:217–225
22. Arnoczky SP (1983) Anatomy of the anterior cruciate ligament. Clin Orthop 172:19–25

23. Girgis FG, Marshall JL, Al Monajem ARS (1975) The cruciate ligaments of the knee joint: anatomical and experimental analysis. Clin Orthop 106:216–231
24. Odensten M, Gillquist J (1985) Functional anatomy of the anterior cruciate ligament and a rationale for reconstruction. J Bone Joint Surg 67A:257–261
25. Woo SLY, Hollis JM, Roux RD, Gomez MA, Inoue M, Kleiner JB, Akeson WH (1987) Effects of knee flexion on the structural properties of the rabbit femur-anterior cruciate ligament-tibia complex. J Biomech 20:557–563
26. Wang CJ, Walker PS, Wolf B (1973) The effects of flexion and rotation on the length patterns of the ligaments of the knee. J Biomech 6:587–596
27. Butler DL, Martin ET, Kaiser AD, Grood ES, Chun KJ, Sodd AN (1988) The effects of flexion and tibial rotation on the 3-D orientations and lengths of human anterior cruciate ligament bundles. Trans ORS 13:59
28. Lew WD, Lewis JL (1978) Technique for calculating in-vivo ligament lengths with application to the human knee joint. J Biomech 11:365–377
29. Hollis MJ, Takai S, Adams DJ, Horibe S, Woo SLY (1991) The effects of knee motion and external loading on the length of the anterior cruciate ligament: a kinematic study. J Biomech Eng 113:208–214
30. Takai S, Adams DJ, Livesay GA, Woo SL-Y (1991) Determination of load in the human anterior cruciate ligament. Trans ORS 16:235
31. Kennedy JC, Hawkins RJ, Wilis RB (1977) Strain gauge analysis of knee ligaments. Clin Orthop 129:225–229
32. Meglan D, Zuelzer W, Buck W, Berme N (1986) Effects of quadriceps force upon strain in the anterior cruciate ligament. Trans ORS 11:55
33. Arms SW, Pope MJ, Johnson RJ, Fischer RA, Arvidsson I, Eriksson E (1984) The biomechanics of anterior cruciate ligament rehabilitation and reconstruction. Am J Sports Med 12(1):8–18
34. Beynnon BD, Fleming BC, Pope MH, Nichols CE, Johnson RJ, Renstrom PA, Stankewich CJ (1993) The measurement of anterior cruciate ligament strain during rehabilitation activities in-vivo. Trans ORS 18:308
35. Beynnon BD, et al. (1989) An in-vivo study of the ACL strain biomechanics in the normal knee, Trans ORS 14:324
36. Fleming B, Beynnon B, Howe J, Pope M (1992) Effect of tension and placement of a prosthetic anterior cruciate ligament on the anteroposterior laxity of the knee. J Orthop Res 10:177–186
37. Fleming B, Beynnon B, Stankewich C, Pope M, Nichols C, Renstrom P, Johnson RJ (1992) The effect of quadriceps and hamstrings activity on strain in the anterior cruciate ligament. N Am Cong Biomech 16:117–118
38. Pope MH, Stankewich CJ, Beynnon BD, Fleming BC (1991) Effect of knee musculature on anterior cruciate ligament strain in-vivo. J Electromyogr Kinesiol 1(3):191–198
39. Ray JD, Feder SM, Bulter DL, Korvick DL (1993) Simultaneous, in vivo measurement of tendon force and strain in the goat model. Trans ORS 18:153
40. Ahmed AM, Burke DL, Duncan NA, Chan KH (1992) Ligament tension pattern in the flexed knee in combined passive anterior tibial translation and axial rotation. J Orthop Res 10:854–867
41. Ahmed AM, Hyder A, Burke DL, Chan KH (1987) In-vivo ligament tension patterns in the flexed knee in passive loading. J Orthop Res 5:217–230
42. Barry D, Ahmed AM (1986) Design and performance of a modified buckle transducer for the measurement of ligament tension. J Biomech Eng 108:149–152

43. Lewis JL, Lew WD, Schmidt J (1982) Note on the application and evaluation of the buckle tranducer for knee ligament force measurement. J Biomech Eng 104: 125–128
44. France PE, Daniels AU, Goble ME, Dunn HK (1983) Simultaneous quantitation of knee ligament force J Biomech 16:553–564
45. Markolf KL, Gorek JF, Kabo JM, Shapiro MS (1990) Direct measurement of resultant forces in the anterior cruciate ligament. J Bonc Joint Surg 72A:557–567
46. Markolf KL, Wascher DC, Finerman GAM (1993) Direct in vitro measurement of forces in cruciate ligaments. Part II: The effect of section of the posterolateral structures. J Bone Joint Surg 75A:387–394
47. Wascher DC, Markolf KL, Shapiro MS, Finerman GAM (1993) Direct in vitro measurement of the forces in cruciate ligaments. Part I: The effect of multiplane loading in the intact knee. J Bone Joint Surg 75A:377–386
48. Holden JP, Korvick DL, Grood ES, Cummings JF, Bylski-Austrow DL (1992) In-vivo forces in the anterior cruciate ligament during walking and trotting in a quadraped. In: The Second North American Congress on Biomechanics, Chicago, IL, 1992
49. Korvick DL, Holden JP, Grood ES, Cummings JF (1993) Relationships between patellar tendon, anterior cruciate ligament and ground reaction forces: an in-vivo study in goats. Trans ORS 18:336
50. Vahey JW, Draganich LF (1991) Tensions in the anterior and posterior cruciate ligaments of the knee during passive loading: predicting the ligament loads from in-situ measurements. J Orthop Res 9:529–538
51. Fujie H, Livesay GA, Kashiwaguchi S, Blomstrom GL, Woo SL-Y (1992) A new methodology for direct, non-contact determination of in-situ forces in soft tissues. N Am Cong Biomech 16:7–8
52. Fujie H, Livesay GA, Woo SL-Y, Kashiwaguchi S, Blomstrom GL (1994) The use of a universal force-moment sensor to determine the in-situ forces in soft tissues: A new methodology. J Biomech Eng (in press)
53. Guan Y, Butler DL, Dormer SG, Feder SM, Mayhan MJ (1993) Anterior cruciate subunit response during anterior drawer. ASME Biomech Symp AMD-113:13–16
54. Livesay GA, Fujie H, Blomstrom GL, Kashiwaguchi S, Woo SL-Y (1993) Determination of the in-situ forces in the human anterior cruciate ligament with a universal force sensor. Trans ORS 18:1
55. Fujie H, Mabuchi K, Woo SL-Y, Livesay GA, Arai S, Tsukamoto Y (1993) The use of robotics technology to study human joint kinematics: a new methodology. J Biomech Eng 115:211–217
56. Fujie H, Mabuchi K, Tsukamoto Y, Yamamoto M, Sasada T (1990) A new method of human knee diagnosis using robotics. World Cong Biomech 1:58
57. Smidt GJ (1973) Biomechanical analysis of knee flexion and extension. J Biomech 6:79–92
58. Grood ES, Suntay WJ, Noyes FR, Butler DL (1984) Biomechanics of the knee-extension exercise. J Bone Joint Surg 66A:725–734
59. O'Conner JJ, et al. (1990) The muscle stabilized knee. In: Daniel DM, Akeson WH, O'Conner JJ (eds) Knee ligaments: structure, function, injury, and repair. Raven, New York
60. Renstrom P, Arms SW, Stanwyck TS, Johnson RJ, Pope MH (1986) Strain within the anterior cruciate ligament during hamstring and quadriceps activity. Am J Sports Med 14:83–87

61. More RC, Karras BT, Neiman R, Fritschy D, Woo SLY, Daniel DM (1993) Hamstrings—an anterior cruciate ligament protagonist: an in-vitro study. Am J Sports Med 21:231–237
62. Biden EN, O'Conner JJ (1990) Experimental methods used to evaluate knee ligament function. In: Daniel DM, Akeson WH, O'Conner JJ (eds) Knee ligaments: structure, function, injury and repair. Raven Press, New York

Variable Response of the Anterior Cruciate Ligament: Nonuniform Deformation and a Simple Model

MASAO TANAKA[1], MASAHIRO INOUE[2], and KEIRO ONO[2]

Summary. The biomechanical behavior of the femur–anterior cruciate ligament–tibia complex depends on knee flexion and the stretching axis. This phenomenon has been anticipated as the result of the difference in the fiber recruitment during stretching caused by knee position. This could be observed partially as the variable response of the ligament tissue with nonuniform deformation. This chapter demonstrates the variation of surface strain of the anterior cruciate ligament from Japanese white rabbits with different knee positions. The deformation was analyzed by using the dot lattice method and paying attention to the reference state. It is experimentally indicated that the anteromedial surface of the ligament does not carry the load under some stretching conditions even at the tension of 98 N on the whole ligament. This nonuniform response is modeled as the variable response model in terms of the area ratio concept, and the recruitment of cross-sectional area as to the stretching is estimated as the increase of the area ratio. This model is simple but provides a tool to understand the influence of knee position on the biomechanical behavior of the ligament.

Key words: Tissue engineering—Anterior cruciate ligament—Surface deformation—Nonuniform deformation—Structural model

Introduction

The bone–ligament–bone complex specimen is commonly used in biomechanical testing of the knee ligament. This is useful for overcoming the difficulties of insufficient tissue size and adjusting the fixing condition [1,2,3]. Several

[1] Department of Mechanical Engineering, Faculty of Engineering Science, Osaka University, Machikaneyama, Toyonaka, Osaka, 560 Japan
[2] Department of Orthopedic Surgery, Osaka University Medical School, Yamadaoka, Suita, Osaka, 565 Japan

techniques have been proposed to obtain tissue deformation from biomechanical testing with a complex specimen, and the photography/videocamera application is the current standard in measuring the surface deformation of the tissue.

By using the video dimensional analyzer system, Woo and associates [1] demonstrated overall tissue strain and variation of strain in the medial collateral ligament. Noyes and coworkers [2] also reported regional strain in the bone–patellar tendon–bone complex by using a high-speed camera system. Butler and associates [4] examined the nonuniform surface strains along the tissue axis of the bone–tissue fascicle–bone complexes from the patellar tendon and cruciate ligaments by using a stereo microscope system.

In the case of cruciate ligaments, variations in fiber bundle lengths within the tissue and less parallel fiber bundle alignment make it difficult to determine tissue deformation of the whole ligament from the deformation of isolated tissue fascicles. This chapter demonstrates surface strain variation of the whole anterior cruciate ligament of the rabbit observed by using the dot lattice method through the stretching testing of the femur–anterior cruciate ligament–tibia complex. A simple variable response model is then discussed, based on the area ratio concept, to provide a phenomenological understanding of the different structural responses of the anterior cruciate ligament with different knee positions.

Testing and Analysis Procedures

Biomechanical testing was conducted for specimens from Japanese white rabbits, and the dot lattice method [5] was applied to examine the strain distribution of the anteromedial surface of the whole anterior cruciate ligament.

Specimen Preparation

The femur–whole knee joint capsule–tibia complexes are acquired from the hind limbs of healthy Japanese white rabbits, cleaned of skin and muscle tissue, and frozen to −30°C until the biomechanical test. The samples are then thawed to room temperature, and the femur and tibia are cut to appropriate length and thickened with bone cement to adjust the shape to fit the attachment of the fixture. These bones are grasped by the attachments with screws, and the relative positions and orientations between the bones are recorded for necessary knee flexion angles at the neutral position of valgus-varus and tibial rotations. Then, all periarticular soft tissue is dissected away, leaving only the anterior cruciate ligament intact, and we have the femur–anterior cruciate ligament–tibia complex specimen.

The medial condylus of the femur is removed to keep the ligament visible to the camera from the medial side, leaving the insertion point of the ligament as is. A number of dots of almost 0.2 mm in diameter are plotted with black India ink on the medial surface of the anteromedial bundle of the ligament. The dots are arranged to be aligned along fibers of the bundle, making a lattice with approximately 1 mm pitch that is used for the observation of surface deformation.

Testing Apparatus

The femur–anterior cruciate ligament–tibia complex specimen grasped by the attachments at femur and tibia is clamped to the fixture and mounted on the electrohydraulic universal testing machine in such a manner that the specified knee position is maintained between the attachments and the stretching axis of the specimen becomes collinear with that of the testing machine. The CCD (Charge Coupled Device)-TV camera is focused on the dot lattice on the ligament surface from the medial side of the specimen, and the movements of the dots are recorded continuously on a standard video cassette recorder with the superimposed clock image as shown in Fig. 1. Stretching is enforced by the specific rate of the piston head movement of the testing machine. The voltage output of the load cell is recorded as the time history triggered by initiation of stretching.

Dot Lattice Processing

A picture frame at the undeformed stage is paged out from the recorded sequence and is captured in the processor. The region of interest for the deformation analysis is defined, and some filtering operations are applied to the gray picture to improve the contrast, to remove the background shading, and so on. The threshold operation then transforms the gray picture to the binary one and extracts the dot lattice. The positions of dots extracted are measured in terms of their centroids. The picture frames at the deformed stage of interest are also paged out and sent to the same image processing procedure, and the positions of dots are obtained again.

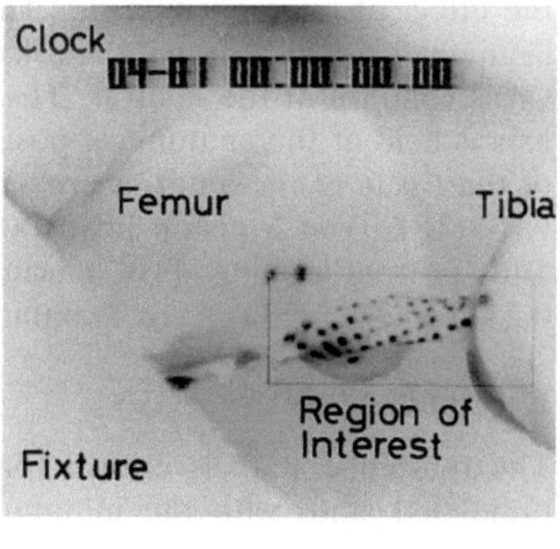

Fig. 1. Dot lattice on medial surface of anteromedial bundle

Deformation Analysis

The undeformed and deformed pictures are compared by referring to the position of the dot placed on the bone. The displacement of each dot of the lattice is then calculated as the difference of the position in the deformed and undeformed pictures. Figure 2a shows an example of extracted dot positions superimposed on the gray pictures at the initial state and at the state of maximum load. The displacements along the tensile and transverse axes are interpolated by using the piecewise polynomial functions defined over the region of interest. The strain distribution is calculated by differentiating the interpolated displacement fields according to the definition. Figure 2b shows an example contour map of the displacement and the strain along the tensile axis. These data are valid only on the surface of the anteromedial bundle contoured by the solid lines in the figure. The coordinate transformation gives us the displacement and the strain along a specific fiber on the ligament surface (Fig. 3), showing the distribution of deformation along the tissue fibers. The example analysis shown in Figs. 2 and 3 is for the case of stretching to failure along the ligament axis with knee flexion at 45° and valgus-varus and tibial rotations at neutral.

Surface Deformation

It has been reported that the structural properties of the femur–anterior cruciate ligament–tibia complex depend on the knee position sensitively. It is considered that the reason results from the difference in recruiting of tissue fibers. This could be visible as the difference in the tissue deformation. In this section, the difference of the surface deformation of the anteromedial bundle of the anterior cruciate ligament is demonstrated for the different knee flexion angles and stretching directions.

Examined Knee Positions

The testing conditions are determined by referring to the previous report on the effect of knee position [6]. That is, the specimens are stretched along the ligament or tibia axis with the knee flexion at 10°, 45°, or 90°, with valgus-varus rotation at the neutral. The rotational position with respect to the tibia axis is held at the anatomical position so as to maintain the helical orientation of the tissue of the anterior cruciate ligament as it normally is. Tensile load to 98 N (10 kgf) at a piston head rate of 50 mm/min is applied to the specimen after preconditioning. Five specimens harvested from rabbits weighing 3.2 ± 0.2 kg (mean ± SD) were examined.

Reference State

The reference state is critical for evaluating the deformation. Although the preloaded state with the minimum load of the apparatus is a conventional

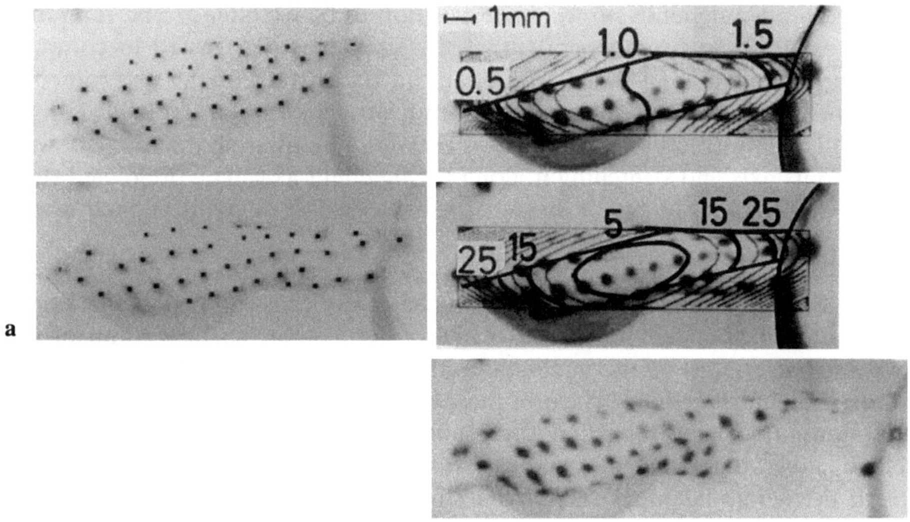

Fig. 2a,b. Analysis of surface deformation. Specimen is stretched along ligament axis with knee flexion at 45° without valgus-varus or tibial rotation. **a** Extracted dot positions at preload of 1.96 N (*top*) and at ultimate load of 167 N (*bottom*). **b** Contour map of displacement (*top*) and strain (*center*) in stretching axis at ultimate load, and failure mode (*bottom*)

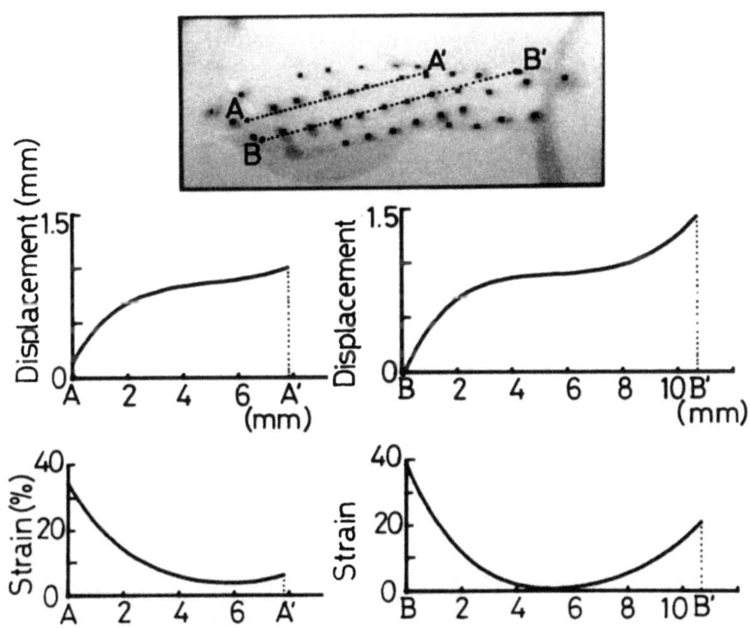

Fig. 3. Surface strain distribution along fiber segment on anteromedial bundle. The deformation is calculated by referring to the preloaded state in this case

reference, it also depends on the knee position to be examined. The reference state is expected to be independent of the knee positions tested to compare the deformations among them. The fibers on the surface of the anteromedial bundle become almost parallel and uniformly stretched when the helical alignment of the anterior cruciate ligament is unwound by rotating the tibia 90° with respect to its axis and stretching it along the ligament axis. The preloaded state with 1.96 N (0.2 kgf) at this position with knee flexion at 10° is defined as the reference state here.

The distance along the stretching axis between the dots at the most femoral and tibial sides among the dot lattice is used as the conventional gauge length of the anteromedial bundle. Figure 4 shows the average surface strain of the anteromedial bundle at the preloaded state for the individual knee position tested. The anteromedial surface of the anterior cruciate ligament is somewhat under tension or neutral when the whole ligament is stretched by a small preload along the ligament axis, but is compressed, that is, relaxed, even though the whole ligament carries the stretching preload along the tibia axis. The preloaded state depends greatly on the stretching direction and also on the knee flexion angle.

Distribution of Surface Strain

This section demonstrates strain distribution along a fiber segment located on the posterior region of the surface of the anteromedial bundle within the view of the camera. Figure 5 shows the distribution of strain along the axis of the fiber segment. The abscissa is the normalized location along the fiber segment

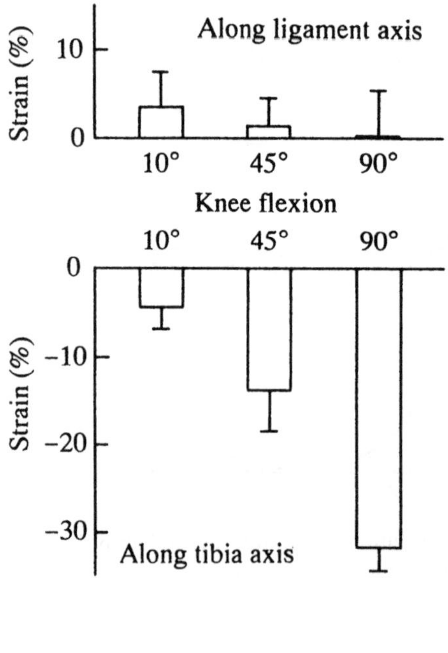

Fig. 4. Average surface strain at the preloaded state. The deformation is calculated on the basis of the reference state independent of knee position

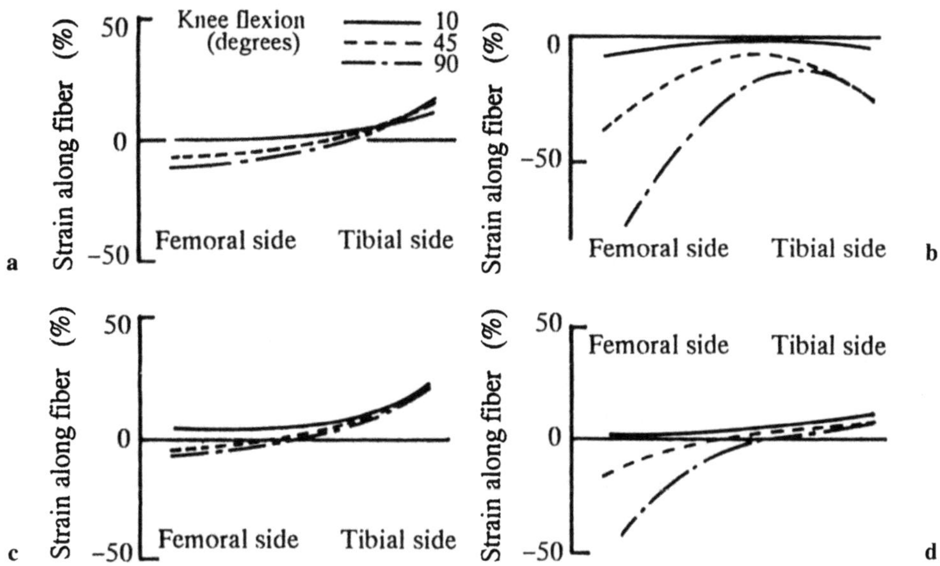

Fig. 5a–d. Surface strain distribution along fiber segment. **a** 1.96 N along ligament axis. **b** 1.96 N along tibia axis. **c** 98 N along ligament axis. **d** 98 N along tibia axis

of interest, and the ordinate is the mean of strain over the specimens. At 1.96 N preload along the ligament axis, the fiber segment is stretched in the tibial side but remains unstretched in the femoral side, showing the nearly zero or negative strain. At preload along the tibia axis, the fiber segment remains relaxed, especially near both the femur and tibia.

At a load of 98 N, the entire region of the fiber segment of interest is stretched when the knee flexion is shallow at 10°. On the other hand, the tibial side is tensioned but the femoral side remains relaxed when the knee flexion is at 45° or 90°. These characteristics are found regardless of the axis of stretching, but the tensile and compressive strains when stretching occurs along the ligament axis are higher than those strains recorded when the stretching is along the tibia axis.

If the preloaded state for an individual testing condition is considered as the reference of the deformation, the increase of strain from the preloaded to the 98 N loaded state looks like the surface strain. This strain is positive regardless of the testing conditions and shows higher tension at both femoral and tibial parts than at the central part [7]. However, this is the apparent strain and is different from the real tissue strain at the surface.

These average strain and strain distributions indicate the nonuniformity of tissue deformation on the surface of the anteromedial bundle of the anterior cruciate ligament under conditions of stretching. It also implies that the surface of the anteromedial bundle is not as active in carrying the tensile load in these testing conditions, and that some other cross-sectional part of the anterior cruciate ligament is strongly responsible for load resistance. This is critical for understanding the mechanical behavior of the anterior cruciate ligament.

Simple Model of Variable Response

The area ratio concept is useful for understanding of nonuniformity over the cross section of the ligament, because some cross-sectional part does not carry the applied load.

Area Ratio and Variable Response Model

The effective cross-sectional area A_e is defined as the area carrying the load. Insofar as no part of the ligament fails, it increases uniformly as to the progress of the stretching that may be represented by the structural elongation x. That is, the area ratio A is then defined as follows:

$$A(x) = A_e(x)/a \tag{1}$$

where a is the physical cross-sectional area of the whole anterior cruciate ligament.

The anterior cruciate ligament is modeled as a collection of tissue fibers of different lengths (Fig. 6) because it is basically a parallel-fibered collagenous tissue in spite of its helical alignment. This modeling is similar to the structural model by Kwan and Woo [8]. Let us consider the ideal uniform response in which all these fibers are stretched uniformly (Fig. 6a), and denote its load versus elongation by $P^*(x)$. The load versus elongation of the various responses (Fig. 6b) is then written in terms of the area ratio A and the uniform response $P^*(x)$:

$$P(x) = \mathrm{P}^*(x)A(0) + \int_o^x P^*(x - u)\frac{dA}{\mathrm{dx}}(u)du \tag{2}$$

where $(dA/dx)(u)du$ denotes the increase of A during the progress of elongation $x = u \ to \ u + du$, and $A(0)$ stands for the area ratio at the preload [9].

In this model, the uniformity of the mechanical properties of fibers is assumed without loss of generality because any mechanical property can be represented as the resultant of fibers with the same mechanical properties but different lengths. It is noted, in this sense, that the effective cross-sectional area does not give the physical area but the equivalent area consisting of the fibers of uniform property.

Estimated Area Ratio

Figure 7 illustrates the load versus structural elongation curves obtained by the stretching test and represented by the exponential toe region and linear region. The area ratio $A(x)$ can be estimated so as to minimize error functional:

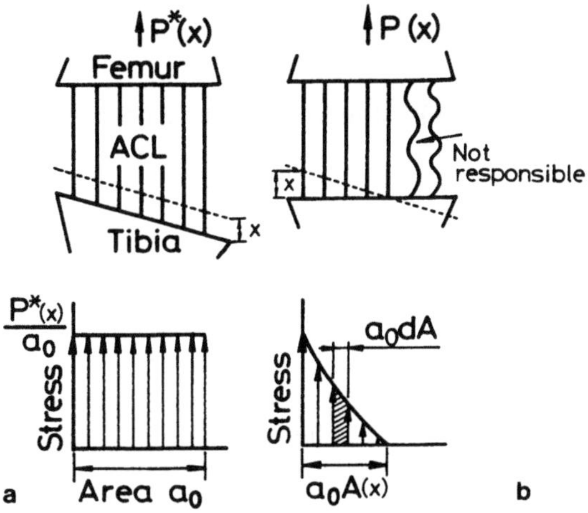

Fig. 6a,b. Simple model of femur–anterior cruciate ligament–tibia complex and area ratio concept. a Ideal uniform response. b Variable response

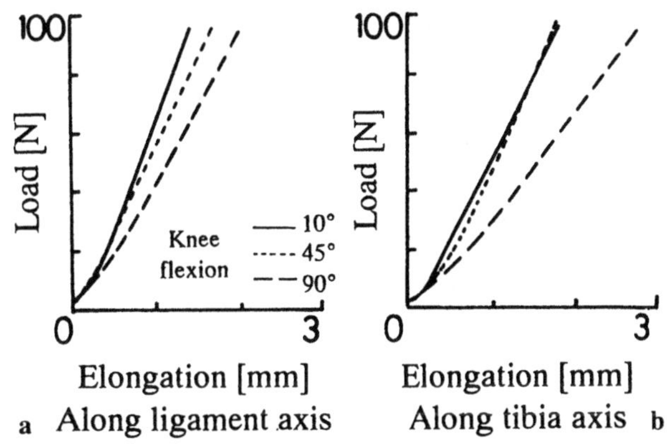

Fig. 7a,b. Load versus structural elongation curve. a Stretched along ligament axis. b Stretched along tibia axis

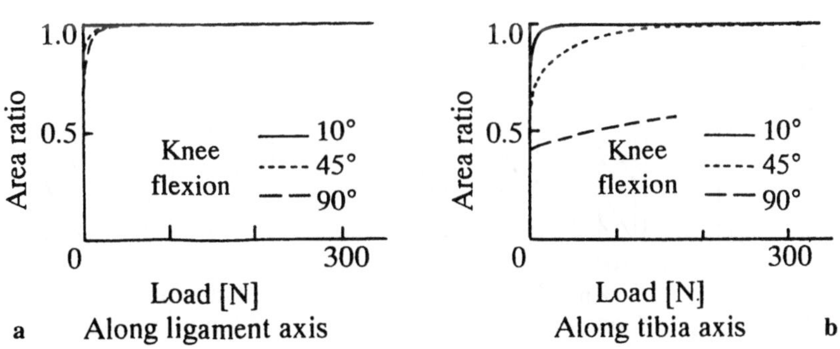

Fig. 8a,b. Estimated area ratio recruitment. a Stretched along ligament axis. b Stretched along tibia axis

$$\int_{o}^{x} [P_{ex}(x) - P(x)]^2 \, dx \tag{3}$$

where $P_{ex}(x)$ is the load versus elongation obtained by stretching test within the load range to $P = P(x)$. Figure 8 shows the estimated area ratio by assuming the exponential form

$$A(x) = 1 - c \exp(dx) \tag{4}$$

In this estimation, the load versus elongation experimentally obtained with the knee flexion at 10° stretched along the ligament axis is used as that of uniform response P^* because it is the stiffest among those in Fig. 7. The area ratio within the load range to $P = 98$ N is the estimated, and that in the remainder is the extrapolation where the ultimate load of the uniform response is assumed to be 343 N (35 kgf). Figure 9 shows the calculated ultimate load based on the estimated area ratio. The calculated ultimate load under stretching conditions along the ligament axis shows no dependency on the knee flexion, but the load when stretching occurs along the tibia axis decreases as to the increase of knee flexion. These qualitative characteristics agree with the experimental results in the literature [6]. The quantitative difference between the calculated and experimental values in the case of stretching along the tibia axis results from the ligament peeling off, which is the failure mode in the experiment.

Conclusions

The nonuniform response of the anterior cruciate ligament was demonstrated by means of surface deformation through the stretching test and dot lattice

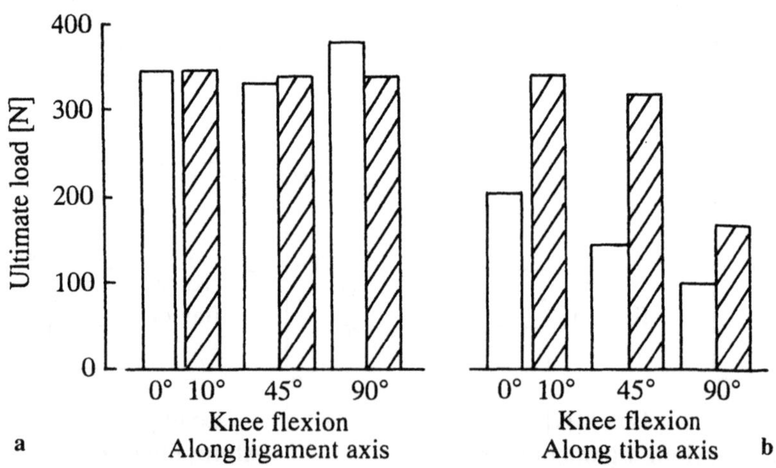

Fig. 9a,b. Calculated ultimate load. **a** Stretched along ligament axis. **b** Stretched along tibia axis. *Open bars*, experiment [6]; *hatched bars*, calculated

analysis. It was observed that the anteromedial surface of the anterior cruciate ligament is not responsible for carrying the load under some stretching conditions. This nonuniform response of the anterior cruciate ligament is modeled as the variable response model in terms of the area ratio. This model is simple but effective in understanding the mechanical behavior of the ligament. The model-based approach has the potential to be extended for further study on ligament mechanics.

Acknowledgment. This work was supported in part by a grant-in-aid for scientific research from the Ministry of Education, Science and Culture of Japan.

References

1. Woo SLY, Gomez MA, Seguchi Y, Endo CM, Akeson WH (1983) Measurement of mechanical properties of ligament substance from a bone-ligament-bone preparation. J Orthop Res 1:22–29
2. Noyes FR, Butler DL, Grood ES, Zernicke RF, Hefzy MS (1984) Biomechanical analysis of human ligament grafts used in knee-ligament repairs and reconstructions. J Bone Joint Surg 66A:344–352
3. Shino K, Kawasaki T, Hirose H, Gotoh I, Inoue M, Ono K (1984) Replacement of the anterior cruciate ligament by an allogenic tendon graft. J Bone Joint Surg 66B:672–681
4. Butler DL, Sheh MY, Stouffer DC, Samaranayake VA, Levy MS (1990) Surface strain variation in human patellar tendon and knee cruciate ligaments. J Biomech Eng 112:38–45
5. Tanaka M, Seguchi Y, Sonoda Y, Inoue M (1988) Feasibility study of in situ measurement of deformation distribution of knee ligament by using image processing. In: Miller GR (ed) (1988) Advances in bioengineering. ASME, New York, pp 61–64
6. Woo SLY, Mollis RD, Roux RD, Gomez MA, Inoue M, Kleiner JB, Skeson WH (1987) Effects of knee flexion on the structural properties of the rabbit femur-anterior cruciate ligament-tibia complex (FATC). J Biomech 20:557–563
7. Tanaka M, Seguchi Y, Mizutani T, Inoue M (1989) A note on the ACL deformation under tensile loading along ligament and/or tibia direction. In: Woo SLY, Seguchi Y (eds) Tissue engineering 1989. ASME, New York, pp 115–118
8. Kwan MK, Woo SLY (1989) A structural model to describe the nonlinear stress-strain behavior for parallel-fibered collagenous tissues. J Biomech Eng 111:361–363
9. Tanaka M, Kannen Y, Inoue M, Maeda A (1993) Model-based identification of variable response of anterior cruciate ligament: effect of knee flexion, valgus-varus rotation and stretching direction. In: Held KD, Brebbia CA, Ciskowski RD, Power H (eds) Computational biomedicine. Computational Mechanics Publications, Southampton, Boston, pp 231–238

Anatomy and Biomechanics of the Human Posterior Cruciate Ligament

Glen A. Livesay, Christopher D. Harner, John W. Xerogeanes, Gregory J. Carlin, Takeshi Kusayama, Hiromichi Fujie, Shinji Kashiwaguchi, and Savio L.-Y. Woo

Summary. With the significant amount of research directed towards the anterior cruciate ligament (ACL) over the last two decades, the functional importance of the posterior cruciate ligament (PCL) has also become evident. As is true for the ACL, PCL insufficiency can lead to progressive laxity of secondary stabilizers of the knee, resulting in localized pain, swelling, and joint instability, and long-term degenerative changes. While surgical reconstruction is often performed in active patients, a better understanding of both the anatomy and biomechanics of the PCL is required in order to ensure the proper selection of appropriate structure and material for the graft, as well as the optimal graft placement. This chapter will attempt to review, quantify, as well as compare the basic anatomical and biomechanical properties of the PCL (and its subportions) and meniscofemoral ligaments (MFLs). In considering PCL anatomy, the midsubstance cross-sectional and insertion site areas of the PCL and MFLs will be compared and contrasted with those of the ACL. The tensile properties of these tissues, i.e., the structural and mechanical properties, will then be addressed. The information reviewed should provide a better understanding of the structure-function relationship of the PCL and supply initial basic science data to assist in guiding PCL reconstruction.

Key words: Posterior cruciate ligament—Meniscofemoral ligament—Anterior cruciate ligament—Anatomy—Biomechanics—Insertion site—Knee—Ligament

---

Musculoskeletal Research Center, Department of Orthopedic Surgery, University of Pittsburgh, 3471 5th Avenue, Pittsburgh, PA 15213, USA

Introduction

During the past 20 years, the study of knee ligament injuries has focused on the anterior cruciate ligament (ACL), primarily because of the predominance of ACL injuries in both the athletic and general population. In the United States alone, as many as one-half million significant ACL injuries occur annually [1]. Because of the poor inherent ability of the ACL to heal [2,3], most of the literature has concentrated on various techniques of ligament reconstruction. Over time, it has become evident that information on the anatomical, biomechanical, and kinematic properties of the ACL is necessary to select the most suitable replacement graft and design the proper surgical reconstruction.

As we learn more about the ACL, and about the knee in general, the functional importance of the posterior cruciate ligament (PCL) becomes increasingly apparent. While the ACL is considered the primary restraint to anterior translation of the tibia, the PCL is the primary restraint to posterior translation of the tibia [4–6]. The PCL is also recognized by many authors as the primary stabilizer of the knee joint [6–8]. It is well documented that PCL insufficiency can lead to progressive laxity of secondary stabilizers of the knee, resulting in localized pain, swelling, and joint instability [8–12]. It has also been shown to lead to changes both in the medial and patellofemoral compartment pressures [13] and in subsequent degenerative osteoarthrosis of these compartments [8,14].

Although the sequelae of PCL injury are similar to those of ACL injuries, interest in the PCL has lagged behind the ACL because PCL injury does not occur as frequently, accounting for between 3.4% and 20% of all knee ligament injuries [15]. The wide range in these reported rates of injury is partly the result of the difficulty associated with the clinical diagnosis of PCL injury. A further difficulty is that some patients have the ability to functionally compensate for PCL deficiency for an extended period of time.

During the past 20 years, methods of evaluating ligament injuries have improved, but questions regarding proper treatment modalities still remain. Surgical reconstruction is often performed for active patients in an attempt to restore knee stability and prevent secondary osteoarthritic changes. The first PCL reconstruction reported was performed by Hey Groves in 1917 [16], and since that time, several reconstructive techniques employing various anatomical alterations and graft materials have been developed [8,17–21]. For these procedures to be successful, however, a better understanding of the anatomy and biomechanics of the PCL is required to ensure the proper selection of appropriate structure and material for the graft, as well as the optimal graft placement.

In this chapter, we attempt to quantify and compare the basic anatomical and biomechanical properties of the PCL and meniscofemoral ligaments (MFLs). When studying the PCL, we chose to divide it into two functionally distinct portions, the anterolateral (AL-PCL) and the posteromedial (PM-PCL) PCLs. In considering PCL anatomy, we compare and contrast the midsubstance cross-sectional and insertion site areas of PCL and MFLs with those of the ACL. The tensile properties of these tissues, that is, the structural and mechanical properties, are then discussed. It is our hope that the data obtained will allow a better understanding of the structure–function relationship of the PCL and provide the necessary basic data to assist in guiding the PCL reconstruction.

Quantitative Anatomy of the PCL

The gross anatomy of the PCL is well described. It arises from the posterior tibia and travels in the anteromedial direction behind the ACL to the lateral surface of the medial femoral condyle. This ligament is intraarticular but extrasynovial; the synovium is reflected from the posterior capsule and covers the ligament on its medial, lateral, and anterior aspects. The posterior portion of the PCL blends with the posterior capsule and periosteum [22,23]. It is composed of two functional portions, the anterolateral (AL-PCL) and posteromedial (PM-PCL). The AL-PCL portion is taut in flexion and the PM-PCL is taut in extension, although the lengths of the two portions appear similar. Using calipers, Girgis et al. [22] determined that the average length of the PCL was 38 mm and the average width was 13 mm.

The meniscofemoral ligaments (MFLs) travel in close proximity to the PCL. They arise from the posterior horn of the lateral meniscus and insert on the lateral aspect of the medial femoral condyle near the PCL insertion site. One runs anteriorly to the PCL (the anterior MFL, or ligament of Humphrey) and one posteriorly (the posterior MFL, or ligament of Wrisberg). Although the location of the ligaments is constant, their incidence is not; individuals may have both, one, or neither of the two MFLs in each knee. Large discrepancies in the incidence of these ligaments have been noted among studies reported in the literature. Published results show both ligaments in 6%–88% of the knees examined, an isolated Humphry in 4%–71%, an isolated Wrisberg in 8%–35%, and neither in 0%–23% [24,25].

Determination of Cross-Sectional Shape and Area

The surgical reproduction of normal cruciate ligament function requires an understanding of ligament geometry in situ. Anatomical studies have documented only qualitative information on ligament size and shape [22,23,25,26]. Quantitative information on the cross-sectional area of these ligaments has been based on width and thickness measurements, assuming a rectangular cross section [27–29]. Others have compressed the ligaments in a slot of known rectangular dimension at a specified pressure [30–33]. Alternatively, some investigators have attempted to preserve ligament shape while determining cross-sectional area, examining soft tissue slices with a planimeter [34], taking molds of the ligament and sectioning [35], or dragging a strain gauge micrometer across the ligament surface to determine ligament shape and cross-sectional area [36].

To avoid possible errors inherent to contact measurements, several non-contact measurement techniques have also been developed for determining the cross-sectional area and shape of a ligament or tendon [37–40]. The majority of these have been loosely based on the concept of shadow projection. In our laboratory, a computer-controlled laser micrometer system has been used to determine ligament cross-sectional shape and area [41,42]. This method requires no contact with the specimen, is compact, and has been shown to be accurate to within 3% for ligaments of variable cross-sectional shape [42].

The cross-sectional shape and area of the PCL, ACL, and MFL from eight knees (mean donor age, 64 years; range, 48–77 years) were measured using the laser micrometer to examine the effects of both knee flexion angle and position along the ligament axis. All epithelial, subcutaneous, and muscular tissue was removed from the knees; the ligaments and joint capsule were preserved and the specimen mounted in a set of clamps that allowed unconstrained knee flexion (Fig. 1). To allow determination of the in situ cross-sectional shapes and areas of the ACLs and PCL, specimens were passively moved to knee flexion positions of 0°, 30°, 60°, and 90° and the relative position and orientation of the tibia and femur recorded. All soft tissues except the cruciate ligaments were removed, and the joint was examined for signs of arthritis or previous ligamentous injury.

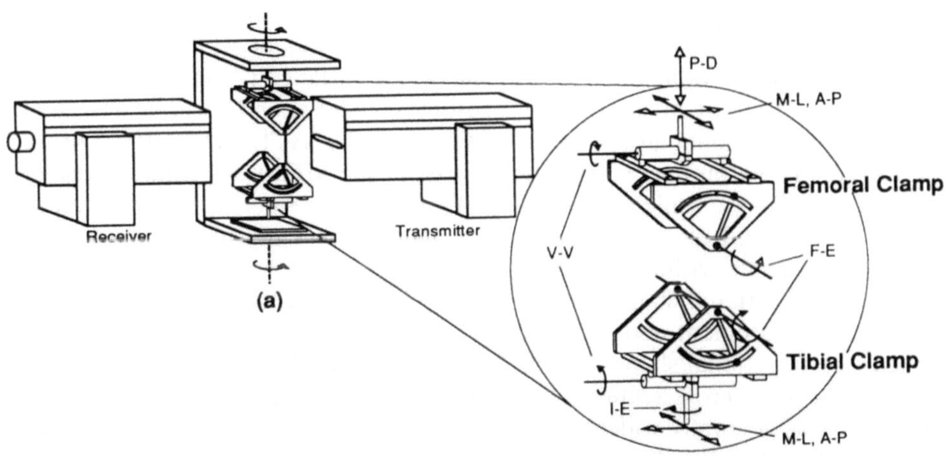

Fig. 1. Schematic diagram detailing the clamps used to fix and reproduce anatomic joint positions, and their relation to the laser micrometer. The degrees of freedom (DOFs) of the clamps are shown in the enlarged region. A collimated laser beam is emitted from the transmitter, passes across the clamped specimen, and is collected by the receiver. A computer-controlled stepper motor rotates the specimen about (*a*) in 3° increments

To clear a path for the 46-mm-wide collimated laser beam to reach the PCL, the ACL, together with the entire lateral femoral condyle and the anterior half of the tibial plateau, was temporarily removed. Bony landmarks were identified and marked for an accurate repositioning of the anterior half of tibial plateau and the lateral femoral condyle. Five equally spaced elastin stain markers were placed on the anterior surface of the ligament, the MFLs (Humphrey and/or Wrisberg) were identified and marked at the same midsubstance levels and then carefully removed.

The position of the entire femur–PCL–tibia assembly was adjusted in the sagittal and coronal planes to ensure that the collimated beam passed perpendicular to the marked PCL cross section. The ligament was minimally tensioned, and PCL profile widths of the proximal level were determined as the specimen was rotated through 180° in 3° increments by the computer-controlled stepper motor. The profile widths were then used in numerical integration to determine cross-sectional area. Subsequently, the entire femur–PCL–tibia assembly was translated with respect to the laser beam to each of the remaining four levels in order from proximal to the most distal level. The ligament was measured twice at each level as a check.

The procedure for measuring cross-sectional shape and area then was then repeated at the previously recorded positions for 30°, 60°, and 90° flexion. The lateral femoral condyle, ACL, and anterior half of the tibial plateau were then repositioned by aligning the previously noted original landmarks. To clear a path for the collimated laser beam to reach the ACL, the entire medial femoral condyle and the posterior half of the tibial plateau, with the PCL, were then removed. The same testing procedure was then followed to obtain shape and area measurements of the ACL as previously utilized for the PCL. MFLs were tested as isolated preparations, precluding evaluation of its shape and area relative to knee flexion. A two-way analysis of variance (ANOVA) was used to test the effects of midsubstance measurement level and knee-flexion angle on the cross-sectional areas of the PCL and the ACL. The effect of the measurement site on the cross-sectional area of the MFL was evaluated using a one-way ANOVA, while paired Student's *t*-tests were performed to compare the PCL and ACL cross-sectional areas at the same midsubstance level and knee-flexion angle within a specimen ($P \leq 0.05$).

The cross-sectional shapes of both the PCL and ACL were irregular at all levels of measurement and changed slightly with flexion angle. Figure 2 shows typical cross-sectional shapes of the ACL, PCL, and MFL with the knee flexed to 30°. The PCL cross section was widest in the mediolateral direction, while the ACL was generally widest in an anteroposterior direction. The MFLs were more circular than the ACL and PCL, with some mediolateral widening similar to the PCL. The cross section of the PCL was more circular approaching the tibial insertion, becoming more elliptical toward the femoral insertion; the ACL was generally more circular at all levels. Changes in flexion angle led to alterations in the cross-sectional shapes of both the ACL and PCL, with the most marked changes in the PCL occurring near its femoral insertion.

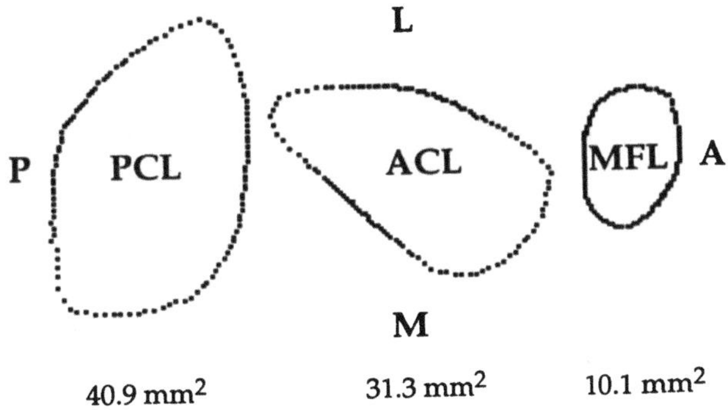

Fig. 2. Typical cross-sectional shapes and areas for the posterior cruciate ligament (*PCL*), anterior cruciate ligament (*ACL*), and menisco femoral ligament (*MFL*) (Wrisberg) at the midsubstance level for a left knee flexed 30° (from Harner et al., to be published)

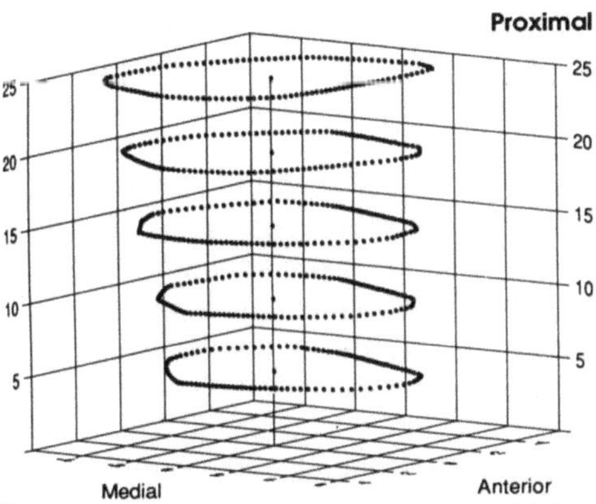

3

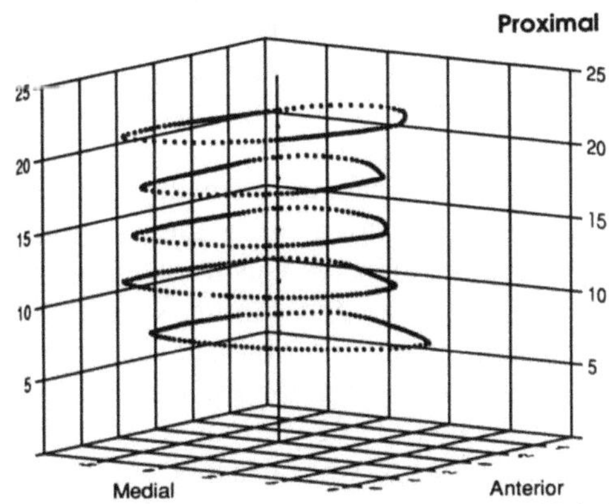

4

Fig. 3. Three-dimensional view of the PCL from the left knee of a typical specimen (68-year-old donor) at 30° flexion, showing the five levels measured along the midsubstance (from Harner et al., to be published)

Fig. 4. Three-dimensional view of the ACL from the left knee of a typical specimen (68-year-old donor) at 30° flexion, showing the five levels measured along the midsubstance (from Harner et al., to be published)

The cross-sectional area of the PCL increased from distal to proximal, whereas the ACL increased in the opposite direction, from proximal to distal. This is shown in Figs. 3 and 4, which represent the five levels of two-dimensional cross-sectional shape measurement reconstructed to show the "three-dimensional" ligament shape. The cross-sectional areas of the PCL and the ACL were significantly different at the midsubstance, midproximal, and proximal levels but not at the middistal and distal levels. The smallest cross-sectional area ratio (PCL/ACL) occurred distally, with the PCL about 120% of the ACL, and the largest proximally, with the PCL about 150% of the ACL. Although kneeflexion changed the cross-sectional shape of the ACL and PCL, it had no significant effect on cross-sectional area.

MFLs were present in all specimens, but with some variation; two knees had Humphrey ligaments only, four had Wrisberg ligaments only, and two had both. In the two specimens with both Wrisberg and Humphrey, one possessed a dominant Wrisberg ligament, while the other had a dominant Humphrey ligament. The mean total cross-sectional area(s) of the MFL were approximately 22% of PCL cross-sectional area.

Evaluation of Ligament Insertions

Successful reconstructions of the human PCL (and ACL) are dependent on not only the accurate reproduction of ligament midsubstance anatomy but also on ligament insertion anatomy. Although many studies have addressed graft properties and placement, little attention has been given to cruciate ligament insertion site geometry. For this reason, we sought to determine the shapes and sizes of the tibial and femoral insertions of the ACL and PCL, and further compare this "insertion area" data with the midsubstance cross-sectional area data for the ligaments.

Five knee specimens (mean age, 63 years; range, 48–77 years) were obtained, and cross-sectional area measurement of the midsubstance of the ACL and PCL was performed using a laser micrometer system as previously discussed [43]. The tibia and femur were then potted in bone cement to allow secure fixation during the digitization of the ligament insertions. A digitizing system (Ascension Technologies, Colchester, VT, USA) was then used to record the coordinates of the periphery of each insertion (accurate to within 0.8 mm). A sharp pointer was attached to the receiver, approximately 30 points were digitized, and the x, y, and z coordinates of each point were calculated.

This three-dimensional data was then loaded into a computer program that allowed the insertion to be mathematically projected to a series of different plane views (two-dimensional). The largest projected area was then taken to be the best planar approximation of the surface area of the insertion. The PCL tibial insertion was found to be 117% ± 16% of the ACL tibial insertion, whereas the femoral insertion was found to be 121 ± 20% of the ACL; however, no statistical difference could be demonstrated in either case.

Typical projected shapes for the ACL and PCL tibial and femoral insertions are shown in Fig. 5. The ACL insertions are both relatively planar, with the femoral insertion being more circular and the tibial insertion being more oval. The PCL femoral insertion is relatively planar and approximates a "half-moon"; however, the tibial insertion is significantly more complex. The "footprint" of this insertion is relatively rectangular, but in three dimensions it is bent across the posterior aspect of the tibial plateau.

The PCL tibial and femoral insertions were both found to be approximately three times greater than its midsubstance, while those for the insertions of the ACL were both more than 3.5-fold greater (Fig. 6). Statistically, all insertions had significantly greater areas than the ligament midsubstance (paired *t* tests, $P < 0.001$).

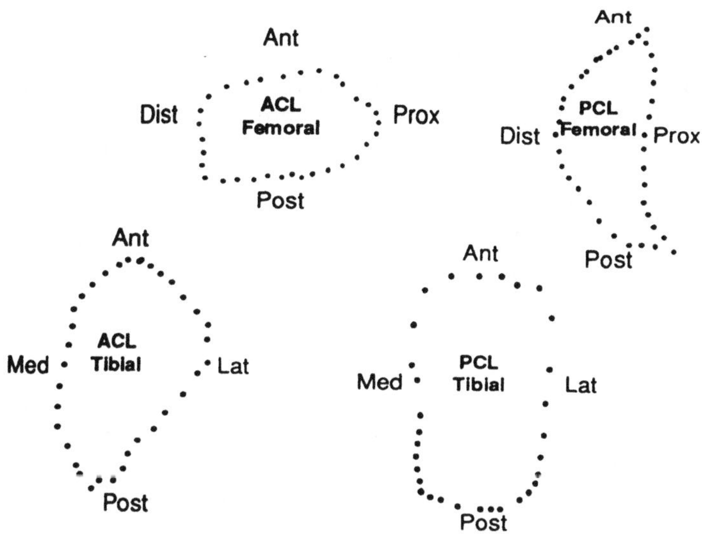

Fig. 5. Typical two-dimensional views of the femoral and tibial insertions of the human ACL and PCL (AL-PCL and PM-PCL portions together) for a right knee

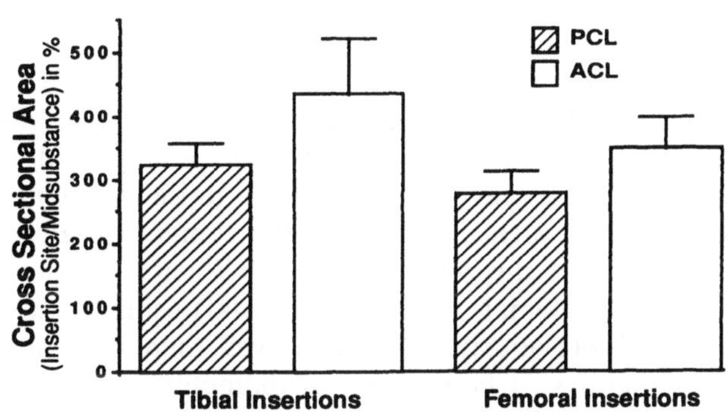

Fig. 6. Bar graph demonstrating the significant difference in cross-sectional area of the insertions of the PCL (*hatched bars*) and ACL (*open bars*) relative to their midsubstance (mean ± SD)

Tensile Properties of the Posterior Cruciate Ligament

Because of the geometric complexity of the PCL, its tensile properties have received little attention compared to those of the ACL [8,44–46]. Furthermore, the tensile properties of the MFLs have not been evaluated.

Kennedy et al. [47] were among the first to measure the tensile strength of the human PCL. They reported that the PCL was approximately twice as strong as the ACL, but were examining excised ligament specimens. Prietto et al. [45] studied PCLs from young cadaveric knees (19–25 years), and tested the entire femur–PCL–tibia complex (FPTC) at 45° of knee flexion, finding linear stiffness and an ultimate load of 204 ± 49 N/mm and 1627 ± 491 N, respectively. Race and Amis recently [46] examined the two bundles of the PCL separately in 10 older knees (53–98 years), determining the AL-PCL to be significantly stronger and stiffer than the PM-PCL.

Fourteen fresh-frozen cadaveric specimens were utilized, and following thawing at room temperature, each specimen was dissected free of skin, subcutaneous tissue, patella, muscle, and other soft tissue. Specimens were then carefully examined for unusual degenerative changes. Several cycles of knee flexion and extension were applied, and the PCL was observed to determine ligament tensioning patterns. The AL-PCL was then identified using the criteria that it is taut during flexion while the PM-PCL is taut toward extension. Final separation was made through the midsubstance using a small pin, and this was tracked to the tibial and femoral insertions and checked with several more cycles of flexion and extension. Component separation was then marked on the femoral and tibial insertion sites, and the MFLs (Wrisberg and/or Humphrey) were also separated and their insertions marked. If both MFLs were present, the smaller was resected at its femoral insertion. The femoral insertions were then separated into AL, PM, and MFL bone blocks with an oscillating bone saw and a small chisel, guided by the prior markings. The tibia was left intact for anatomic positioning during the tensile test. The menisci were dissected away from the tibia, the lateral meniscus separated from the medial meniscus, and left intact for distal MFL clamping. All the femoral bone blocks were then wrapped twice with wire for improved fixation and potted in polymethylmethacrylate (PMMA). To prevent heat damage to the ligament insertions, the bone blocks and PMMA were chilled while the cement cured.

Cross-sectional areas of the AL-PCL, PM-PCL, and MFL were then measured at three levels (proximal, middle, and distal) using the laser micrometer system, as described previously. The tibia was then fixed in an aluminum cylinder with five diaphyseal bolts 6 mm in diameter, and then mounted in a custom five-degree-of-freedom (5-DOF) clamp to allow "anatomic" positioning during testing [48]. The femoral cement–bone block of the PM-PCL was clamped, and the tibial insertion was fixed in an alignment that allowed uniaxial tension to be applied along the component axis. Four elastin stain markers were placed longitudinally on the surface of the ligament to allow tissue strain measurement with a motion analysis system (Motion Analysis, Santa Rosa, CA, USA) [49]. The specimen was preconditioned (10 cycles of 0–2 mm extension at 20 mm/min) and then tested to failure at 200 mm/min while tensile force and elongation data were recorded. Similar testing was then performed on the AL-PCL and the MFL components.

Tissue strain was determined by analyzing the collected video data with the motion analysis system [49]. Stress was calculated using force data and the cross-sectional area measurements. The resulting stress–strain curves were plotted and the elastic moduli calculated. Statistical analysis of the linear stiffness, ultimate loads, and elastic moduli was then performed using paired t tests with a significance level of $P < 0.05$.

The data were normalized within each specimen to allow for direct comparison of the structural and mechanical properties between specimens. The linear stiffness of the AL-PCL component was found to be significantly higher than that of the PM-PCL component (2.6 fold; $P < 0.001$) and MFL (2.7 fold; $P < 0.01$), while the linear stiffness of the PM component and MFL were not significantly different ($P > 0.5$). The ultimate loads of the AL-PCL component were significantly greater than the PM-PCL component (3.4 fold; $P < 0.0001$) and MFL (5.0 fold; $P < 0.0001$) within the same knee. Typical load-extension plots of the AL-PCL, PM-PCL, and MFL are shown in Fig. 7.

For younger specimens (49 years of age of less), 75% failed in the midsubstance, while 67% of older specimens failed through the insertion. All but one of the insertional failures for the AL-PCL and PM-PCL component occurred on the femoral side. The MFL component was shown to possess a slightly greater elastic modulus than the AL-PCL, although statistical significance could not be demonstrated. The elastic moduli of both the MFL and AL-PCL were found to be significantly greater than the PM component (2.2 fold; $P < 0.05$; and 1.9 times; $P < 0.05$ respectively) (Fig. 8).

Discussion

Quantitative data regarding the anatomy of the PCL and MFLs, including determination of the cross-sectional areas along the ligament substance and characterization of insertion site areas, were obtained in this study. Further, the tensile properties of these structures were also determined.

We observed the cross-sectional shapes of the PCL and ACL to be irregular, neither circular nor rectangular, and to vary from femoral to tibial insertions. The PCL and ACL were also noted to have different cross-sectional shapes. These shapes will be useful when applying mathematical modeling to either of

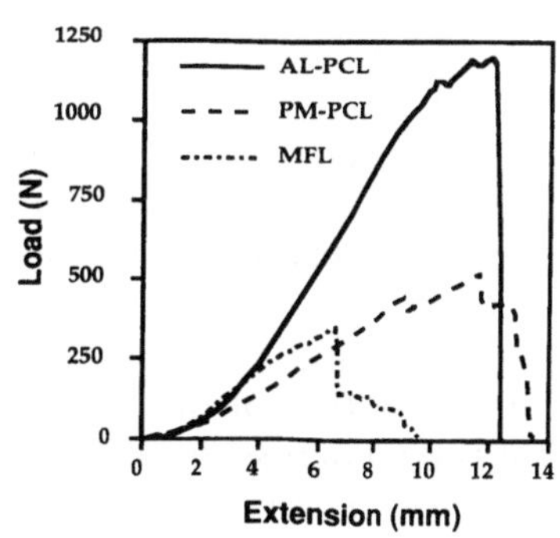

Fig. 7. Typical load-extension curves for the AL-PCL, PM-PCL, and MFL components from the same knee

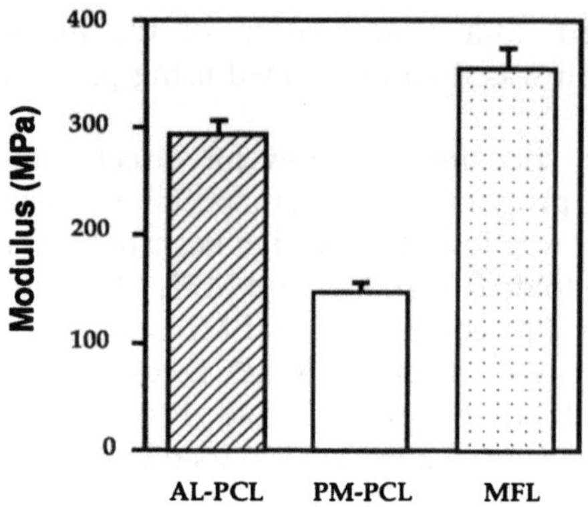

Fig. 8. Bar graph showing the elastic moduli of the AL-PCL, PM-PCL, and MFL components, demonstrating the superior tissue quality of the MFL (mean ± SEM)

these ligaments. Additionally, the three-dimensional complex shapes of these structures indicate that fully 6-DOF assessment of their function is necessary.

We have demonstrated that the PCL cross-sectional area is between 120% and 150% that of the ACL. While this is smaller than previously believed, it still gives some support to those authors who believe that the PCL is the most important ligament within the knee [6,50,51]. In addition, the large size and complexity of the PCL may reflect a more distinct separation of the AL-PCL and PM-PCL components [23,46,52,53].

Changes in the cross-sectional area along the length of the cruciate ligaments in this study were not found to be significant because of the large variability; however, overall trends were shown to be consistent for all specimens. The PCL was shown to increase in cross-sectional area proximally, and the ACL increased distally. Although kneeflexion altered the cross-sectional shape of the PCL and ACL (at the same level), the fact that it did not significantly affect the cross-sectional area indicates that some movement of parallel fibers occurs within the cruciate ligaments with knee flexion. This phenomenon may have practical implications as to which portions of these ligaments are loaded in different knee orientations.

The current finding is that MFL cross-sectional areas are approximately 22% of the PCL, and this value is larger than previous reports [25]. Given that MFLs were present in all specimens considered, and that their size is not negligible, their functional importance in the intact knee may be underestimated.

Determination of the sizes and shapes of the insertions of the human PCL and ACL demonstrated that insertion areas are several times larger than the respective ligament substance. We have also determined that the insertions of the ACL are generally circular or oval, as noted by Odensten and Gillquist [26], and that the PCL femoral insertion approximates a "half-moon." The tibial insertion of the PCL is rectangular when considered in two dimensions, but is bent across the posterior aspect of the tibia, making for complex insertion anatomy. The significant differences between midsubstance area and insertion area may indicate that ultrastructural changes occur in ligament composition within the cruciate ligaments near their insertions.

It has been shown previously that the force in the PCL increases with knee flexion [54]. Observations of functional anatomy further demonstrate that the AL-PCL component becomes increasingly taut while the PM-PCL component becomes lax during this motion. Therefore, it is reasonable that both the stiffness and ultimate load in the AL-PCL component will be significantly greater than that in the PM-PCL component and MFL. Our structural data are also in general agreement with previously published work, although we have found slightly higher elastic moduli for the portions of the PCL [46].

The PM-PCL component and MFL possess similar linear stiffnesses, while the inherent tissue quality of the MFL (i.e., elastic modulus) is superior. This indicates that the MFL is a significant anatomical and biomechanical structure that may have clinical significance. For this reason, our results should be helpful for both graft selection and graft placement during the surgical treatment of PCL insufficiency.

Many questions still remain with respect to the surgical treatment of PCL injuries. From this work we have provided some fundamental data, but have also rasised new questions regarding the role of the PCL and MFLs in knee kinematics. In the next phase of our research, we will utilize our recently developed robotic technology [55] together with a universal force-moment sensor approach [56,57] to elucidate the contributions of the individual components of the PCL in situ during motion and external loading of the joint.

Acknowledgments. This work was supported by the Orthopedic Research and Education Foundation, the Whitaker Foundation, NIH grant #AR 39683, and the University of Pittsburgh Medical Center. The technical assistance of Duane Morrow is gratefully acknowledged.

References

1. Jackson D (ed) (1993) The anterior cruciate ligament: current and future concepts. 1st edn. Raven Press, New York, 462
2. O'Donoghue D, Frank C, Jeter G, Johnson W, Zeiders J, Kenyon R (1971) Repair and reconstruction of the anterior cruciate ligament in dogs. Factors influencing long-term results. J Bone Joint Surg 53A:710–718
3. O'Donghue D, Rockwod C (1966) Repair of the anterior cruciate ligament in dogs. J Bone Joint Surg 48A:503–519
4. Butler DL, Noyes FR, Grood ES (1980) Ligamentous restraints to anterior-posterior drawer in the human knee. J Bone Joint Surg 62A:259–270
5. Grood ES, Noyes FR, Stowers SF (1988) Limits of movement in the human knee: the effects of sectioning the PCL and posterolateral structures. J Bone Joint Surg 70A:88–97
6. Kennedy JC, Grainger RW (1967) The posterior cruciate ligament. J Trauma 7:367–377

7. Cain TE, Schwab GH (1981) Performance of an athlete with straight posterior knee instability. Am J Sports Med 9:203–208
8. Hughston JC, Bowden JA, Andrews JR, Norwood LA (1980) Acute tears of the posterior cruciate ligament: results of operative treatment. J Bone Joint Surg 62A:438–450
9. Brantigan OC, Voshell AF (1941) The mechanics of ligaments and menisci of the joint. J Bone Joint Surg 23A:44–66
10. Hughston JC, Degenhardt TC (1982) Reconstruction of the posterior cruciate ligament. Clin Orthop 164:59–77
11. Hughston JC, Norwood LA (1980) The posterolateral drawer test and external rotational recurvatum test for posterolateral rotatory instability of the knee. Clin Orthop 147:82–87
12. Torg JS, Barton TM, Pavlov H (1982) Natural history of the posterior cruciate ligament-deficient knee. Clin Orthop 164:59–77
13. Shyhar M, Warren R, Ortiz G, Schwartz E, Otis J (1993) The effects of sectioning of the posterior cruciate ligament and the posterolateral complex on the articular contact pressures within the knee. J Bone Joint Surg 75A:694–99
14. Parolie JM, Bergfeld JA (1986) Long-term results of nonoperative treatment of isolated posterior cruciate ligament injuries in the athlete. Am J Sports Med 14:35–38
15. Miyasaka KC, Daniel DM, Stone ML, Hirshman P (1991) The incidence of knee ligament injuries in the general population. Am J Knee Surg, 4(1):3–8
16. Hey Groves E (1917) Operation for the repair of the cruciate ligaments. Lancet 2:674–675
17. Fleming RE Jr, Blatz DJ, McCarroll JR (1981) Posterior problems in the knee: posterior cruciate insufficiency and posterolateral rotary insufficiency. Am J Sports Med 9:107–113
18. Insall JN, Hood RW (1982) Bone-block transfer of the medial head of the gastrocnemius for posterior cruciate insufficiency. J Bone Joint Surg 64A:691–699
19. Kennedy JC, Galpin RD (1982) The use of the medial head of the gastrocnemius muscle in the posterior cruciate-deficient knee. Am J Sports Med 10:63–74
20. McCormick WC, Bagg RJ, Kennedy CW, Leukens CA (1976) Reconstruction of the posterior cruciate ligament. Clin Orthop 118:30–34
21. Tilberg B (1977) The late repair of torn cruciate ligaments using menisci. J Bone Joint Surg 59B:15–19
22. Girgis FG, Marshall JL, Al Monajem ARS (1975) The cruciate ligaments of the knee joint: anatomical and experimental analysis. Clin Orthop 106:216–231
23. Van Dommelen BA, Fowler PJ (1989) Anatomy of the posterior cruciate ligament. Am J Sports Med 14:24–29
24. Friederich N, O'Brien W (1990) Functional anatomy of the meniscofemoral ligaments. In: Fourth Congress of the European Society for Knee Surgery and Arthroscopy
25. Heller L, Langman J (1964) The menisco-femoral ligaments of the human knee. J Bone Joint Surg 46B:307–313
26. Odensten M, Gillquist J (1985) Functional anatomy of the anterior cruciate ligament and a rationale for reconstruction. J Bone Joint Surg 67A:257–261
27. Gratz CM (1931) Tensile strength and elasticity tests on human fascia lata. J Bone Joint Surg 13:334–340
28. Nunley RL (1958) The ligamenta flava of the dog. A study of tensile and physical properties. Am J Phys Med 37:256–268

29. Wright DG, Renneis DC (1964) A study of the elastic properties of plantar fascia. J Bone Joint Surg 46A:482–492
30. Allard P, Thirty PS, Bourgault A, Drouin G (1979) Pressure dependence of the area micrometer method in evaluation of cruciate ligament cross-section. J Biomed Eng 1:265–267
31. Butler DL, Grood ES, Noyes FR, Zernicke RF, Brackett K (1984) Effects of structure and strain measurement technique on the material properties of young human tendons and fascia. J Biomech 17:579–596
32. Ellis DG (1969) Cross-sectional area measurements for tendon specimens: a comparison of several methods. J Biomech 2:175–186
33. Walker LB, Harris EH, Benedict JV (1964) Stress-strain relations in human cadaveric plantaris tendon: A preliminary study. Med Electron Biol Eng 2:31–38
34. Cronkite AE (1936) The tensile strength of human tendons. Anat Rec 64:173–186
35. Race A, Amis AA (1994) A molding method to find cross-sections of soft tissue bundles with complex shapes. Trans ORS 19:783
36. Shrive NG, Lam TC, Damson E, Frank CB (1988) A new method for measuring the cross-sectional area of connective tissue structures. J Biomech Eng 110:104–109
37. Ellis DG (1986) A shadow amplitude method for measuring cross sectional area of biological specimens. In Annu Conf Eng Med Biol 51:6
38. Gupta BN, Subramanian KN, Brinker WO, Gupta AN (1971) Tensile strength of canine cranial cruciate ligaments. Am J Vet Res 32:183–190
39. Iaconis F, Steindler R, Marinozzi G (1987) Measurements of cross-sectional area of collagen structures (knee ligaments) by means of an optical method. J Biomech 20:1003–1010
40. Njus GO, Njus NM (1986) A noncontact method for determining cross-sectional area of soft tissues. Trans ORS 11:126
41. Lee TQ, Woo SL-Y (1988) A new method for determining cross-sectional shape and area of soft tissues. J Biomech Eng 110:110–114
42. Woo SL-Y, Danto MI, Ohland KJ, Lee TQ, Newton PO (1990) The use of a laser micrometer system to determine the cross-sectional shape and area of ligaments: A comparative study with two existing methods. J Biomech Eng 112:426–431
43. Harner CD, Livesay GA, Choi NY, Fujie H, Fu FH, Woo SL-Y (1992) Evaluation of the sizes and shapes of the human anterior and posterior cruciate ligaments: a comparative study. Trans ORS 17:123
44. Butler DL, Kay MD, Stouffer DC (1986) Comparison of material properties in fascicle-bone units from human patellar tendon and knee ligaments. J Biomech 19(6):425–432
45. Prietto MP, Bain JR, Stonebrook SN, Settlage RA (1988) Tensile strength of the human posterior cruciate ligament (PCL). Trans ORS 13:195
46. Race A, Amis AA (1994) The mechanical properties of the two bundles of the human posterior cruciate ligament. J Biomech 27(1):13–24
47. Kennedy JC, Hawkins RJ, Willis RB, Danylchuk KD (1976) Tension studies of human knee ligaments, yield point, ultimate failure, and disruption of the cruciate and tibial collateral ligaments. J Bone Joint Surg 58A:350–355
48. Woo SL-Y, Hollis JM, Adams DJ, Lyon RM, Takai S (1991) Tensile properties of the human femur-anterior cruciate ligament-tibia complex: the effect of specimen age and orientation. Am J Sports Med 19:217–225
49. Lee TQ, Danto MI (1992) Application of a continuous video digitizing system for tensile testing of bone-soft tissue-bone complex. ASME Adv Bioeng 22:87–90

50. Hughston JC, Andrews JR, Cross MJ, Moschi A (1976) Classification of knee ligament instabilities: Part I. The medial compartment and cruciate ligaments. J Bone Joint Surg 58A(2):159–72
51. Hughston JC, Andrews JR, Cross MJ, Moschi A (1976) Classification of knee ligament instabilities: Part II. The lateral compartment. J Bone Joint Surg 58A: 173–179
52. Barton TM, Torg JS (1984) Posterior cruciate ligament insufficiency. A review of the literature. Am J Sports Med 12:419–430
53. Dorlot JM, Christel P, Sedel L, Witvoet J (1983) The displacement of the bony insertion sites of the cruciate ligaments during flexion of the knee. Trans ORS 8:328
54. Wascher DC, Markolf KL, Shapiro MS, Finerman GAM (1993) Direct in vitro measurement of the forces in cruciate ligaments. Part I: The effect of multiplane loading in the intact knee. J Bone Joint Surg 75A:377–386
55. Fujie H, Mabuchi K, Woo SL-Y, Livesay GA, Arai S, Tsukamoto Y (1993) The use of robotics technology to study human joint kinematics: A new methodology. J Biomech Eng 115:211–217
56. Fujie H, Livesay GA, Kashiwaguchi S, Blomstrom GL, Woo SL-Y (1992) A new methodology for direct, non-contact determination of in-situ forces in soft tissues. N Am Cong Biomech 16:7–8
57. Fujie H, Livesay GA, Woo SL-Y, Kashiwaguchi S, Blomstrom GL (1994) The use of a universal force-moment sensor to determine the in-situ forces in soft tissues: a new methodology. J Biomech Eng (in press)

Part 6. Muscle

Modulation and Adaptability of Mechanical Properties of Mammalian Skeletal Muscle

Kenzo Akazawa

Summary. Basic mechanical properties of the isolated skeletal muscle fiber are outlined in this chapter. Behavior of the motor units and modulation of the mechanical properties in voluntary contraction are then explained. Finally, findings concerning the morphological and metabolic changes resulting from nerve-induced activities or exercise training are described. (1) Viscoelasticity of the whole muscle becomes greater almost linearly with an increase in force of the voluntary contraction. This is attributed to increase in the number of active muscle fibers and in their activities; that is, when the force of the muscle is increased, the motor units are recruited in order of increasing size and their firing rates are increased. (2) Exercise training is accompanied by an increase in fiber size of the muscle (hypertrophy), and possibly by an increase in fiber number (hyperplasia). The transformation of the muscle phenotype occurs in animal experiments; fast muscle fibers that are made to perform a more tonic activity become slower-contracting.

Key words: Skeletal muscle—Mechanical property—Viscoelasticity—Motor unit—Fiber type—Force–velocity relation—Adaptability

Introduction

While human beings are incapable of generating such a strong force as robots, they adapt skillfully to changes in objects or external environments and execute actions smoothly. Such skill in movement is attributed not only to harmony between the muscles and the neural control system, but to the flexibility or adaptability of these systems as well.

Department of Computer and Systems Engineering, Kobe University, 1-1 Rokkodai, Nada, Kobe, 657 Japan

The muscle is not a simple force generator. The dependence of muscle force on change in length and its rate of change gives rise to a rubber-like or viscoelastic behavior that would play an important role in maintaining posture or controlling limb movements as well as in absorbing mechanical impacts [1,2]. The muscle, however, does not behave like a simple rubber. Basic mechanical properties of the contracting muscle fiber can be explained by a two-component model [3–5]: the contractile component and the series elastic component. The viscoelasticity of the in vivo muscle is not fixed, but is regulated by the brain [6,7]. Adjusting effectively our muscle viscoelasticity, we can easily handle breakable objects, open a door, or smoothly carry out biped walking.

A muscle is composed of many motor units, the properties of which vary widely even within the muscle [8]. Changes of the muscle viscoelasticity in voluntary contraction are closely related to activities of the motor units; change in the muscle force is accompanied by a change in the firing rate and by recruitment of motor units [9,10].

It is of interest that mechanical activity (exercise) and neural activity are responsible for adaptive changes in morphology or metabolism of individual muscle fibers of the adult muscle [11–13].

This chapter outlines mechanical properties of the skeletal muscle, behavior of the motor units, modulation of the viscoelasticity in voluntary contraction, and adaptive changes (plasticity) of the muscle.

Mechanical Properties of Isolated Muscle Fibers

Force–Length Relation

It is well known that force developed by muscular contraction is a function of the length of the muscle. Gordon et al. [14] and Edman et al. [15] observed isometric force by varying the length of isolated muscle fibers. When the sarcomere length (roughly proportional to the muscle length) is less than 2.2 μm the isometric tetanus tension increases with an increase in the muscle length; this is an ascending limb of the force–length relation. However, as the sarcomere length becomes longer than 2.2 μm, the isometric tetanus tension decreases because of the decrease in the overlap region of the thin and thick filaments; this is the descending limb of the curve.

Effects of both muscle length and stimulation frequency on the isometric tension of cat soleus muscle are shown in Fig. 1; a method of subdividing ventral roots and supplying stimulation pulses to different five motor units in rotation was used [16]. Isometric tension increases with an increase of stimulation frequency as well as with an increase of muscle length over a certain range of the length.

The static force–length relation of the ascending limb seems to represent that of a parallel elastic component (PEC). If so, PEC should have the negative spring constant in the descending limb; the system becomes unstable. In the

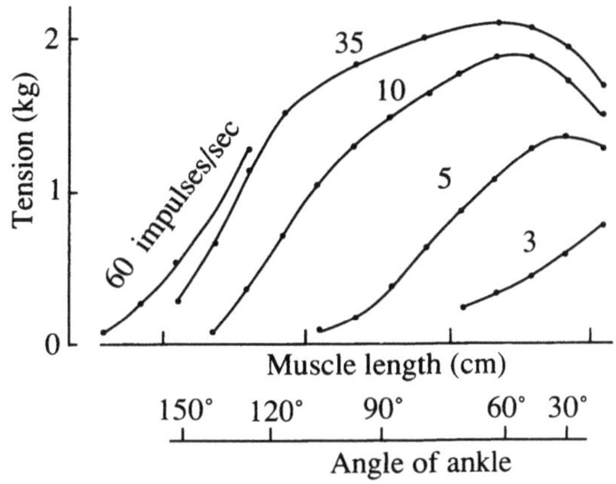

Fig. 1. The effect of length on active tension at different stimulus rates of distributed stimulation (five stimulating channels) in cat soleus muscle [16]

descending limb, however, by examining dynamic responses to stretch, we could identify another type of PEC; developed offset force of the contracting muscle in response to stretch increases with the amount of stretch, and the offset force is much larger than the isometric force [15]. In any case, this behavior seems to represent the existence of PEC, not as a real component but as an analogy.

Series Elastic Component

Existence of the series elastic component (SEC) can be showed by the quick release experiment shown in Fig. 2a. When one end of the muscle is released quickly during isometric tetanic contraction, a small amount of quick shortening occurs just after the release, followed by a steady shortening at a velocity that depends on the applied load. This quick shortening just after the release is attributed to the activity of SEC, and the steady shortening corresponds to the contractile component (CC).

This phenomenon is schematically illustrated in Fig. 2b. The muscle during isometric contraction in (b) in the figure is quickly release in (c) and allowed to shorten with a load P in (d). The force–extension relation of SEC can be estimated as the relation between the force change and the instantaneous shortening (S). The relation was estimated from frog skeletal muscle [17]; the slope of the force–extension curve is steeper at greater tension (Fig. 3c); this means that the elastic coefficient (stiffness) is greater at the larger tension. This property was also reported in cat soleus muscle [18] and human upper arm muscle [19].

Force–Load–Velocity Relation

One of the most important features that characterize the dynamic properties of muscular contraction is the relation between the load, P, and the velocity, v.

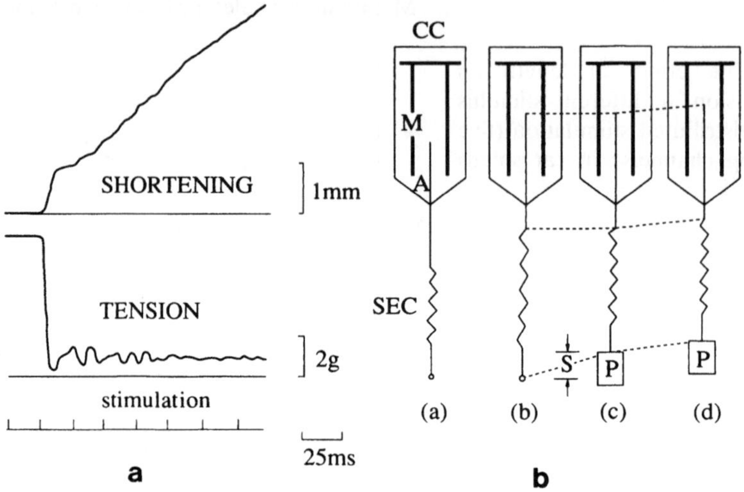

Fig. 2. a Experimental result of the isotonic quick release in frog semitendinosus muscle. Intrinsic shortening of the series elastic component (*SEC*) appeared just after the quick release. **b** Diagram for estimating the force-extension relation of SEC: (a) resting state; (b) isometric contraction; (c) immediately after quick release with applied load *P*; (d) following isotonic contraction

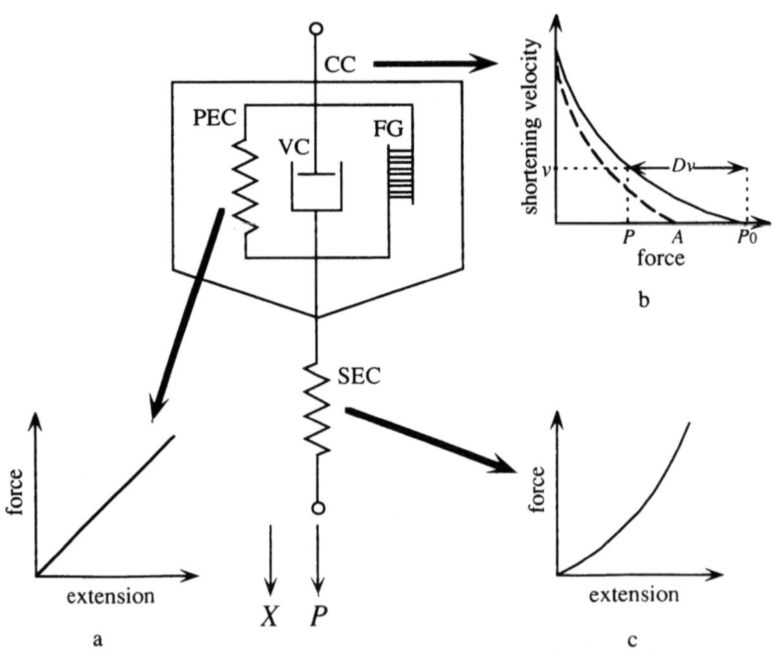

Fig. 3. Mechanical model of muscle and basic properties of each component. *CC*, contractile component; *SEC*, series elastic component; *FG*, force generator; *VC*, viscous component; *PEC*, parallel elastic component

Examining heat production, Hill [3] expressed the load–velocity relation of the tetanus muscle by a hyperbolic equation:

$$(P + a)(v + b) = b(P_O + a) \tag{1}$$

where P_O is the maximum isometric force, a is the heat constant and b is the rate constant of energy liberation. This equation holds for almost all skeletal muscles in animals [4,19].

We determined load–velocity relations at different levels of the contraction [20]. The relations were also hyperbolic (see Fig. 3b) and are expressed by a generalized form of Hill's equation:

$$(P + a')(v + b) = b(A + a')$$
$$a' = a(A/P_O) \tag{2}$$

where A is the contractile force (intensity of the active state). This relation is also expressed by

$$P = A - Dv \tag{3}$$

$$D = (A/P_O)(P_O + a)/(v + b) \tag{4}$$

The second term of the right-hand side of Equation 3 corresponds to the velocity–dependent force loss or a viscous-like force of the viscous coefficient D; the coefficient D is not constant but increases linearly with an increase of A. On the basis of Equation 3, the contractile component can be divided into two components, a force generator FG (the contractile force, A) and a viscous component VC (Fig. 3).

Model of Muscle Fiber

On the basis of these properties, the mechanical model of muscle fiber can be shown by Fig. 3, where CC is divided into three components, FG, VC and PEC. The model does not imply their real existence, but is given only as a mechanical analogy for the sake of easy understanding.

For simplicity, the coefficient of each component is assumed to be constant. The force applied to the muscle is denoted by $P(t)$ as a function of time t, the contractile force by $A(t)$, the change of the muscle length by $X(t)$, elastic constants of SEC and PEC by Es and Ep, respectively and viscous constant by D. By using the Laplace transform, the force $P(s)$ can be expressed as

$$P(s) = \frac{K_1}{1 + T_1 s}A(s) + \frac{K(1 + T_2 s)}{1 + T_1 s}X(s) \tag{5}$$

where $T_1 = \mathrm{D}/(Es + Ep)$, $T_2 = \mathrm{D/Ep}$, $K_1 = Es/(Es + Ep)$, and $K = Es\ Ep/(Es + Ep)$. The first term on the right-hand side represents isometric force. The second term is the force caused by stretching the viscous and elastic components of the muscle. The gain K corresponds to the elastic constant of the whole muscle.

Muscle Property in Voluntary Contraction

Neural Mechanism of Force Production

During voluntary movements, muscular contraction takes place in sequences as follows. Command impulses from the brain reach an α-motoneuron of the spinal cord, exciting the α-motoneuron. When a single α-motoneuron is excited, all muscle fibers it supplies contract simultaneously (Fig. 4). A set of the α-motoneuron plus the muscle fibers is called a motor unit, which is the smallest basic unit of force production of the muscle. Muscle fibers belonging to the same motor unit have a similar histochemical appearance; they have nearly identical contents of enzymes of the oxidative pathway and contain identical isomyosins.

Motor units vary widely in their properties even within the muscle, and can be classified into three groups, which are given various names. Burke and Edgerton [8] referred to them as FF, FR, and S. The FF type is fast contracting, fatigable, and can develop large twitch tension; muscle fibers of the FF type are fast-twitch glycolytic (FG, type IIA). The S type are slow contracting with small twitch tension, and are less subjective to fatigue because of their slow oxidative metabolism; muscle fibers of the S type are slow-twitch oxidative (SO, type I). The FR type is between the these two types; it is essentially the same as the fast-contracting, glycolytic type, except that it is fatigue-resistant. Muscle fibers of the FR type are fast-twitch, oxidative, and glycolytic (FOG, type IIB).

The muscle is composed of several tens to several hundreds of motor units. When the muscle is exerting force, not all the motor units within the muscle are contracting but are working to a certain rule (size principle [21]). The force

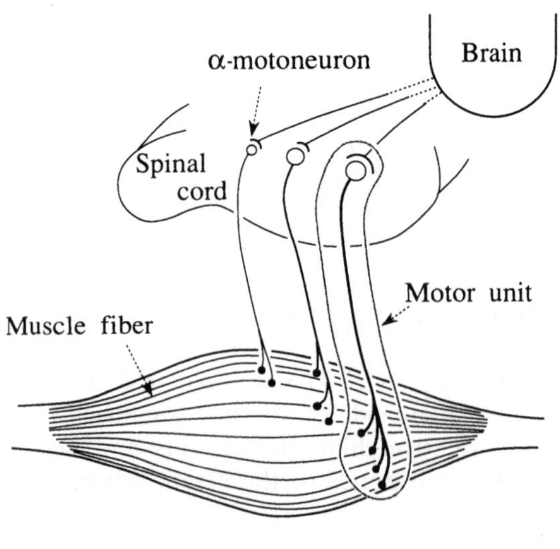

Fig. 4. A motor unit, which is a set of α-motoneurons plus the muscle fibers it supplies, is a basic unit for generating the force

of the whole muscle is the sum of the force generated per motor unit, so that the force is controlled by three factors:

1. The firing rates of the motor units
2. The size of the active motor units (larger sized motor unit can produce larger force)
3. The number of the active motor unit

When the force of the muscle increases gradually, motor units become activated one after the other, and the firing rates of the active motor units also increase [9,10]. The smallest motor units are recruited first, and larger motor units are then recruited systematically according to size. In general, the smaller motor units are always active during contraction. We investigated quantitatively these three factors of human muscles [10], and proposed a neural network model based on the size principle [22].

Modulation of Muscle Viscoelasticity

Dynamic properties of the human skeletal muscle have been studied by many workers; for example, the force–velocity relation for upper arm muscles [19], and stretch-evoked stiffness for the elbow [23] and the ankle [24]. However, the intrinsic viscoelastic property of the muscle itself has not been precisely measured, taking account of its dependence on the contraction level. We have examined the mechanical properties of flexor pollicis longus muscle at the different levels of voluntary isometric contraction [25]. The results are briefly explained next.

A quick length perturbation of a small amount was applied to the thumb at the resting state and during maintenance of the constant isometric force. These experiments were repeated varying the level of contraction and using the same length of perturbation. We estimated the stretch-induced force Fv to be attributed solely to the muscle viscoelasticity. When the larger isometric force was generated, the greater increases in Fv were observed (Fig. 5). A larger change in Fv means larger viscoelasticity because of the same length of perturbation. It was shown that the muscle viscoelasticity increases linearly with the isometric force.

Simultaneous changes of isometric force and muscle viscoelasticity can easily be understood by considering the behavior of motor units. As described earlier, both the number of active muscle fibers (active motor units) and the firing rates increase with increasing the isometric force. Both increases are responsible for an increase in the total viscoelasticity of the muscle.

A whole muscle is composed of many muscle fibers, and not all the muscle fibers are active during contraction, so that the whole muscle cannot be represented by a single two-component model. In general, such a model seems to be appropriate in that two-component models are connected in parallel. However, as far as the isometric contraction is concerned, the whole muscle can be

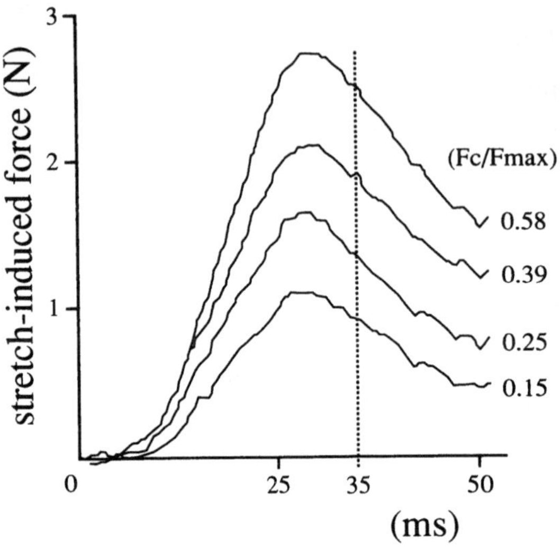

Fig. 5. Effect of isometric force on the stretch-induced force, Fv [25]. Number at right of each trace is the isometric force (*Fc*), normalized by the maximum voluntary isometric force (Fmax); the abscissa is the time after the onset of ramp stretch in milliseconds (ms)

represented by the single two-component model in Fig. 3 because the muscle viscoelasticity varies linearly with the isometric force.

The movements of the limb are actually controlled by contractions of a pair of antagonistic muscles, a flexor muscle and an extensor muscle. The driving torque around the joint is the difference between the torque generated by the flexor and that produced by the extensor muscle. On the contrary, the viscoelasticity around the joint is the sum of the viscoelasticity of both muscles. This implies that the driving torque (or the limb position) and viscoelasticity around the joint could be controlled separately by two independent neural commands to the flexor and extensor muscles. Consider how effectively the muscle viscoelasticity works in real movements. When we jump down from a high place or change from the swing phase to the stance phase during walking, we can make a soft landing on the ground by using the viscoelasticity of our muscles. An example of this modulation of muscle viscoelasticity was shown in a walking cat [7].

Adaptive Changes of Contractile and Morphological Properties

The contractile and metabolic characteristics of muscles are not predetermined but can be altered. This was firmly established by the cross-innervation experiments of Eccles and colleagues [26]. They sutured the motor nerve from the slow soleus muscle of the cat into the fast flexor digitorum longus and vice versa. Some time after reinnervation by the alien nerve, it was found the fast muscle, now supplied by the slow nerve, became slow-contracting and the slow muscle, now supplied by a fast nerve, became fast-contracting. Such factors as exercise, muscle activity, specific changes of impulse frequency, hormones, and change of environmental temperature are known to be capable of causing qualitative changes in a muscle [11,12].

Goldspink [12] indicated that exercise training is accompanied by hypertrophy (increase in muscle fiber size of the adult muscle). Figure 6 shows the effects of exercise on the size of muscle fibers of athletes. The greatest hypertrophy is seen in training for weight lifting or throwing (shot, discus), which requires the production of a lot of force. In these activities, obvious increase in size is found in the fast fibers. On the other hand, endurance activities (running or bicycling) do not give rise to extensive hypertrophy, but rather increase the number of mitochondria and the levels of oxidative enzymes; these are necessary for producing sufficient energy and preventing fatigue. Effects of exercise training on metabolic properties of different muscle fiber types were examined by measuring the glycolytic and oxidative enzyme activities in single muscle fibers of rat soleus and extensor digitorum longus (EDL) muscles [27]. Endurance training resulted in hypertrophy of slow-type fibers (SO) and in an increase of oxidative enzymes activities of SO, FOG, and FG fibers. Sprint training resulted in hypertrophy of fast-twitch fibers (FG and FOG), reduced distribution of the SO fibers in EDL muscles, and significant increase of glycolytic enzyme activity of FOG and FG fibers.

Goldspink [12] reported that exercise was not accompanied by an increase in the number of muscle fibers; while special situations in which the muscle was

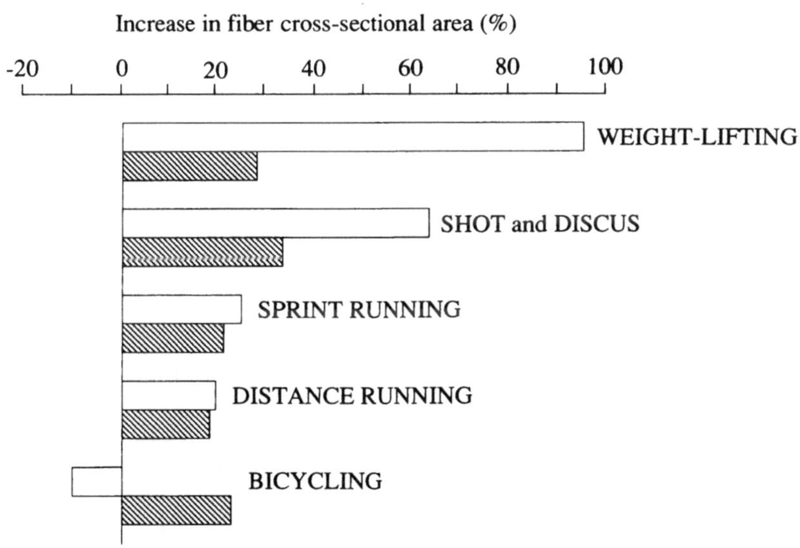

Fig. 6. Effects of different types of exercise on human skeletal muscle [12]. Muscle samples were obtained by needle biopsy. *Open bars*, fast glycolytic fibers; *hatched bars*, slow oxidative fibers

greatly overloaded (e.g., surgical removal of a synergetic muscle) resulted in the splitting of muscle fibers, normal types of exercise (such as running and weight lifting) resulted in no obvious changes in the number of muscle fibers. However, recent evidence from several laboratories strongly suggests the occurrence of muscle fiber hyperplasia (new fiber formation) in adult animals and possibly in human athletes [28]. Direct counts of muscle fibers using nitric acid digestion techniques show that both exercise and stretch overload result in significant increases (range, 9%–52%) in fiber number. Also, satellite cells have been shown to be involved in muscle fiber hyperplasia in stretch and exercise.

Vrbová and colleagues [11,13] have examined effects of electrical activity and shown that not only the nerve-induced activity but also the electrically elicited activity can alter the contractile property of the adult muscle. For example, when rabbit fast muscles were electrically stimulated chronically by an activity pattern resembling that typical of slow motor units, these muscles were transformed to slow-contracting muscle. The fast-to-slow transition of the muscle is seen in both morphological and metabolic aspects; for example, the size of larger muscle fibers decreases. Thus, the diffusion distance of oxygen to the center of the fiber becomes shorter, thus enhancing aerobic oxidative capacity and allowing increase in resistance to fatigue.

When the continuous low-frequency activity to the slow soleus muscle was reduced or abolished by tenotomy, the soleus became a faster-contracting muscle. However, there is less detailed information regarding the conversion of a slow muscle to a fast muscle.

References

1. Winters JM, Woo SL-Y (1990) Multiple muscle systems. Springer, Berlin Heidelberg New York, pp 121–716
2. Special issue on "Modeling the control of upper limb movement" (1993) J Motor Behav 25:131–250
3. Hill AV (1938) The heat of shortening and the dynamic constants of muscle. Proc R Soc Lond B126:136–195
4. Stein RB (1980) Nerve and muscle; membrane, cells, and systems. Plenum, New York, pp 117–205
5. Akazawa K, Fujii K (1989) Mathematical model of the frog skeletal muscle—analysis of nonlinear mechanical properties. Front Med Biol Eng 1:331–340
6. Akazawa K, Milner TE, Stein RB (1983) Modulation of reflex EMG and stiffness in response to stretch of human finger muscle. J Neurophysiol 49:16–27
7. Akazawa K, Aldridge JW, Steeves JD, Stein RB (1982) Modulation of stretch reflexes during locomotion in the mesencephalic cat. J Physiol 329:553–567
8. Burke RE, Edgerton VR (1975) Motor unit properties and selective involvement in movement. Exercise Sport Sci Rev 3:31–81
9. Milner-Brown HS, Stein RB, Yemm R (1973) The orderly recruitment of human motor units during voluntary isometric contractions. J Physiol 230:359–370

10. Kanosue K, Yoshida M, Akazawa K, Fujii K (1979) The number of active motor units and their firing rates in voluntary contraction of human brachialis muscle. Jpn J Physiol 29:427–443
11. Vrbová G, Gordon T, Jones R (1978) Nerve–muscle interaction. Chapman and Hall, London, pp 119–165
12. Goldspink G (1980) Growth of muscle. Development and specialization of skeletal muscle. In: Goldspink DF (ed) Development and specialization of skeletal muscle. Cambridge University Press, Cambridge, pp 19–35
13. Peter D, Vrbová G (1985) Neural control of phenotypic expression in mammalian muscle fibres. Muscle & Nerve 8:676–689
14. Gordon AM, Huxley AF, Julian FJ (1966) The variation in isometric tension with sarcomere length in vertebrate muscle fibers. J Physiol 184:170–192
15. Edman KAP, Elzinga G, Noble MIM (1976) Force enhancement induced by stretch of contracting single isolated muscle fibers of the frog. J Physiol 258:95–96
16. Rack PMH, Westbury DR (1969) The effects of length and stimulus rate on tension in the isometric cat soleus muscle. J Physiol 204:443–460
17. Akazawa K (1990) Theoretical and experimental study on the series elastic component in frog skeletal muscle. Tech Rep Osaka Univ 40:41–48
18. Joyce GC, Rack PMH (1969) Isotonic lengthening and shortening movements of cat soleus muscle. J Physiol 204:475–491
19. Wilkie DR (1950) The relation between force and velocity in human muscle. J Physiol 110:249–280
20. Mashima H, Akazawa K, Kushima K, Fujii K (1972) The force-load-velocity relation and the viscous-like force in the frog skeletal muscle. Jpn J Physiol 22:103–120
21. Henneman E (1974) Peripheral mechanisms involved in the control of muscle. In: Mountcastle VB (ed) Medical physiology. Mosby, St Louis, pp 617–635
22. Akazawa K, Kato K (1990) Neural network model for control of muscle force based on the size principle of motor unit. Proc IEEE 78:1531–1535
23. Joyce GC, Rack PMH, Ross HF (1974) The force generated at the human elbow joint in response to imposed sinusoidal movements of the forearm. J Physiol 240:351–374
24. Agarwal GC, Gottlieb GL (1977) Compliance of the human ankle joint. Trans ASME 99:166–170
25. Akazawa K, Kusumoto H (1992) Cybernetic actuators—muscle mechanics and hardware model. Proc 1992 Int Biomed Eng Days 48–57
26. Buller AJ, Eccles JC, Eccles RM (1960) Interactions between motoneurones and muscles in respect to the characteristic speeds of their responses. J Physiol 150:417–439
27. Takekura H, Yoshioka T (1990) Different metabolic responses to exercise training programmes in single rat muscle fibers. J Muscle Res Cell Motil 11:105–113
28. Antonio J, Gonyea WJ (1993) Skeletal muscle hyperplasia. Med Sci Sports Exercise 25:1333–1345

Part 7. Vascular System

Biomechanical Response of Aortic Wall to Hypertension in the Rat

KOZABURO HAYASHI[1] and TAKEO MATSUMOTO[2]

Summary. Effects of hypertension on the morphometry and elastic properties of thoracic aortas were studied in rats from the standpoint of mechanical adaptation of biological soft tissues. The wall thickness and outer diameter increased with the elevation of blood pressure. The wall hypertrophy was primarily caused by the thickening of lamellar units in the inner media. However, the hoop stress calculated for the systolic blood pressure before killing, P_{sys}, was independent of P_{sys} and was kept at a normal level throughout the experimental period of 2–16 weeks. On the other hand, the elastic modulus of wall was linearly correlated with P_{sys} until 8 weeks; however, it also became independent of P_{sys} at 16 weeks and had a normal value. These results indicated that the aortic wall adapts well to hypertension structurally and mechanically, and that morphometrical adaptation proceeds material adaptation.

Key words: Aortic wall—Hypertension—Remodeling—Functional adaptation—Morphometry—Elastic modulus—Wall stress

Introduction

Living tissues and organs grow or resorb in response to change in mechanical environment as a phenomenon of functional adaptation and optimal operation. They have the ability to homeostatically respond to mechanical demands. Such a biomechanical phenomenon is called tissue remodeling. Although Wolff's law [1] on bone remodeling is widely known, relatively little is known about stress

[1] Department of Mechanical Engineering, Faculty of Engineering Science, Osaka University, Toyonaka, Osaka, 560 Japan
[2] Department of Mechatronics and Precision Engineering, Faculty of Engineering, Tohoku University, Sendai, 980 Japan

effects on the homeostasis in biological soft tissues. However, several recent studies have shown that soft tissues also have the ability of remodeling.

It has been reported that the thickness of aortic walls in hypertensive rats increased with the elevation of blood pressure [2–4]. In addition, many studies of large arteries from hypertensive subjects indicated that these vessels had an increased slope of the stress–strain curve in the circumferential direction [5]. However, Vaishnav et al. [4] showed that the stress–extension ratio relationship of the canine thoracic aorta was very similar in the hypertensive and normotensive animals in the circumferential direction. According to Berry and Greenwald [6], initial large difference in the elastic modulus of the aorta between hypertensive and normotensive rats disappeared within 16 weeks. If living tissues and organs operate optimally, these morphometrical and mechanical changes occurring in response to hypertension could be attributable to the adaptation of vascular walls to mechanical demands.

To obtain more quantitative knowledge on the mechanical response of blood vessels to hypertension, the authors studied temporal changes of wall dimensions, elastic properties, and histology of the thoracic aortas in hypertensive rats [7–9].

Method

Experimental

Goldblatt hypertension was induced in male Wistar rats aged 8–9 weeks (body weight, 200–300 g) by constricting their renal arteries with small silver U-shaped clips. Systolic blood pressure, P_{sys}, was measured once a week with a tail plethysmography under conscious condition.

Tubular segments of thoracic aortas 15 mm long were resected 2, 4, 8, or 16 weeks after surgery. After being attached to a test device with the in vivo length, each segment was immersed in a bath filled with Krebs-Ringer solution of 37°C. The segment was inflated and deflated five to eight times between 0 and 250 mmHg with Krebs-Ringer solution at the pressure change rate of about 3 mmHg/s. Final pressure–diameter data were recorded, where intraluminal pressure, P_i, and outer diameter, D_o, were measured with a fluid-filled pressure transducer (MPU-0.5-290-0-III, Orientec, Tokyo, Japan) and a TV system consisting of a CCD camera (TK-66, Micro Technica, Tokyo, Japan) and an image analyzer (E-1000, Elcow, Hakodate, Japan), respectively.

After the pressure–diameter testing, each blood vessel was sliced with scissors at the position where the outer diameter was measured and its inner and outer diameter was measured with a low-magnification microscope. The inner diameter of a pressurized segment, D_i, was calculated assuming incompressibility of the wall material [5]. Nontreated rats of the same age as the operated animals were used to obtain control data.

In addition to these experiments, from five 8-week-old hypertensive rats we also obtained 20-mm-long aortic segments from the portion distal to the segments used for the pressure–diameter testing. Each segment was fixed with 5% formaldehyde solution overnight while applying its in vivo intraluminal pressure (P_{sys}) and axial extension ratio. Thin sections parallel to the aortic axis were then cut out from the segment, stained with Azan, and observed with a light microscope.

Details of the experimental procedures have been reported elsewhere [8,9].

Data Analysis

Circumferential hoop stress, $\bar{\sigma}_\theta$, was calculated by the law of Laplace:

$$\bar{\sigma}_\theta = P_i D_i/(D_o - D_i) \tag{1}$$

To represent the elastic properties of wall material, the incremental elastic modulus [5,10], $H_{\theta\theta}$, was used:

$$H_{\theta\theta} = 2D_o(\Delta P D_i^2/\Delta D_o + P D_o)/(D_o^2 - D_i^2) \tag{2}$$

where ΔD_o is the increment of D_o produced by ΔP_i, which is the increment of P_i.

Results

Blood Pressure and Wall Dimensions

In the present study, if the systolic blood pressure of a rat was equal to or higher than 160 mmHg, we classified the animal as a hypertensive rat. According to this definition, hypertension was induced in about 79% of the operated animals. As shown in Fig. 1, in hypertensive rats blood pressure started increasing within a week, and reached a plateau about 6 weeks after the operation. Although the data are not shown in this article, body weight of the control animals increased rather rapidly for the first 8 weeks but somewhat gradually thereafter. The hypertensive rats gained weight for 6 weeks in a

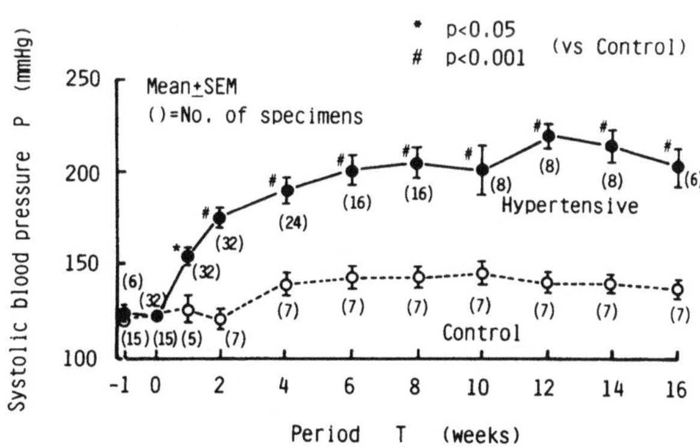

Fig. 1. Temporal change of systolic blood pressure of hypertensive and age-matched control rats [8]. Week "−1", 1 week before operation; P_{sys}, systolic blood pressure immediately before killing

fashion similar to the control animals, although their weight reached a plateau thereafter [8]. Body weight was significantly lower in the hypertensive rats than in the control ones between 6 and 14 weeks after the operation.

Figure 2 shows examples of the relationship between P_{sys} measured immediately before sacrifice and wall thickness and those between P_{sys} and outer and inner diameters of unloaded aortas [8]. Because body weight was significantly different between the hypertensive and control rats during the period of 6–14 weeks, as stated, these wall dimensions were normalized by the cube root of body weight. A good correlation was observed between wall thickness and P_{sys}. Outer diameter correlated well with P_{sys}, although inner diameter did not. Similar relationships were observed for nonnormalized data.

Wall Stress

The relationships between P_{sys} and the circumferential wall stress calculated for this pressure are shown in Fig. 3 [8]. The wall stress was almost independent of P_{sys}, namely, the extent of hypertension, at each period. Although data are not shown in this article, the wall stresses calculated for 100 and 200 mmHg had negative correlation with P_{sys} [8].

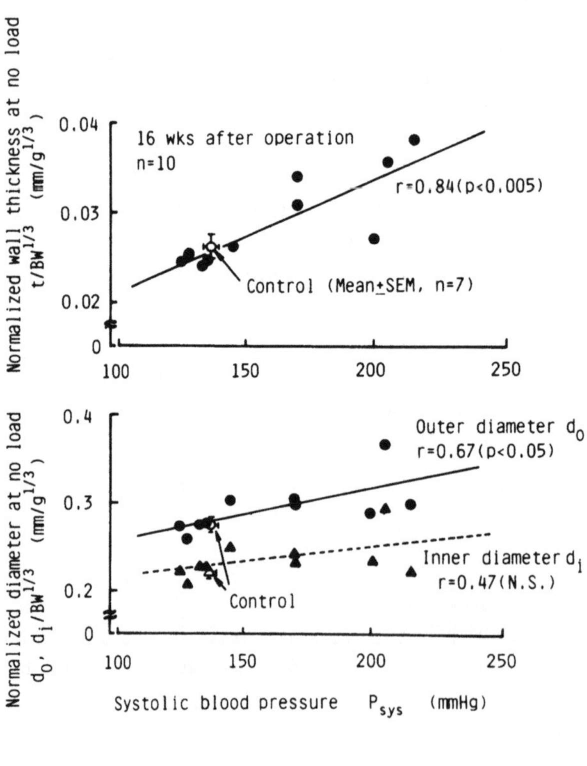

Fig. 2. Systolic blood pressure before killing versus wall thickness (t), outer diameter (d_o), and inner diameter (d_i) of unloaded aortas from 16-week-old rats [8]. These wall dimensions were normalized by the cube root of body weight (BW) in each animal

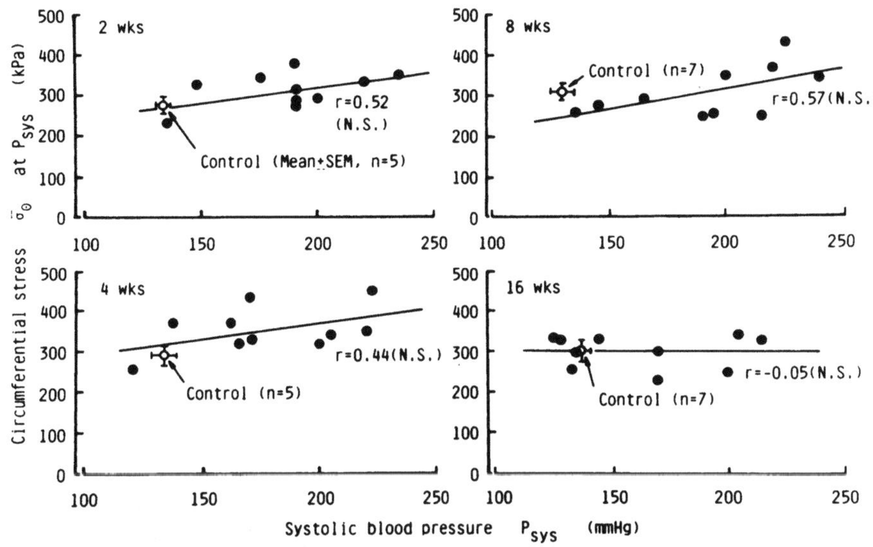

Fig. 3. Systolic blood pressure before killing versus circumferential wall stress calculated for this pressure [8]

Elastic Modulus

The incremental elastic modulus at P_{sys} had significant correlations with P_{sys} until 8 weeks after the operation (Fig. 4) [8]. However, the correlation disappeared at 16 weeks, and the elastic modulus was almost constant regardless of the blood pressure.

Histology

Figure 5 shows micrographs of the aortic sections of a control (P_{sys} = 140 mmHg), hypertensive (P_{sys} = 200 mmHg), and severely hypertensive (P_{sys} = 240 mmHg) rats. As stated previously, wall thickness increased with the elevation of blood pressure. The thickness of lamellar units was almost uniform throughout the wall thickness in the control rat, while in the severely hypertensive rat, the units were much thicker near the inner surface than near the outer one.

Using the photomicrograph of each specimen, total medial thickness, number of lamellar units, and thickness of each lamellar unit were measured at five randomely selected locations. The results are shown in Table 1, where subintimal and subadventitial units indicate the lamellar units nearest to the intima and the adventitia, respectively [9]. The number of lamellar units and the relative thickness of subadventitial lamellar units were not significantly different between the control and hypertensive animals. However, the total

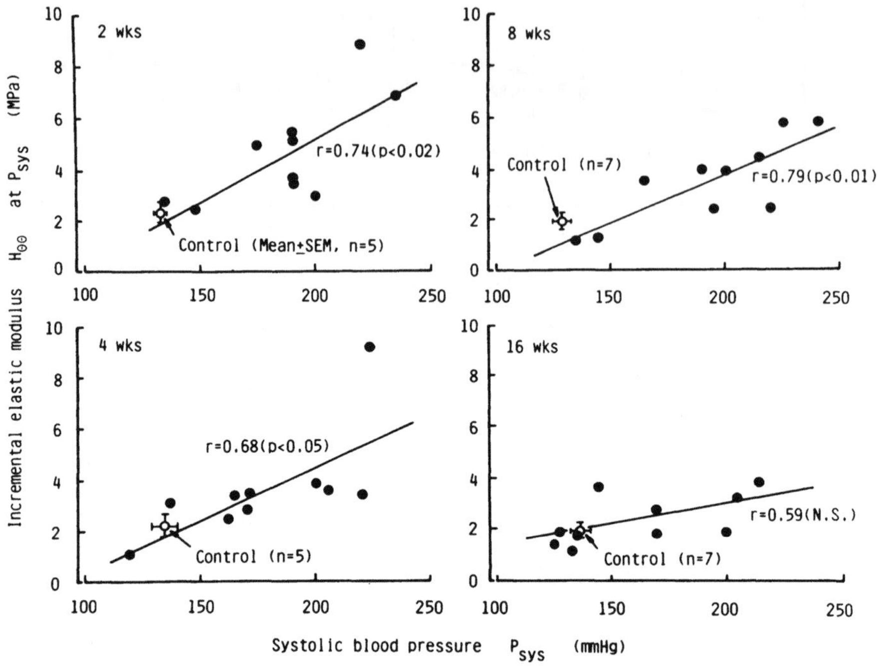

Fig. 4. Systolic blood pressure before killing versus incremental elastic modulus at this pressure [8]

medial thickness and the relative thickness of subintimal lamellar units were significantly larger in the hypertensive rats than in the control ones. These results indicate that the hypertrophy in the hypertensive animals is primarily attributable to the wall thickening in the subintimal regions.

Discussion

Many studies have shown that the thickness and stiffness of arterial walls are increased by hypertension [2,3,6,11–21]. Most of these studies, however, did not present full data on the temporal change of blood vessels induced by hypertension. In particular, no studies have been made of the mechanical response of arterial walls from the standpoint of functional adaptation.

Wolinsky [2] reported that the circumferential wall stress in the thoracic aorta in systole was not different between hypertensive rats and age-matched controls at 20 weeks after the initiation of hypertension. In the present study, we obtained similar results, although the wall stress was almost independent of P_{sys} at all periods for 16 weeks after operation, whereas P_{sys} continued to increase until 6 weeks. However, the wall stresses calculated for 100 and 200 mmHg varied depending on P_{sys}. These results indicate that aortic hyper-

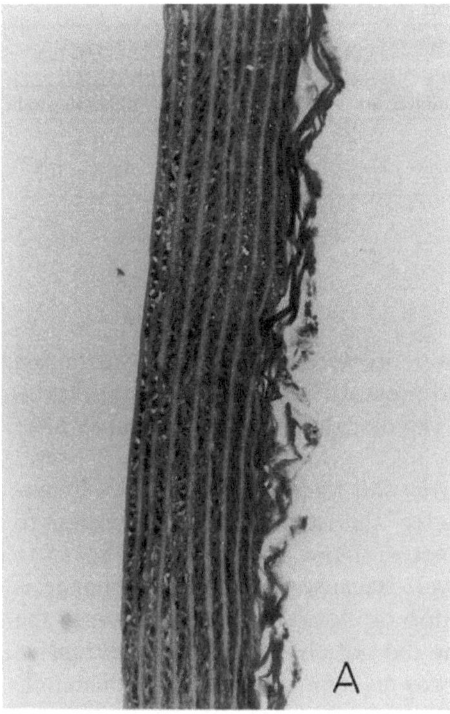

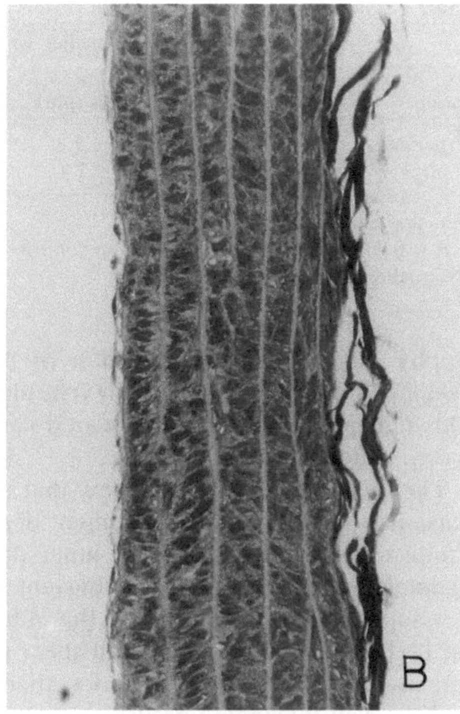

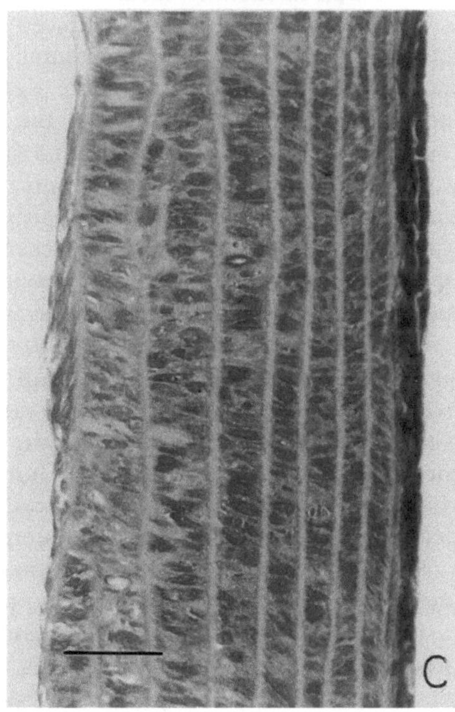

Fig. 5. Photomicrographs of longitudinal sections of the aortas fixed under in vivo load condition and stained with Azan [9]. **A** Control rat (P_{sys} = 140 mmHg). **B** Hypertensive rat (P_{sys} = 200 mmHg). **C** Severely hypertensive rat (P_{sys} = 240 mmHg). The intraluminal surface faces left. Length marker (50 μm) in **C** applies to all parts of figure

Table 1. Summary of morphological observation (from [9])

Group	Number of rats	Total medial thickness (μm)	Number of lamellar units	Relative thickness of lamellar units (%)[a]	
				Subintimal	Subadventitial
Hypertensive	5	109.7 ± 3.8*	7.38 ± 0.25	17.0 ± 0.9**	9.7 ± 0.6
Control	5	57.3 ± 2.4	7.52 ± 0.53	12.9 ± 1.2	11.0 ± 0.8

Data are mean ± SE.
*, $P < 0.001$ *vs* control; **, $P < 0.05$ *vs* control.
[a] Normalized by total medial thickness.

trophy occurs with the elevation of blood pressure so that the wall stress developed inside the body by in vivo blood pressure is kept at a constant level. This functional adaptation appeared very rapidly, that is, within 2 weeks after the elevation of blood pressure.

The data shown in Fig. 2 show that aortic wall thickening caused by hypertension occurs, keeping the inner diameter almost constant. According to Kamiya and Togawa [22], the inner diameter of the artery changes so as to maintain wall shear rate at a constant level. Because no significant change is developed in cardiac output by the induction of Goldblatt hypertension in the rat [21], it is presumed that wall shear rate did not change in the hypertensive rats used for this study. Therefore, there was no change in the inner diameter, and the wall thickening occurred toward the outer (adventitial) direction.

Our histological obervation showed that wall hypertrophy induced by hypertension is mainly caused by thickening of medial lamellar units, in particular, of those in subintimal region. There were almost no changes in the number of lamellar units. The thickening of the lamellar units in subintimal region corresponds to a larger circumferential stress in the inner wall than in the outer wall [9], which indicates that the morphometrical changes occur depending on the magnitude of the mechanical stress applied to it. The mechanisms for the wall response are not known at present. Arterial smooth muscle cells seem to be capable of a wide range of biosynthetic responses, and the responses are induced and modulated by the stress imposed [9].

Unlike the circumferential wall stress, the incremental elastic modulus had significant correlations with P_{sys} until 8 weeks after operation (Fig. 4). However, the elastic modulus became independent of the blood pressure at 16 weeks. These results indicate that the aortic walls in hypertensive rats restore the in vivo elastic properties to a normal level in 16 weeks. In addition, the change of the material properties caused by the change of applied force appears much later than the morphometrical change of the wall.

The results obtained from the current study showed that rapid thickening and relatively slow change of elastic properties occur in the rat aortic wall in response to hypertension, which indicates that mophometrical adaptation proceeds material adaptation. That is, the aotic wall responds to the change of mechanical load by two processes: first, it changes the wall thickness to keep stress in an optimal range, and second, it changes the wall elasticity to restore the wall function to a normal level.

Acknowledgments. This work was supported financially in part by the Grants-in-Aid for International Scientific Research Program (no. 01044009), for General Scientific Research, Category B (no. 02452096), and for Scientific Research on Priority Areas (Biomechanics, nos. 04237102 and 04237103), from the Ministry of Education, Science and Culture, Japan, all with K. Hayashi as the principal investigator.

References

1. Wolff J (1986) The law of bone remodeling. Springer, Berlin
2. Wolinsky H (1971) Effects of hypertension and its reversal on the thoracic aorta of male and female rats. Circ Res 28:622–637
3. Wolinsky H (1972) Long-term effects of hypertension on the rat aortic wall and their relation to concurrent aging changes. Circ Res 30:301–309
4. Vaishnav RN, Vossoughi J, Patel DJ, Cothran LN, Coleman BR, Ison-Franklin EL (1990) Effects of hypertension on elasticity and geometry of aortic tissue from dogs. Trans ASME J Biomech Eng 112:70–74
5. Hayashi K (1993) Experimental approaches on measuring the mechanical properties and constitutive laws of arterial walls. Trans ASME J Biomech Eng 115:481–488
6. Berry CL, Greenwald SE (1976) Effects of hypertension on the static mechanical properties and chemical composition of the rat aorta. Circ Res 10:437–451
7. Matsumoto T, Hayashi K (1991) Mechanical response of rat aorta to hypertension. In: Vanderby R Jr (ed) 1991 Advances in bioengineering. American Society of Mechanical Engineering, New York, pp 631–634
8. Matsumoto T, Hayashi K (1994) Mechanical and dimensional adaptation of rat aorta to hypertension. Trans ASME J Biomech Eng (in press)
9. Matsumoto T, Hayashi K (1994) Analysis of stress and strain distribution in hypertensive and normotensive rat aorta considering residual strain. Trans ASME J Biomech Eng (in press)
10. Hudetz AG (1979) Incremental elastic modulus for orthotropic incompressible arteries. J Biomech 12:651–655
11. Feigel EO, Peterson LH, Jones AW (1963) Mechanical and chemical properties of arteries in experimental hypertension. J Clin Invest 42:1640–1647
12. Greene MA, Friedlander R, Boltax AJ (1966) Distensibility of arteries in human hypertension. Proc Soc Exp Biol Med 121:580–585
13. Aars H (1968) Static load-length characteristics of aortic strips from hypertensive rabbits. Acta Physiol Scand 73:101–110
14. Bandik N, Sparks H (1970) Viscoelastic properties of the aorta of hypertensive rats. Proc Soc Exp Biol Med 134:56–60
15. Wolinsky H (1970) Response of the rat media to hypertension. Circ Res 26:507–522
16. Sharma MG, Hollis TM (1974) Rheological properties of arteries under normal and experimental hypertension conditions. J Biomech 4:293–300

17. Cox RH (1979) Comparison of arterial wall mechanics in normotensive and spontaneously hypertensive rats. Am J Physiol 237:H159–H167
18. Michelini LC, Krieger EM (1986) Aortic caliber changes during development of hypertension in freely moving rats. Am J Physiol 259:H662–H671
19. Cox RH, Bagshaw RJ (1988) Effects of hypertension and its reversal on canine arterial wall properties. Hypertension 12:301–309
20. Imura T, Yamamoto K, Satoh T, Mikami T, Yasuda H (1988) Arteriosclerotic change in the human abdominal aorta in vivo in relation to coronary heart disease and risk factors. Atherosclerosis 73:149–155
21. Stacy DL, Prewitt RL (1989) Effects of chronic hypertension and its reversal on arteries and arterioles. Circ Res 65:869–879
22. Kamiya A, Togawa T (1980) Adaptive regulation of wall shear stress to flow change in the canine carotid artery. Am J Physiol 239:H14–H21

Biomechanics of Atherosclerosis and the Blood Flow

T. Yamaguchi and Tad W. Taylor

Summary. It is widely accepted that computational fluid mechanics can provide a comprehensive view into the delicate structures of blood flow. In this article, we discuss several aspects of the application of the computational fluid mechanics to biofluid mechanics. First, we introduce the basic methods of computational fluid mechanics. We next discuss several methods for the formation of realistic mathematical models using casts of some cardiovascular structures, such as rabbit carotid arteries, a canine left ventricle, and others. The three-dimensional (3-D) coordinate values of the cast surfaces were measured and input into a computer, and 3-D models were constructed using curve-fitting algorithms including a cubic spline surface. Surface volume smoothing and filling was performed by a system developed for a body-fitted-coordinate, finite-volume-type flow simulation. The flow field in the models were calculated by solving the 3-D unsteady Navier–Stokes equation using a finite-volume method from a commercially available simulation package. The velocity and pressure field obtained from the commercial solver allowed us to calculate various physiologically significant parameters such as the wall shear stresses. Unsteady spatial distribution of these fluid mechanical quantities were visualized in the heart left ventricle (LV) models, and our graphics routines allowed us to "travel" through the vessels. The realistic model of the LV constructed from a canine heart allowed us to understand some fluid mechanical effects such as cardiac infarction and showed complex flow field formation.

Key words: Atherosclerosis—Hemodynamics—Computational fluid mechanics—Vascular cast—Blood flow—Flow visualization—Wall shear stress—Left ventricle

Department of Bio-Medical Engineering, School of High Technology for Human Welfare, Tokai University, 317 Nishino, Numazu, Shizuoka, 410-03 Japan

Introduction

Cardiovascular disease is one of the leading causes of mortality in industrialized countries. Vascular diseases such as atherosclerosis, stroke, and thrombosis are known to be strongly influenced by blood flow in terms of their onset and advancement [1–3]. Atherosclerosis is known to start at vessel locations where the wall shear stress is expected to be low [4]. These so-called low wall shear stress regions of the artery, not yet exactly located for a three-dimensional (3-D) flow field, are thought to be specific locations prone to atherosclerosis. It is highly conceivable that the wall shear stress distribution shows markedly different patterns when the configuration of the artery varies [5,6]. The vascular system has a complex structure, including branching, curvature, constriction, and dilatation. Blood flow in the artery is, moreover, unsteady with a complex time course. Direct measurement methods of blood flow, such as ultrasound Doppler velocimetry and laser Doppler velocimetry, have serious limitations when they are applied to real arterial systems. For example, an ultrasound velocimeter has poor time and space resolution, although it can be used with a minimum degree of invasion. In contrast, although the laser velocimeter is accurate, it cannot be applied to in vivo measurements. Even though the magnetic resonance imaging (MRI) method is one of the most promising technologies for the total flow field detection in the cardiovascular system, it is not readily available for experimental purposes. No other measurement method currently available can deal directly with the total flow field. In particular, these techniques cannot measure near-wall phenomena, such as fluid shear stress on the vascular endothelium, with sufficient spatial resolution to be able to compare it with pathophysiological observations. Moreover, the data interpretation is sometimes difficult because the flow field in such places is essentially three-dimensional and unsteady. Analogies from two-dimensional (2-D) flow fields may be seriously misleading.

As we previously reported [7–17], computational fluid mechanics can be a powerful tool for investigating complex structures of blood flow. Mathematical models also offer an aid to interpret noninvasive monitoring techniques. Unfortunately, most arterial and cardiac flow models are based on simple geometries with only one- or two-dimensional modeling methodologies. Models reported so far were based on rather imaginary shapes of arteries [18]. Because the most significant factor affecting the flow structure is known to be the 3-D configuration, the construction of a realistic model with its configuration based on real arteries is mandatory for correlating fluid mechanics to physiological and pathophysiological phenomena such as the atherogenetic process. To overcome this limitation, we are developing models based on realistic vascular chamber structures.

Recent advances in computer technology and implementation techniques now make more realistic modeling methodologies possible. Vascular casts can be useful for the study of complex vascular structures, for example, in the geometry of the aortic leaflet [19,20], and these casts provide a real-world representation of the actual vascular structure. Using these casts as a starting point allows a more accurate comparison of blood flow measurement techniques and computational fluid mechanical results.

In this article, we discuss several methods for obtaining realistic mathematical models of arteries of a wide variety of sizes. These are useful for 3-D computational fluid mechanics calculations, computational visualization of the flow, and understanding flow-related mechanical phenomena based on these computations.

Methods

The equations solved were the 3-D, unsteady Navier–Stokes equations of incompressible viscous flow as shown next [21]:

$$\rho \frac{\mathbf{Dv}}{\mathbf{Dt}} = -\nabla \mathrm{r} + \mu \nabla^2 \mathbf{v} \tag{1}$$

$$(\nabla \cdot \mathbf{v}) = 0 \tag{2}$$

where $\mathbf{v}$ is the velocity vector, π is the density, p is the pressure, and μ is the viscosity. As is frequently mentioned, there are several basic problems in applying the Navier–Stokes equations to blood flow. First, it is assumed that the fluid is Newtonian, i.e., the viscosity does not vary with the flow condition. This is not true for blood, because it is known to show shear-dependent viscosity, particularly when the shear rate is very low. However, the viscosity becomes relatively constant if the shear rate exceeds a certain limiting value, which the shear rate usually exceeds in the flow in larger arteries. Therefore, we neglect the non-Newtonian viscosity when we consider the blood flow in large arteries. The next frequently asked question is related to the elasticity or viscoelasticity of the arterial wall. The arterial wall deforms from the inside pressure. This pressure changes primarily because of the pulsatile nature of the blood flow and secondarily because of fluctuations of the velocity field caused by disturbances such as vortices and turbulence. The changes are unfortunately very difficult to incorporate into currently available computational methods, and intensive studies have been performed in the field of computational fluid mechanics. We therefore neglect this problem at the moment, expecting to include it in our future studies.

Third, we must consider what methods to use in the numerical analysis. In general, currently available techniques fall into three or four classes. They are the finite-difference or the finite-volume type method, the finite-element approach, the boundary-element method, and so forth. The finite-difference or the finite-volume method was first developed in this field and has been available for many years. It can give stable results with relatively little computation time required. However, the flexibility of the model construction is less than the later developed finite-element method. Although the finite-element method allows greater flexibility to build the computational domain, the computation generally requires much more time. The boundary-element method is mainly used for a class of flow problems (the potential flow) that are not relevant to blood flow. We employed the finite-volume method because a number of systems that have been validated are commercially available. As pointed out, the computational time required to obtain stable results is relatively small and this point is important because we are using workstation class computers. Although they have been enormously improved in terms of the computation speed and have reached the point where they are feasible to use in the field of computational fluid mechanics, they are still far behind the state-of-the-art supercomputers and the time factor can be a very important, even a determinant, issue at times.

In our studies, equations 1 and 2 were integrated by a finite-volume method with time matching by a SIMPLE-like algorithm. The computational grid was staggered, and the computation was performed using a body-fitted (or boundary-fitted) coordinate system [22]. Convective terms were discretized by first-order upwind differencing, and unsteady terms were discretized by central differencing. The solver code used was SCRYU (Software Cradle, Japan) V1.4. The blood was assumed to be a Newtonian, homogeneous, and incompressible fluid.

Computations were performed on a variety of workstation class computers. Velocities and pressure values were computed using the SCRYU package, which allowed shear stress values to be computed from our own user-developed programs. The SCRYU package has a graphical interface for directly viewing the input geometrical structure and also the computed flow and pressure fields.

The Silicon Graphics GL library routines were used with our program system to allow advanced 3-D color graphics visualization. Our system has the capability of moving the point of the observer and world coordinates according to input values, which is a useful feature for making color videos.

Small-Scale Modeling

Animal Experiments and Measurements

Three male Japanese white rabbits (body weight, 2.3–2.9 kg) were anesthetized by the intravenous injection of 25 mg/kg body weight of Nembutal (sodium pentobarbital). Both carotid arteries were exposed through median incisions. After ligation of the inferior thyroidal arteries, both common carotid arteries were constricted using an 8-0 nylon suture to give an approximately 75% reduction in the cross-sectional area. The same surgical operation was carried out on both arteries, and the contralateral artery was fixed and examined microscopically to investigate tissue reactions.

After 1 h of constriction, both carotid arteries were cannulated and the rabbits were killed by injection of excess Nembutal. The carotid arteries were rinsed using Ringer's saline solution injected through cannulae, and fixed by 2.5% glutaraldehyde solution at 100–120 mmHg for 30 min in situ. The left carotid artery was usually excised after fixation and microscopically examined. After fixation, methacrylate resin (Mercox CL, Dainippon Ink Co., Ltd., Japan) was injected through the cannula into the right carotid artery at 100–120 mmHg, and the pressure was maintained until the resin hardened (approximately 10 min). Finally, the cast was prepared by digesting the arterial tissue with sodium hypochlorite (Fig. 1).

Three-dimensional coordinate values were obtained from the cast using a 3-D measurement microscope (Nikon UM-3, Tokyo, Japan). The microscope was equipped with a 3-D digital position sensor. The resolution was 1.0 mm in the 3-D space. A vascular cast was held on the X–Y stage of the microscope using a holding device originally designed to align a fiberoptic cable. By using this device, the rotation angle of the cast could be precisely determined. A point on the axis of rotation of the holding device was chosen as a reference point, and coordinate values were taken relative to this point.

As the first step, the cast was fixed at an arbitrary angle, and a pair of coordinate values representing the maximum width of the cast were measured at some distance away from the constriction. Similar measurements were taken at a longitudinal interval of 0.1 mm. Next, the cast was rotated 90° around the holding device axis, and similar measurements were made at the exact locations where the previous pair of coordinate values were taken. These locations were identified by the distance from the reference point on the holding device.

Thus, two pairs of 3-D coordinate values were obtained for each cross-sectional slice separated by 100 mm along the rotational axis. Taking the center of gravity these four points as one of the foci, an ellipsoid curve was used to

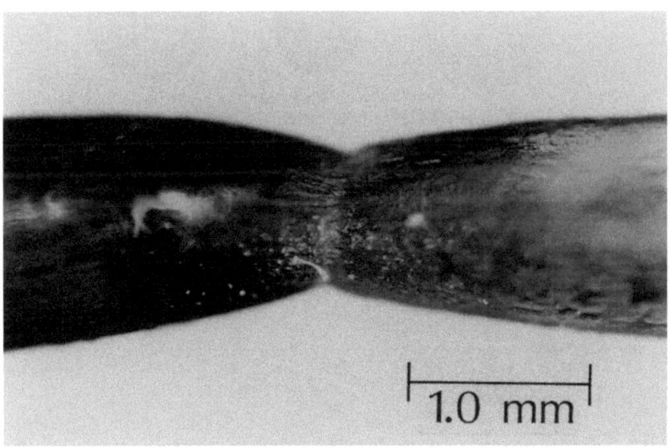

Fig. 1. A cast made of methacrylate resin injected into a constricted rabbit carotid artery. Nylon suture left after digesting the arterial wall was removed when measurements were taken. *B*ar, 1.0 mm

interpolate values between a pair of neighboring points. Finally, computational grids of 19 × 19 were formed using the Poisson equation on smooth the crossing points of the grids. Because the computational method is a finite-volume method, we used a generalized coordinate system, or a so-called boundary-fitted coordinate system. The total grid dimension was 19 × 19 × 35=12 365 for X-, Y-, and Z- axes (Fig. 2), respectively.

Results Obtained from Small-Scale Models

In Fig. 3, results are shown of the computational fluid mechanical calculation using a realistic model formed by the present method. Because rather simplified flow conditions were assumed, we could not discuss the detailed correlation between the flow field and the atherogenetic process or any other related biological or pathological phenomena. However, it can be stressed that the flow nature must be very complicated in real arteries. The 3-D configuration of the model shown here could be regarded as a rather simple one, that is, an axisymmetric constriction. The shape reconstructed from the actual measurements of the arterial cast was, however, much more complicated than we expected. As seen in Fig. 2, cross-sections at various longitudinal points were not symmetric. The computed velocity distribution downstream to the stenosis showed remarkable asymmetry, reflecting such an irregular configuration of the model. The flow separation appeared to be totally different from that of an ideal axisymmetric stenosis model. This and other types of complex natures of

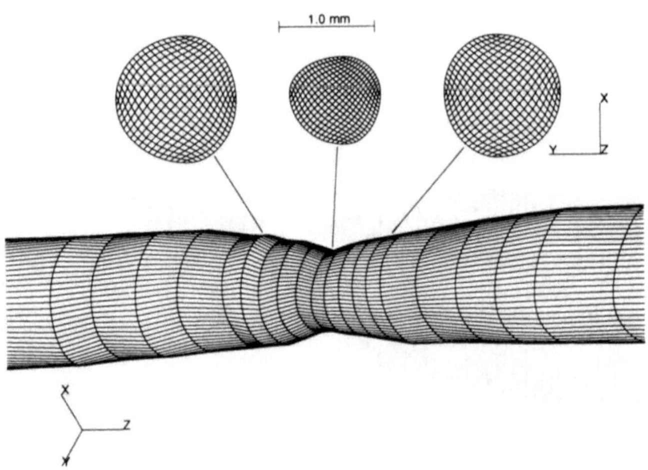

Fig. 2. Examples of computational grids. The surface at the vicinity of the stenosis neck is shown with several cross-sectional grids. The outer surface was made first on the basis of the measured data and by interpolating with an ellipsoid. The internal grid points were assigned using a method that utilizes the Poisson equation

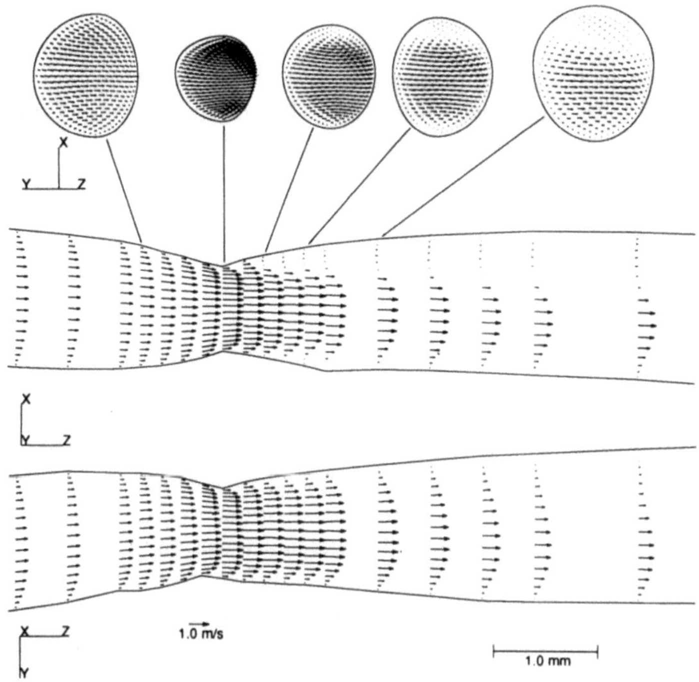

Fig. 3. Velocity profile near the stenosis throat of the cast. Significant asymmetry of flow is seen because of asymmetry of the actual stenosis. Sizes and positions of the 3-D separation bubbles are considerably different from those expected from 2-D axisymmetric configurations

blood flow should be taken into account when cell biological or pathological reastions of blood vessels are considered with respect to hemodynamic events.

Remarks on the Small-Scale Model Studies

The computational method makes it possible to reconstruct as complex a model as we wish, provided that the computer power is sufficient. Applications of the computational fluid mechanics to the arterial flow so far reported were based on rather idealistic models. When we try to correlate biological or pathological findings to fluid mechanical phenomena within a certain blood vessel, the fluid mechanical environment should ideally be estimated using a blood vessel model based on the actual vessel from which the biological or pathological data were obtained. The method reported was not fully 3-D because of limitations of the measurement device. Only four points were measured, and the cross section of the model was interpolated assuming a partial ellipsoid. To reconstruct the true 3-D configuration it is necessary to obtain the true 3-D coordinate values. From the measuring microscope used in this study, we can measure X–Y positions by the movement of the stage. The Z-direction was based on the focus of the microscope, and the view position cannot be as precisely determined as the X–Y positions. However, the present method is suitable to an object whose size is within a millimeter range, and it is useful for the interpretation of the morphological data related to atherogenesis.

Large-Scale Realistic Modeling

Cast Preparation

Casts were made from three important large-scale blood flow structures in the left ventricular chamber of the heart. Figure 4 shows the canine left ventricle cast. This cast was made by injecting a methacrylate resin into the left ventricle (LV) of a canine heart that was kept in diastole. The heart was obtained from a mongrel dog weighing approximately 15 kg that was used in a cardiac energetics experiment. The chest wall was opened, the connecting arteries and veins were severed, and the heart was removed. The heart was then washed and placed partly in a water bath and connected with sutures to obtain a diastolic shape. A syringe was used to inject the methacrylate resin. After the resin solidified, the heart was placed in a 10% KOH solution and allowed to remain for approximately 3 weeks, after which the ventricular muscle was dissolved, leaving the cast intact.

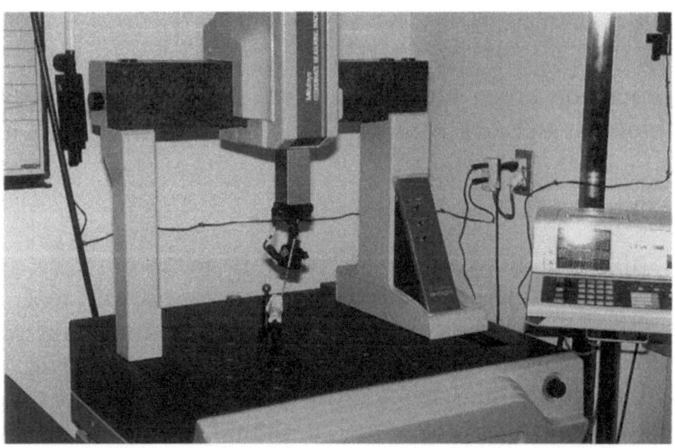

Fig. 4. The Mitutoyo BH303 Coordinate measuring machine and analog to digital interface. The canine left ventricle (LV) is shown attached to the mounting pin cast-holding device. The canine LV cast was made from methacrylate resin injected into an excised canine LV in diastole. Physical dimensions: approximately 60 mm from the base to the apex and approximately 30 mm in diameter. Circular grid lines were drawn approximately every 2 mm

Coordinate Measurement

To reconstruct the true 3-D configuration, it is necessary to obtain the 3-D coordinate values. A coordinate measuring machine from (Mitutoyo, Tokyo, Japan) was used to measure and digitize the coordinates of the cast. This machine was employed by first zeroing the coordinate position with a fixed stainless steel ball used for calibration of the coordinate position. After that, the measurement pin was moved to the cast circular grid markings and the pin was touched to the cast. A small displacement of the measurement pin caused a beep that signified the sample position had been properly recorded. The signal was digitally sent to a Silicon Graphics IRIS 4D/310 VGX workstation where the data values were recorded in a data file. Approximately 4000 points were measured in the canine LV cast. The measurement accuracy was approximately 10 μm.

Results and Remarks of the Large-Scale Modeling

Surface volume smoothing and filling were performed first by our own system and then by spline function from the preprocessor of the SCRYU system. These coordinate values are directly input into the package for flow simulation.

Figure 5 shows the canine LV representation. The smoothed volume representation is illustrated, again being in the surface rendering mode. The coordinate values of the vascular structure geometric patterns shown in Figs. 3 through 5 have been inserted into the SCRYU three-dimensional flow simu-

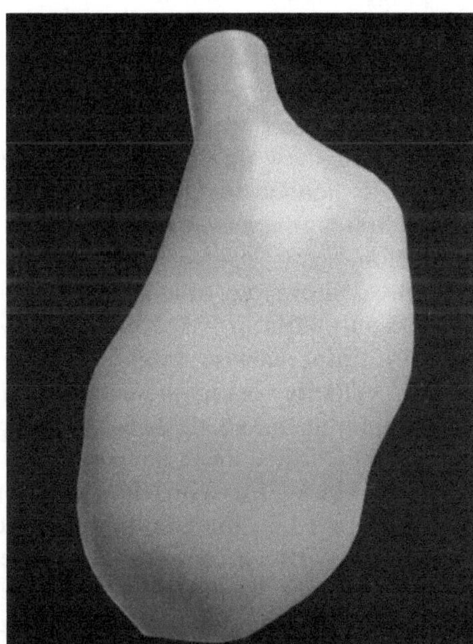

Fig. 5. The canine systolic LV representations. Gouraud shaded rectangular surface patch representation

lation package and will be used to compute complex flow patterns, wall shear stress, and pressure contours.

The construction of a realistic model whose configuration is based on real vascular structures is necessary for correlating fluid mechanics to the atherogenic and other vascular disease processes. The cast measurement technique coupled with our computational method should make it possible to compare simulation results with important measurement techniques such as ultrasound and MRI, and should allow a finer detail of flow understanding than is presently available using either a modeling or imaging method alone.

Computational Flow Visualization

The interpretation of various flow monitoring techniques patterns can be difficult because the visualization modes are usually 2-D views of 3-D phenomena, and our analysis should aid this interpretation. We see an example of 3-D representations of our computational flow visualization studies in the following section.

In Fig. 6, SCRYU heart LV visualization graphs with grid construction and velocities are presented. In A, the LV diastolic side view shows the realistic shape taken from the canine LV cast. In the model there are 30 lateral grids, 40 longitudianl grids, and 10 sets of radial grids. The size is approximately 30 mm in diameter and 70 mm from the aortic outlet to the apex, and the computed volume is 25.8 ml.

When ventricular septal wall infarction is present, the centerline velocity vectors computed at the end of contraction (t = 0.25 sec) do not show that vortices are forming (Fig. 4B). However, the cross-sectional velocity vectors computed at the end of contraction (t = 0.25 sec) with ventricular wall infarction present show that complex flow field formation is present, and in this case the flow is headed toward the infarcted LV wall at the top of the figure. This demonstrates that a 2-D simulation can be misleading with regards to flow field prediction.

Figure 7 shows our Silicon Graphics GL side view velocity vectors computed at end contraction (t = 0.25 sec) with ventricular wall infarction present. Here the wire frame viewing mode is shown and the ends of the golden arrows are brown. Velocity vectors are shown at six levels of the LV, and the flow reaches a maximum at the aortic outlet. The figures presented here represent only one part of the cardiac cycle and only one point of view. Videos have been made that show the entire ventricular ejection cycle from any vantage point.

In the heart LV, the results presented in Figs. 5–7 indicate that there should be areas of vortex formation in infarcted hearts with reduced wall motion. The vortices formed should be observable with sophisticated flow measurement techniques. The model and our visualization techniques also show that there is a pressure gradient between the apex of the LV and the aorta. This pressure

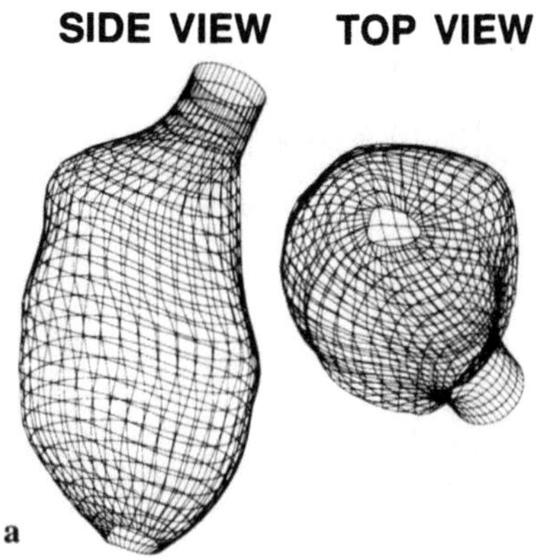

a

Fig. 6a–c. Heart LV visualization graphs, grid construction and velocities. **a** LV diastolic side and top view. There are 30 lateral grids and 40 longitudinal grids in the model. Size is approximately 30 mm in diameter and 70 mm from the aortic outlet to the apex. The computed volume is 25.8 ml. **b** Centerline velocity vectors computed at the end of contraction (t = 0.25 s velocity = 50 cm/s) with ventricular wall infarction present. **c** Cross-sectional velocity vectors computed at end contraction (t = 0.25 sec) with ventricular wall infarction present. *Arrows*, infarction centers; x, infarction boundaries

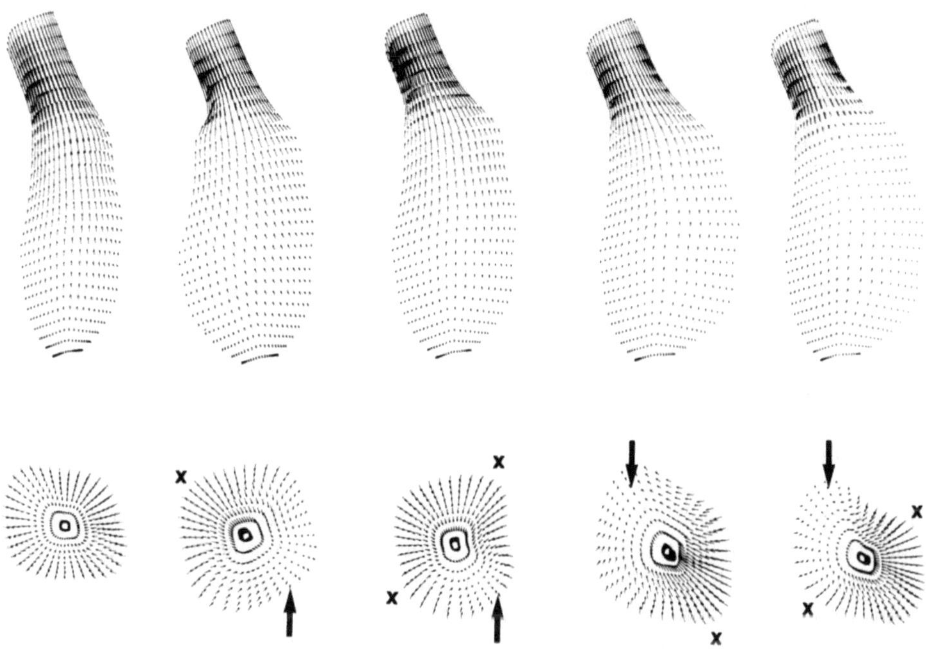

b

c

rise was experimentally observed in humans by Pasipoularides et al. [23]; they found that the pressure could reach 20 mmHg even in normal patients with no ventricular or valve abnormalities.

General Remarks

It should be noted that the available computer power was unsatisfactory when we conducted the present study to examine very complicated flow fields in

Fig. 7. Silicon Graphics GL centerline velocity vectors computed at end contraction, t = 0.25 sec, with ventricular wall infarction present. Here the wire frame viewing mode is shown

the cardiovascular system. However, the rapid advancement of computer technology has made it possible to compute fairly complicated flow fields even by a workstation. Therefore the use of a realistic model such as the one reported in this study should become a mandatory part of an investigation relating fluid dynamic factors to normal and pathological behavior of vascular tissues and cells.

Using physiological casts and inputting the exact dimensions into our 3-D flow simulations will allow a better understanding of the flow processes in important vascular structures. This will be useful to elucidate the possible contribution of physical factors to the pathogenesis of vascular diseases.

In conclusion, the 3-D computational and visualization techniques show a finer detail of the flow structure than that observed in the 2-D computational case as is seen by comparing the cross-sectional and centerline velocity profile simulation results; cross-sectional velocity profiles are difficult to observe experimentally. With the 3-D abdominal aortic aneurysm and heart LV flow solution and visualization possible, this should make it easier to interpret measurements such as ultrasound and MRI.

Acknowledgments. Part of this study was supported by the Grant-in-Aid for Scientific Research "Biomechanics of Structure and Function of Living Cells, Tissues, and Organs" (#04237101) and Grant #04454537 from the Ministry of Education, Science and Culture of Japan. This work was also partially supported by the Suzuken Memorial Foundation.

References

1. Caro CG, Fitz-Gerald JM, Schroter RC (1971) Atheroma and arterial wall shear observation, correlation, and proposal of a shear dependent mass transfer mechanism for atherogenesis. Proc R Soc Lond (Biol) 177:109–159
2. Fry DL (1972) Localizing factors in arteriosclerosis. In: Likoff W, Segal BL, Insull W Jr (eds) Atherosclerosis and coronary heart disease. Grune and Stratton, New York, pp 85–104
3. Yoshida Y, Yamaguchi T, Caro CG, Glagov S, Nerem RM (eds) (1988) Role of blood flow in atherogenesis. Springer, Berlin Heidelberg New York Tokyo
4. Yoshida Y, Sue W, Yamane T, Okano M, Oyama T, Mitsumata M, Suda K, Yamaguchi T, Ooneda G (1990) Structural differences of arterial walls which are either vulnerable or resistant to atherosclerosis. Acta Med Biol (Jpn) 38(suppl):1–19
5. Yamaguchi T, Hanai S (1988) To what extent does a minimal atherosclerotic plaque alter the arterial wall shear stress distribution? A model study by an electrochemical method. Biorheology 25:31–36
6. Yamaguchi T, Hanai S (1990) Atherosclerotic plaques and the three dimensional distribution of wall shear stress. A numerical model study. In: Liepsch D (ed) Biofluid mechanics blood flow in large vessels. Springer, Berlin, pp 513–519
7. Yamaguchi T, Nakano A, Hanai S (1991) Three dimensional shear stress distribution around small atherosclerotic plaques with steady and unsteady flow. In: Mosora F, Caro CG, Krause E, Schmid-Shönbein H, Baquay C, Pelissier R (eds) Biomechanical transport process. Plenum, New York, pp 173–182
8. Yamaguchi T, Taylor TW (1992) A parametrically defincd computational fluid mechanical model for the study of the flow in arterial bifurcations. Adv Bioeng 22:237–240
9. Yamaguchi T, Taylor TW (1992) Computational fluid mechanical study of the coronary spasm. Adv Bioeng 22:333–340
10. Yamaguchi T, Taylor TW (1993) Computational fluid mechanical study of the blood flow with moving walls in the cardiovascular system. In: Hori M, Kobayashi S (eds) Theoretical and applied mechanics, vol 42. University of Tokyo Press, Tokyo, pp 331–338
11. Yamaguchi T (1993) A computational fluid mechanical study of blood flow in a variety of asymmetric arterial bifurcations. Front Med Biol Eng 5:135–141
12. Sakurai A, Yamaguchi T, Okino H, Hanai S, Masuda M (1993) A method for formulating realistic mathematical models based on arterial casts for the computational fluid mechanical studies on arterial flow and atherosclerosis. J Phys III(3):1551–1556
13. Taylor TW, Yamaguchi T (1994) Three-dimensional simulation of blood flow in an abdominal aortic aneurysm—steady and unsteady flow cases. J Biomech Eng 116:89–97
14. Taylor TW, Okino H, Yamaguchi T (1993) The effects of supravalvular aortic stenosis on realistic three-dimensional left ventricular blood ejection. Biorheology 30:429–434
15. Taylor TW, Okino H, Yamaguchi T (1994) Three-dimensional analysis of left ventricular ejection using computational fluid mechanics. J Biomech Eng 116: 127
16. Taylor TW, Okino H, Yamaguchi T (2993) Three dimensional analysis of left ventricular ejection using computational fluid dynamics. 1993 Bioengineering Conference ASME, BED 24:136–139

17. Taylor TW, Okino H, Yamaguchi T (1993) Realistic three-dimensional left ventricular ejection determined from computational fluid dynamics. Adv Bioeng 26: 119–122
18. O'Brien V, Ehrlich LW (1980) Pulsatile flow through a constricted artery. In: Schneck DJ (ed) Biofluid mechanics, vol 2. Plenum, New York, p 497
19. Mercer JL, Benedicty M, Bahnson AT (1973) The geometry and construction of the aortic leaflet. J Thorac Cardiovasc Surg 65:511–518
20. Vesely I, Eickmeier B, Campbell G (1991) Automated 3-D reconstruction of vascular structures from high definition casts. IEEE Trans Biomed Eng 38:1123–1129
21. Bird RB, Stewart WE, Lightfoot EN (1960) Transport phenomena. Wiley, New York, pp 80–81
22. Thompson JF, Warsi ZUA, Mastin CW (1982) Boundary-fitted coordinate systems for numerical solutions of partial differential equations—A review. J Comp Physics 47:1–108
23. Pasipoularides A, Murgo JP, Miller JW, Craig WE (1987) Nonobstructive left ventricular ejection pressure gradients in man. Circ Res 61:220–227

Fluid Shear Stress and Vascular Endothelial Cell Biomechanics

Akira Kamiya[1] and Joji Ando[2]

Summary. Vascular endothelium is now recognized as a very active cell layer that plays an important role in maintaining the vessel wall functions. The endothelial cells (ECs) have a variety of proliferative, synthetic, secretory, and self-adaptive capabilities that regulate vascular tonus, morphology, or adhesiveness by interacting with other cells such as vascular smooth muscle cells, platelets and white cells. Recent evidence suggests that these functions are all affected by fluid shear stress on the endothelial wall, which is a rheological force shearing the luminal surface of the blood vessel when blood flows over it. Wall shear stress is a regulating factor of adaptive vessel growth and angiogenesis, and might be a local risk factor in the pathogenesis of atherosclerosis. Shear stress also modulates the production of vasoactive substances such as endothelium-derived relaxing factor, prostacyclin, histamine, and endothelin, and regulates macromolecule permeability and endocytosis. More recent studies have shown that shear stress exerts an influence on the expression of mRNA of various EC-producing substances such as tissue plasminogen activator mRNA. These facts suggest that ECs have mechanoreceptors to detect changes in shear stress, the signal of which is transmitted to intracellular organelles. It is established that the intracellular Ca^{2+} response is involved in this shear-sensing mechanism of EC as the second messenger of the internal signaling system.

Key words: Vascular diameter—Wall shear stress—Adaptive remodeling—Endothelial cell—Proliferation—Protein synthesis—Signal transduction—Cytoplasmic calcium ion

[1] Institute of Medical Electronics and [2] Department of Cardiovascular Biomechanics, Faculty of Medicine, University of Tokyo, 7-3-1 Hongo, Bunkyo-ku, Tokyo, 113 Japan

Introduction

In the past 15 years, much attention has been paid to the function and biomechanics of the vascular endothelial cells (ECs). The reason is that many studies have successively offered many pieces of evidence indicating various important roles of ECs in maintaining the physiological functions of the cardiovascular system. It is now well recognized that the EC monolayer is not merely an anticoagulant or antipermeable passive barrier between flowing blood and vascular tissue, but an active layer controlling vascular growth and remodeling, and also regulating vascular tonus via secretion of both relaxant and contractile factors. ECs also interact with other cells by means of various cell growth factors, adhesion proteins, and other chemical mediators. In addition, many reports have revealed that ECs respond not only to chemical stimuli such as hormones, autacoids, and neurotransmitters, but also to mechanical stimuli such as shearing stress elicited by blood flow or tensile stress generated by blood pressure. Investigation of the influence of mechanical stresses on EC functions, that is, EC biomechanics, is expected to give new insight into the physiology of vascular growth and remodeling and the pathogenesis of various vascular diseases including atherosclerosis. This category of research, extending interdisciplinarily over biology, medicine, and engineering, is forming a new major field in biomechanics as well as in biorheology. We outline here recent findings concerning the effects of wall shear stress on EC functions, mainly focusing on our own studies. More detailed reviews are found elsewhere [1,2].

Adaptive Responses of the Vascular Wall to Fluid Shear Stress

Statistical analyses of vascular branching systems [3,4] suggest that when blood flow (f) through a vessel is related to its internal radius (r) as

$$f/r^n = \text{constant} \tag{1}$$

the value of n is nearly equal to 3. The mechanism inducing such a relationship has long been suspected to be the adaptation of vessel wall to flow rate. Nearly 100 years ago, Thoma [5] documented that vascular sprouting (angiogenesis) and growth are rapid in vessels with high flow in the chick embryo. Later Rodbard [6] put forward the hypothesis that rheological viscous drag generated by blood flow, which is now commonly called wall shear stress, may regulate the vascular caliber. His hypothesis was based on many clinical findings, such as vessel dilatation at the arteriovenous fistula or dilatation and atrophy of the maternal uterine artery during pregnancy and after delivery respectively, etc.

Kamiya and Togawa [7] tried to verify this mechanism experimentally, because they noticed that this can also be a persuasive regulating mechanism of the foregoing statistical flow–radius relationship; wall shear stress (τ) to the Poiseulle flow is written as follows:

$$\tau = \mu\dot{\gamma} = 4\,\mu f/(\pi r^3) \qquad (2)$$

where μ and $\dot{\gamma}$ are blood viscosity and wall shear rate (the vertical blood velocity gradient at the wall surface), respectively. Equation 2 indicates that when flow is increased or decreased and if the concomitant changes in shear stress induces enlargement or reduction of radius, the stress τ is eventually kept constant, because the third power of the radius (r^3) may exert a potent negative feedback effect on τ. A constant τ in Equation 2 is consistent with the flow–radius relationship in Equation 1 for n = 3. The animal experiments in the Kamiya and Togawa study [7] were performed, as depicted in Fig. 1a, by constructing an arteriovenous shunt at the canine carotid artery and then measuring changes in arterial blood flow and internal diameter at varied times after the shunt operation. Figure 1b and c shows the results obtained: chronic changes in the ratio of shunted/control arterial radius and those of wall shear stress against the flow ratio revealed that the vascular walls respond to sustained change in flow and remodel their radii to regulate the wall shear stress constant, as predicted.

These results were confirmed later by several studies [8–10]. These studies examined the adaptive responses of vessel diameter to the shear stress at several arteries in various animals, by increasing blood flow with a similar shunt or by reducing flow with a flow-restricting clamp. Langille and O'Donnel [9] showed that these vessel diameter responses to flow changes are abolished when the endothelium is removed from the vessel, indicating that the vascular wall response to shear stress is endothelium dependent.

It has also been well documented that blood flow or shear stress plays an important physiological role in capillary angiogenesis [11], because sustained increase in peripheral blood flow in muscles caused by repeated electrical nerve stimulation or by chronic vasodilator administration stimulated capillary angiogenesis and increased capillary density, while reduced peripheral blood flow caused the size and number of arterioles to decrease [12].

These in vivo investigations suggested that the adaptive responses of large vessels and capillary angiogenesis are regulated by wall shear stress through the endothelium. Meanwhile, development of cell culture techniques and molecular biological approaches to cell activities have enabled us to examine the effect of fluid shear stress on EC functions in vitro by using cultured ECs and flow-loading devices. Some of the results are summarized here, together with in vivo data.

Effects of Fluid Shear Stress on EC Migration and Proliferation

One of the most striking characteristics of EC functions is that ECs always construct a confluent monolayer. If a part of such a monolayer is denuded, surrounding ECs migrate and proliferate to cover the denuded area. Ando et al. [13,14] determined in vitro that EC regeneration in this repairing process is

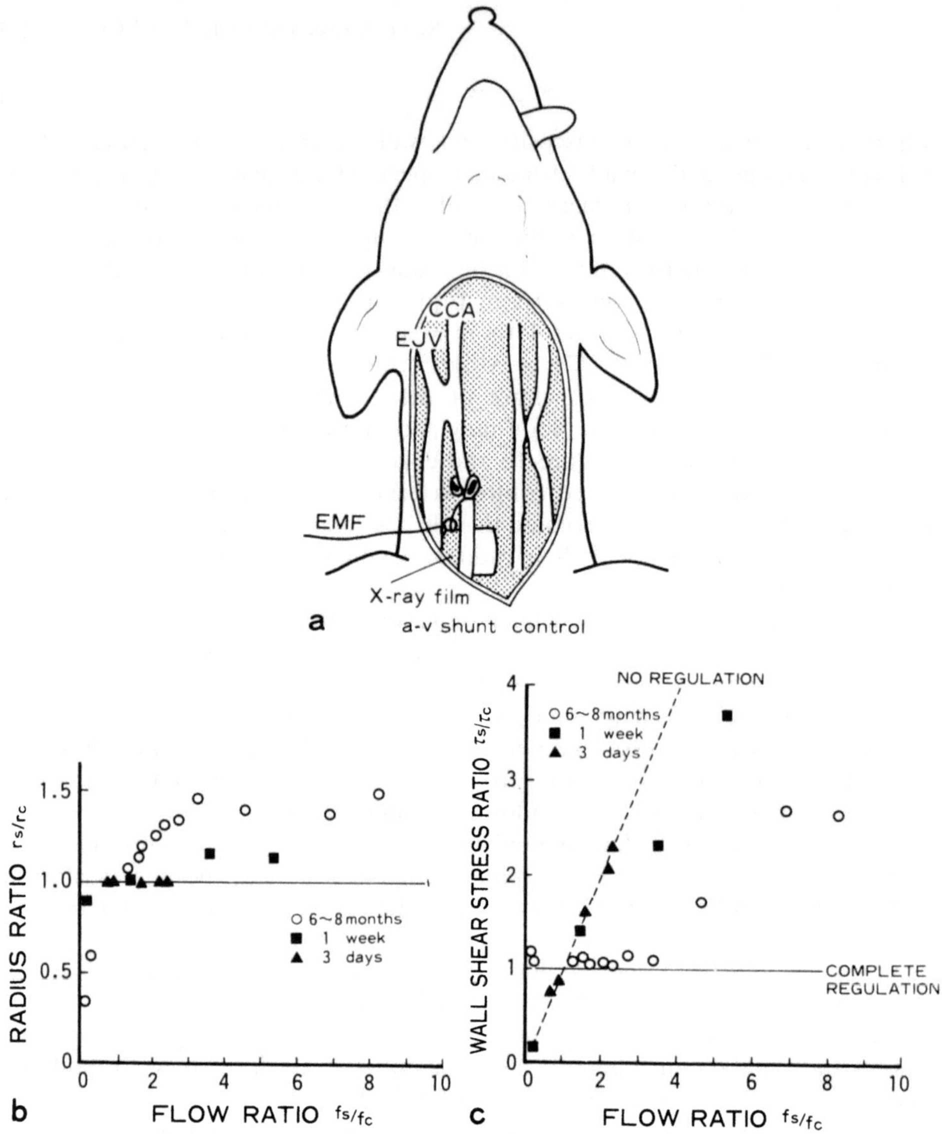

Fig. 1a–c. The arteriovenous (*a-v*) shunt experiment and its results, showing the adaptive responses of vascular internal radius to flow regulated by wall shear stress [7]. **a** The a-v shunt was constructed between the canine common carotid artery (*CCA*) and the external juglar vein (*EJV*). After some flow-loading periods (3 days, 1 week, or 6–8 months), the blood flow rate (f_s) and the internal radius (r_s) of the shunted artery were measured using electromagnetic flowmetry (*EMF*) and angiography (X-ray film) to compare them with those (f_c and r_c) of the contralateral control artery. **b** The obtained radius ratio (r_s/r_c) plotted against flow ratio (f_s/f_c). **c** Relationships of shear stress ratio (τ_s/τ_c), calculated from radius ratio (r_s/r_c) by Equation 2 against flow ratio (f_s/f_c). Note that for the chronic data (6–8 months) with a flow ratio less than 4, the values of shear stress ratio have shifted close to the line of unity, indicating complete regulation of radius by wall shear stress

affected by fluid shear stress application. Following mechanical denudation of a part of a cultured confluent monolayer [13], they exposed ECs to shear stress (0.4–1.4 dyn/cm^2) for 24 h in a flow-loading device (a rotating disc type) and measured the number of cells that migrated to and proliferated in the denuded area and the migration distance. The number of cells in the denuded area and the migration distance were greater under shear stress than under the static condition, demonstrating that the shear stress stimulates EC regeneration in the repairing process. In addition, the area covered by ECs in the downstream site of the flow was larger than that in the upstream site, suggesting a vector effect of shear stress on EC. In the long-term observation of a single EC motility using a video recording, Masuda and Fujiwara [15] found a similar vector effect; that is, the individual cell moves at the same speed in all directions immediately after flow application but thereafter turns to go downstream.

Figure 2 shows the effects of shear stress on cell proliferation during the repair of mechanical denudation observed in the DNA histogram [14]. It shows that the cells subjected to fluid shear stress contain a relatively low proportion of cells in the resting state (left peak; G_1/G_0 phases) and a relatively high proportion of cells in the mitotic state (right peak; S_1, G_2, and M phases of the cell cycle). This indicates that shear stress enhances DNA synthesis and stimulates ECs to divide.

In the confluent monolayer, however, the effects of shear stress on cell proliferation are still controversial. Eskin et al. [16] applied shear stress of 3.3 dyn/cm^2 to cultured bovine ECs for 1–2 weeks and observed a 1.5 to 5-fold increase in cell density compared with static controls, whereas Dewey [17]

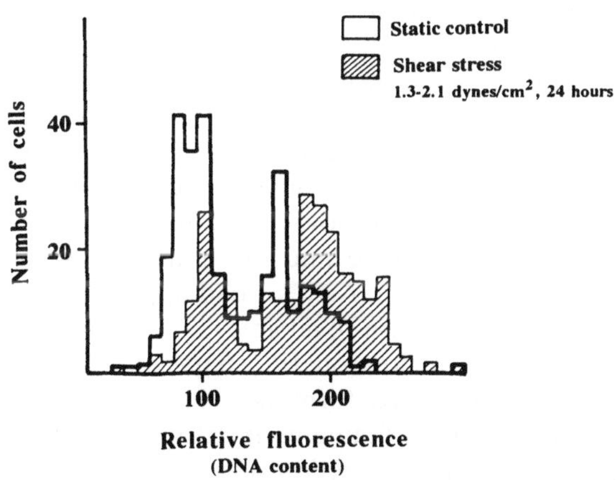

Fig. 2. The shear effects on endothelial cell (EC) DNA synthesis during repair of mechanical denudation [14]. Cell nuclei were stained with propidium iodide and their DNA histogram was obtained by photometric fluorescence microscopy. In the shear-loaded cultures, the number of cells decreased in the G_0 and G_1 phases of the cell cycle (left peak), while it increased in the G_2 and M phases, which suggests a stimulatory effect of shear stress on DNA synthesis in ECs

reported neither a change in cell density after a 1-week application of shear stress of 5 dyn/cm^2 nor an increase in [^{3}H]thymidine uptake under the shear stress application of 1–8 dyn/cm^2 for 24 h. On the other hand, Davies et al. [18] showed that turbulent shear stress as low as 1.5 dyn/cm^2 stimulated substantial DNA synthesis of ECs, although laminar shear stress (range, 8–15 dyn/cm^2) revealed no effect. They suggested that the unstable characteristics of shear stress, rather than its magnitude, are the major determinant of the effect of shear on EC growth. In an in vivo study concerning shear effects on EC growth, Masuda et al. [19] showed that EC density increased twofold in a month when the wall shear stress had been increased up to fivefold by making an arteriovenous shunt in a canine carotid artery.

In addition to shear stress, ECs in vivo are subjected to stretch tension generated by blood pressure. This mechanical stress seems to stimulate EC growth as well. The uptake of [^{3}H]thymidine by ECs cultured on an elastic membrane was reported to increase markedly following cyclic stretching [20].

Effects of Fluid Shearing Force on EC Morphology

The shape and orientation of EC have long been known to be influenced by blood flow in vivo. In regions where blood flows unidirectionally at high speed, ECs are ellipsoidal in shape, with their long axis aligned in the direction of flow, while in regions where blood flows at low speed or stagnates, they are rounded and show no definite orientation. Flaherty et al. [21] surgically removed a segment of canine thoracic aorta where ECs were aligning along the flow stream and reimplanted it in a direction vertical to its original orientation. They observed that the nuclear pattern of the ECs was realigned along the flow direction within 10 dyas after the surgery. In in vitro experiments, Dewey et al. [22] applied an 8 dyn/cm^2 shear stress to cultured calf ECs in a flow-loading device of a cone plate type. No significant cell reorientation was observed until 24 h, but by 48 h the cells had uniformly assumed an ellipsoidal configuration with their major axis aligned in the flow direction. It has now been confirmed that the onset of cell shape change and orientation is sensitive to both the duration and magnitude of the shear stress load applied, and that these shear-induced changes in morphology are reversible.

From the sensitive effect of flow on EC shape and orientation, it is easily anticipated that the cytoskeletons forming cell shape and structure can be affected by flow. Actually, actin filaments in ECs exposed to a high flow rate are aligned parallel to the flow direction and form striking rich stress fibers [23,24]. When shear stress was experimentally increased in a canine carotid artery by an arteriovenous shunt, the bundles of 6- to 7-nm microfilaments in the ECs were increased [25]. Kim et al. [26,27] observed in experimentally coarctated aortas that the stress fibers in a high-shear region upstream of the coarctation became markedly thicker and longer than in control aortas. Similar findings were obtained in vitro. Franke et al. [28] applied shear stress to cultured human umbilical vein ECs (HUVEC) in a cone plate apparatus and observed that stress fibers that were not seen in static controls were formed in the shear-loaded ECs. Wechezak et al. [29,30] also reported that under the influence of shear stress, F-actin stress fibers in cultured bovine ECs were aligned with flow direction, that the actin reorganization was accompanied by redistribution of the underlying fibronectin matrix, and that following the stress application focal contacts which coincide with the termini of stress fibers at the basal plasma membrane were concentrated in the downstream regions of the cell.

Effects of Shear Stress on Protein Synthesis and Secretion

Shear stress modulates various protein synthesis and secretion by ECs. Concerning histamine, which increases endothelial permeability, it was reported that when cultured bovine ECs were subjected to shear stress of 2.8–6.2 dyn/cm^2 for 1.5 h in a flow chamber, the activity of histidine decarboxylase in ECs increased 2.8 to 3.7 times more than the static control [31]. By applying mean shear stress ranging from 22 to 109 dyn/cm^2 by pulsatile perfusion of rabbit aortas with platelet-free blood, DeForrest and Hollis [32] found a high correlation between the shear stress and the EC histamine-forming capacity.

The shear stress effect on production of prostacyclin, a potent vasodilator and endogenous inhibitor of platelet aggregation, was first reported by Reeves et al. [33]. When the perfusion flow to an isolated rat lung was enhanced, the minute production of prostacyclin measured from its stable metabolite, 6-keto-prostaglandin $F_1\alpha$, increased linearly with the flow rate. Frangos et al. [34] applied a shear stress of 0.016–10 dyn/cm^2 to HUVEC and observed that the onset of shear stress led to a sudden burst of prostacyclin production, which decreased to a steady state within several minutes. The rate of steady-state production by cells subjected to pulsatile shear stress (frequency, 1 Hz; range, 12–8 dyn/cm^2) was more than twice that of cells exposed to steady shear stress (10 dyn/cm^2) and 16 fold greater than that of cells in stationary culture. Grabowski et al. [35] reported, concerning the prostacyclin production induced by shear stress in cultured bovine ECs, that only the peak rate of the burst increased in a stress-dependent manner, and that once stressed, cells did not produce additional prostacyclin in response to repeated stimulation.

The production and release of endothelium-derived relaxing factor (EDRF), a potent vasodilating and antiplatelet-aggregating substance like prostacyclin, are also found to be modulated by shear stress. In a blood-perfused coronary artery, Holtz et al. [36] found that EDRF but not prostacyclin was the sustained mediator of the flow-dependent, endothelium-mediated dilation of the artery. This finding was confirmed by Rubanyi et al. [37] and by Miller and Vanhoutte [38]. In an in vitro study using cultured ECs, Cooke et al. [39] showed that when bovine aortic ECs cultured on microcarrier beads were subjected to vortical flow in a bath by magnetic stirring, relaxation of vascular rings in the same bath occurred in a flow-rate-dependent manner. The authors also indicated that the mediator of this response EDRF is nitric oxide (NO) or NO-related substances, because the flow-rate-dependent relaxations were inhibited by methylene blue and hemoglobin, which interfere with biosynthesis of NO.

More recently, Korenaga et al. [40] clearly proved the shear stress-dependent NO production in ECs by employing concomitant responses of cyclic guanosine monophosphate (cGMP) as its monitor. The working hypothesis underlying this study was that if the cGMP level in ECs, which is known to be regulated by NO in the cell in a dose-dependent manner through cyclic guanylate activation, responds to flow load depending on stress and if these cGMP reponses are completely abolished by a selective inhibitor of NO (e.g., N^G-monomethyl-L-arginine (L-NMA)), then the cGMP level reflects NO production rate in ECs. The results of their in vitro experiments, as shown in Fig. 3, revealed the validity of their hypothesis. They demonstrated that the amplitude of NO response depends on both shear stress and extracellular adenosine triphosphate (ATP) concentration and is reversible.

The effects of shear on endothelin (ET) release by ECs remain poorly defined. Yoshizumi et al. [41] measured ET released from cultured porcine ECs by using an enzyme-linked immunosorbent assay (ELISA) following the

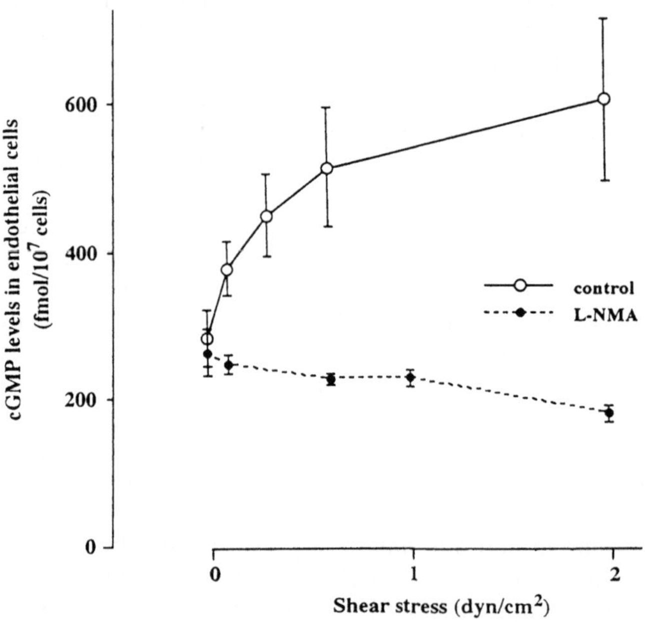

Fig. 3. The effects of shear stress on nitric oxide (NO) production in ECs [40]. NO production rate was monitored from cyclic guanosine monophosphate (*cGMP*) levels, induced through guanilate cyclase activation by NO and complelely abolished by a specific NO inhibitor, L-NMA. Shear stress dependency of NO production in ECs was confirmed

application of a shear stress of 5 dyn/cm^2 for 24 or 48 h in a cone plate apparatus. The amount of ET increased from 0 for the control to 3.5 after 24 h to 5 ng/ml after 48 h, and the level of ET mRNA showed a peak increase 2–4 h after the initiation of shear application. The shear-induced increase in the production of ET was confirmed by Milner et al. [42] in rabbit ECs. However, the opposite results were shown by Sharefkin et al. [43]. They tested the effect of a shear stress of 25 dyn/cm^2 on ET-1 precursor mRNA and ET peptide release in HUVEC in parallel plate flow chambers and found that application of shear stress for 24 h reduced the levels of this mRNA and also decreased ET-1 peptide release. The reason for the discrepancy is not clear, but it may be the difference in the intensity of shear stress applied or the phase-dependent expression of ET to the load.

Shear stress also modulates the production of tissue plasminogen activator (tPA) in ECs. Diamond et al. [44] observed that low (venous) shear stress (4 dyn/cm^2) had no effect on tPA secretion but that at high (arterial) shear stresses of 15 and 25 dyn/cm^2, the tPA secretion rate was 2.1 and 3.0 times greater than that of the basal tPA secretion, respectively. The increase in tPA production was not observed in 4–6 h but occurred 10–30 h after shear application, indicating that the tPA response to shear is very slow. tPA mRNA levels were also found to increase in ECs exposed to shear stress [45].

The production of platelet-derived growth factor (PDGF), which stimulates smooth muscle cell (SMC) growth and constricts vessels, is affected by shear stress as well. The levels of both PDGF A and B mRNA in HUVEC were shown to increase shear dependently [46]. Ono et al. [47] observed that SMC migration factors were released from ECs by shear stress and surmised that PDGF might be related to the phenomenon. They also showed the enhanced synthesis of collagen and an extracellular matrix produced by ECs as a result of shear application.

In some cases, however, shear stress downregulates synthetic activities of EC. For instance, application of shear stress to ECs decreases the production of fibronectin, one of the typical adhesive proteins [48]. The production of proteoheparan sulfate, a glycoprotein, has also been shown to be downregulated by shear stress [49]. By utilizing immunostain with monoclonal antibodies and flow cytometry, Ohtsuka et al. [50] recently found that flow application to mouse lymph node ECs markedly reduced the expression of vascular cell adhesion molecule-1 (VCAM-1) but induced no change in CD44. This VCAM-1 response to flow was shear stress and loading timedependent and also reversible.

Effects of Shear Stress on Other EC Functions

It has long been known, since the work of Fry [51], that shear stress exerts an influence on endothelial permeability. Jo et al. [52] reported data showing that EC permeability is sensitive to shear stress. Bovine aortic ECs cultured on a filter membrane were subjected to flow in an in vitro flow system, and simultaneous changes in permeability for albumin were measured. The results showed that permeability increased 4.3 and 10.3 times in 1 h at 1 and 10 dyn/cm^2 of shear stress, respectively. After removal of the shear stress, permeability returned to preshear levels within 2 h. By using Cr-ethylene diamine tetra acetic acid (Cr-EDTA) as a test tracer having a molecular weight comparable with sugar, Shibata and Kamiya [53] observed about a twofold increase in transcapillary permeability in the rabbit tenuissimus muscle against capillary flow increased from 500 to 2500 μm/s and suggested that the increased capillary wall shear stress induced this permeability enhancement.

Pinocytosis, as measured by uptake of horseradish peroxidase by cultured bovine ECs, was shown to increase in response to shear stress time and amplitude dependently [54]. Shear stress also enhances receptor-mediated low-density lipoprotein internalization and degradation by ECs [55].

Adhesive interaction between blood cells and ECs is modulated by shear stress as well. Shear stress of even less than 2 dyn/cm^2 decreases the rate of adherence of neutrophils to ECs [56], suggesting that shear stress might be an important determinant of the number of neutrophils which adhere to and migrate under the endothelium in postcapillary venules or inflammatory sites. Further studies are required to clarify whether shear stress acts as a physical force to detach neutrophils or modulates EC adhesiveness by changing the expression of adhesive proteins on their surface, as suggested by VCAM-1 downregulation described earlier [50].

Changes in mechanical properties of ECs caused by shear stress have been reported by Sato et al. [57]. After cultured bovine aortic ECs were exposed to shear stress, the authors measured cell stiffness by aspirating cells with a micropipette. Their results showed that cells exposed to shear stress showed mechanical stiffness significantly greater than that of control cells.

Shear Stress Signal Transduction

As described so far, ECs change their morphology and function in response to shear stress. This strongly suggests the presence of a mechanism by which ECs recognize shear stress and mediate the signal to intracellular organelles. Recent findings concerning this mechanism are now summarized.

It has been generally accepted that a wide variety of cell types have an internal signaling system. When external stimuli arrive at the cell membrane, they are recognized by specific receptors on the membrane and are transferred into changes in second messengers such as G-protein, inositol triphosphate, diacylglycerol, cyclic adenosinc monophosphate (cAMP), cGMP, and Ca^{2+}. These second messengers are known to regulate cell functions via many biochemical processes. Ca^{2+} response in ECs to shear stress was first reported by Ando et al. [58]. Nollert et al. [59] also found flow-induced inositol triphosphate transients in ECs. In addition, it is reported that shear stress increases the frequency of opening of the K^+ channel stress-dependently [60], leading to hyperpolarization of the cell membrane [61]. This response of the K^+ channel might be associated with the cytoplasmic Ca^{2+} increase, because ECs are known to have Ca^{2+}-dependent K^+ channels.

The responses of intracellular Ca^{2+} have been used in our studies [62] to analyze how ECs perceive fluid shear stress. Cultured monolayers of bovine aortic ECs loaded with the fluorescent Ca^{2+}-probe, fura-2, were exposed to different flow rates in a flow chamber, and simultaneous changes in cytoplasmic Ca^{2+} concentration were monitored by a photometric fluorescence microscope. In the presence of 500 nM extracellular ATP, the onset of flow induced an immediate increase in Ca^{2+} concentration, and when the flow stopped, Ca^{2+} concentration returned to the control level. The intracellular Ca^{2+} level increased corresponding to the stepwise increase in flow rate (Fig. 4a). These Ca^{2+} responses disappeared, when extracellular Ca^{2+} concentration was reduced to zero by adding 2 mM ethyleneglycolbis(2-aminoethyl ether)-tetraacetic acid (EGTA) to the medium, indicating that the observed stepwise responses were caused by Ca^{2+} influx from the external medium. Such flow-induced Ca^{2+} responses were strongly modified by the extracellular ATP level. When ATP concentration in the external medium was reduced to 100 nM, no noticeable responses were observed. For ATP levels greater than 1 mM, however, every step response became large and unstable, associated with a sharp initial peak that did not disappear by adding EGTA to the medium. This implies that the initial transient at high ATP level resulted from Ca^{2+} release from the intracellular stores such as the endoplasmic reticulum or mitochondria.

Such high sensitivity of flow-induced Ca^{2+} responses to ATP provoked an argument about the cellular sensing mechanism of EC to shear [63–65]. A conventional concept of the mechanism is, as suggested from the vector effects of flow on the cell morphology, that EC has some mechanoceptors which can directly detect mechanical deformation of the cell induced by wall shear stress (the shear stress hypothesis). The sensitivity of the receptors may be modulated by chemical agonists like ATP. Another plausible mechanism later proposed takes into account the diffusional accumulation of ATP to the cell surface. This concept is based on the fact that ATP binding to P_{2y}-purinoceptors on the EC surface has been known to elicit Ca^{2+} transients dose-dependently [66], and the fact that the diffusional accumulation of a substance to the surface from the external medium is hindered by the diffusional boundary layer, being built up near the surface and modulated by wall shear rate. Hence, this mechanism can also interpret the characteristic behaviors of flow-induced Ca^{2+} responses (the shear rate hypothesis). According to Caro and Nerem [67], the thickness of the diffusion boundary layer (δ) and the accumulation rate to the surface ($\dot{m}$) are related to the shear rate ($\dot{\gamma}$) as

$$\delta = \alpha(D/\dot{\gamma})^{1/3}$$
$$\dot{m} = (D/\delta)\,\Delta\,C/0.595 \qquad (3)$$

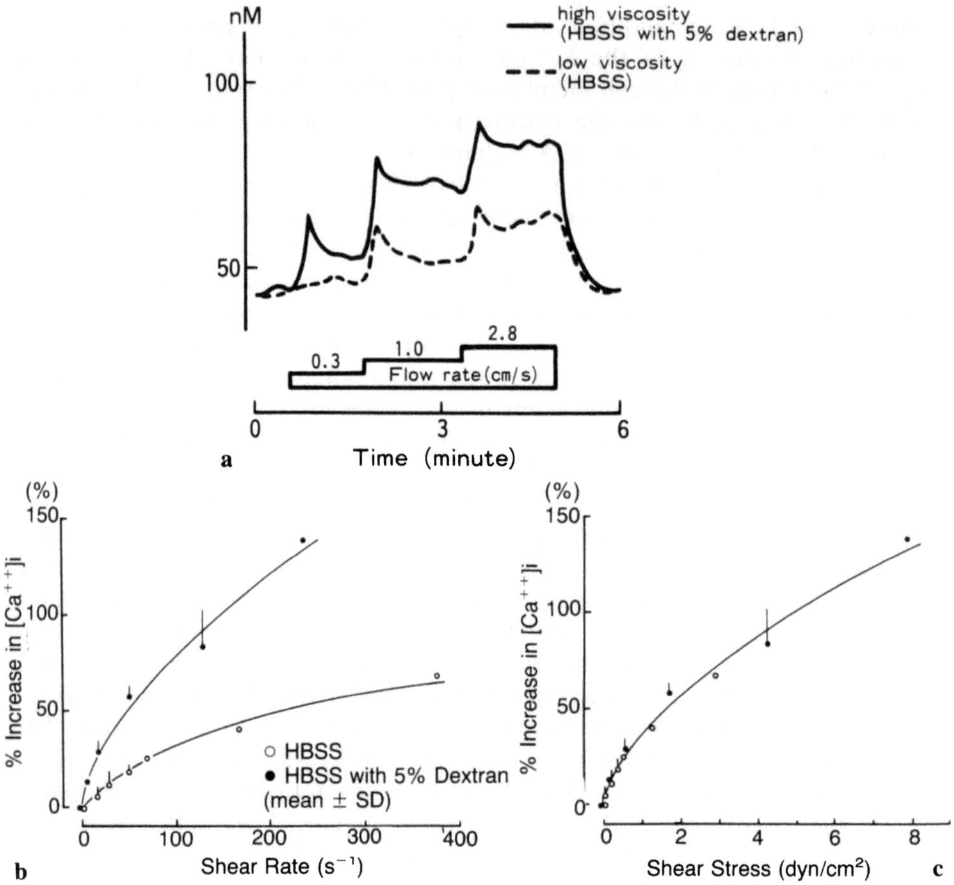

Fig. 4a–c. a Flow-induced Ca^{2+} responses obtained by perfusing a low-viscosity medium (Hanks' solution, HBSS) and a high-viscosity medium (HBSS with 5% dextran) [68]. The response magnitudes are plotted against wall shear rate **b** and wall shear stress **c**. Note in **b** that the levels of Ca^{2+} response to the high-viscosity medium are always greater than those in the low-viscosity medium at any shear rate level; **c**, both high- and low-viscosity data are clustered around a single curve against the shear stress. These results demonstrated the validity of the shear stress hypothesis rather than that of the shear rate hypothesis

where D and ΔC indicate ATP diffusivity and its concentration difference beyond the diffusion baundary layer (α, a constant).

As a matter of practice, it is not easy to prove which hypothesis is true by discriminating the effect of shear stress (τ) and that of shear rate ($\dot{\gamma}$), because they are closely related ($\tau = \mu\dot{\gamma}$), as seen in Equation 2. One possible way is to change viscosity (μ) and to see; (1) whether the same τ at different μ and $\dot{\gamma}$ induces the same Ca^{2+} response (to confirm the shear stress hypothesis), or (2) whether Ca^{2+} response diminishes when μ is increased at the same γ (to support the shear rate hypothesis), because the accumulation rate ($\dot{m}$) of ATP is, from Equation 3 and the Stokes–Einstein equation, expressed as

$$\dot{m} = \kappa\dot{\gamma}^{1/3}\mu^{-2/3} \tag{4}$$

where κ is a constant relevant to neither $\dot{\gamma}$ nor μ. Figure 4a shows a typical example of Ca^{2+} response in our in vitro experiments recently performed on these considerations, in which the medium viscosity was varied from 0.765 cP to 3.372 cP by adding 5% dextran [68]. The summary of results in Fig. 4b disproves the shear rate hypothesis, because the Ca^{2+} response to higher viscosity was always larger than that to lower viscosity at any shear rate level. On the contrary, clear one-to-one correspondence between Ca^{2+} response and shear stress regardless of viscosity difference in Fig. 4c verified the shear stress hypothesis. It was confirmed that flow-induced Ca^{2+} responses in ECs are elicited by mechanical deformation of the cell caused by wall shear stress rather than ATP accumulation to the cell modulated by wall shear rate. This conclusion is supported by the fact that when ECs are rubbed lightly with a latex balloon and exposed to frictional force, a transient Ca^{2+} increase occurs in a ATP-free medium and even in the air [69]. ATP (or other chemical ligands) may be regarded as the sensitivity modulator of EC responses to fluid shear stress.

The intracellular or extracellular mechanism of EC for sensing its own shear deformation has not been well identified, even for where the sensor exists. Besides shear stress, pulsatile stretching of ECs cultured on a flexible membrane also induces an immediate increase in intracellular Ca^{2+} concentration, probably through stretch-activated Ca^{2+} channels found by Lansman et al. [70], and stimulates adenylate cyclase activity [71]. Because physiological vascular responses to shearing stress and to tangential stress are apparently different, it is of interest to see how ECs discriminate these stresses. Ingber [72] has suggested that integrins, transmembrane molecules mediating cell attachment by physically interlinking the extracellular matrix with the intracellular cytoskeleton, may act as mechanochemical transducers. Including such a possibility, further analyses are necessary to visualize the sensing mechanisms in EC to physical stresses.

References

1. Ando J, Kamiya A (1993) Blood flow and vascular endothelial cell function. Front Med Biol Eng 5:245–264
2. Davies PF, Tripatchi S (1993) Mechanical stress mechanism and the cell: an endothelial paradigm. Circ Res 72:239–245
3. Suwa N, Takahashi T (1971) Morphological and morphometrical analysis of circulation in hypertension and ischemic kidney. Urban and Schwarzenberg, München
4. Hutchins GM, Milner MM, Boitnott JK (1976) Vessel caliber and branch angle of human coronary arterial branch-points. Circ Res 38:572–576

5. Thoma R (1911) Über die Histomechanik des Gefäßsystems und die Pathogenese der Angiosklerose. Pathol Anat Physiol 204:1–74
6. Rodbard S (1959) Physical forces and the vascular lining. Ann Int Med 50:1339–1351
7. Kamiya A, Togawa T (1980) Adaptive regulation of wall shear stress to flow change in the canine carotid artery. Am J Physiol 239:H14–H21
8. Guyton JR, Hartley CJ (1985) Flow restriction of one carotid artery in juvenile rats inhibits growth of arterial diameter. Am J Physiol 17:H540–H546
9. Langille BL, O'Donnell F (1986) Reductions in arterial diameter produced by chronic decreases in blood flow are endothelium-dependent. Science 231:405–407
10. Zarins CK, Zatina MA, Giddens DP, Ku DN, Glagov S (1987) Shear regulation of arterial lumen diameter in experimental atherogenesis. J Vasc Surg 5:413–420
11. Hudlicka O, Brown M, Egginton S (1992) Angiogenesis in skeletal and cardiac muscle. Physiol Rev 72(2):369–417
12. Wang DW, Prewitt RL (1991) Microvascular development during normal growth and reduced blood flow: introduction of a new model. Am J Physiol 260:H1966–H1972
13. Ando J, Nomura H, Kamiya A (1987) The effect of fluid shear stress on the migration and proliferation of cultured endothelial cells. Microvasc Res 33:62–70
14. Ando J, Komatsuda T, Ishikawa C, Kamiya A (1990) Fluid shear stress enhanced DNA synthesis in cultured endothelial cells during repair of mechanical denudation. Biorheology 27:675–684
15. Masuda M, Fujiwara K (1991) Morphological responses of single endothelial cells exposed to physiological levels of fluid shear stress. Med Biol Eng Comput 29[suppl]:268
16. Eskin SG, Sybers HD, O'Bannon W, Navarro LT (1982) Performance of tissue cultured endothelial cells in a mock circulatory loop. Artery 10:159–171
17. Dewey CF Jr (1984) Effects of fluid flow on living vascular cells. J Biomech Eng 106:31–35
18. Davies PF, Remuzzi A, Gordon EJ, Dewey CF Jr, Gimbrone MA Jr (1986) Turbulent fluid shear stress induces vascular endothelial cell turnover in vitro. Proc Natl Acad Sci USA 83:2114–2117
19. Masuda H, Kawamura K, Tohda K, Shozawa T, Sageshima M, Kamiya A (1989) Increase in endothelial cell density before artery enlargement in flow-loaded canine carotid artery. Arteriosclerosis 9:812–823
20. Sumpio BE, Banes AJ, Levin LG, Johnson G Jr (1987) Mechanical stress stimulates aortic endothelial cells to proliferate. J Vasc Surg 6:252–256
21. Flaherty JT, Pierce JE, Ferrans VJ, Patel DJ, Tucker WK, Fry DL (1972) Endothelial nuclear patterns in the canine arterial tree with particular reference to hemodynamic events. Circ Res 30:23–33
22. Dewey CF Jr, Bussolari SR, Gimbrone MA Jr, Davies PF (1981) The dynamic response of vascular endothelial cells to fluid shear stress. J Biomech Eng 103:177–184
23. Wong AJ, Pollard TD, Herman IM (1983) Actin filament stress fibers in vascular endothelial cells in vivo. Science 219:867–869
24. White GE, Fujiwara K (1986) Expression and intracellular distribution of stress fibers in aortic endothelium. J Cell Biol 103:63–70
25. Masuda H, Shozawa T, Hosoda S, Kanda M, Kamiya A (1985) Cytoplasmic microfilaments in endothelial cells of flow loaded canine carotid arteries. Heart Vessels 1:65–69

26. Kim DW, Langille BL, Wong MKK, Gotlieb AI (1989) Patterns of endothelial microfilament distribution in the rabbit aorta in situ. Circ Res 64:21–31
27. Kim DW, Gotlieb AI, Langille BL (1989) In vivo modulation of endothelial F-actin microfilaments by experimental alterations in shear stress. Arteriosclerosis 9: 439–445
28. Franke RP, Grafe M, Schnittler H, Seiffge D, Mittermayer C, Drenckhahn D (1984) Induction of human vascular endothelial stress fibres by fluid shear stress. Nature 307:648–649
29. Weckezak AR, Viggers RF, Sauvage LR (1985) Fibronectin and F-actin redistribution in cultured endothelial cells exposed to shear stress. Lab Invest 53(6): 639–647
30. Wechezak AR, Wight TN, Viggers RF, Sauvage LR (1989) Endothelial adherence under shear stress is dependent upon microfilament reorganization. J Cell Physiol 139:136–146
31. Rosen LA, Hollis TM, Sharma MG (1974) Alterations in bovine endothelial histidine decarboxylase activity following exposure to shear stresses. Exp Mol Pathol 20:329–343
32. DeForrest JM, Hollis TM (1978) Shear stress and aortic histamine synthesis. Am J Physiol 234:H701–H705
33. Reeves JT, Grondelle AV, Voelkel NF, Walker B, Lindenfeld J, Worthen S, Mathias M (1983) Prostacyclin production and lung endothelial cell shear stress. In: Hypoxia, exercise, and altitude: proceedings of the third Banff international hypoxia symposium. Liss, New York, pp 125–131
34. Frangos JA, Eskin SG, McIntire LV, Ives CL (1985) Flow effects on prostacyclin production by cultured human endothelial cells. Science 227:1477–1479
35. Grabowski EF, Jaffe EA, Weksler BB (1985) Prostacyclin production by cultured endothelial cell monolayers exposed to step increases in shear stress. J Lab Clin Med 105:36–43
36. Holtz J, Forstermann U, Pohl U, Giesler M, Bassenge E (1984) Flow-dependent, endothelium-mediated dilation of epicardial coronary arteries in conscious dogs: effects of cyclooxygenase inhibition. J Cardiovasc Pharmacol 6:1161–1169
37. Rubanyi GM, Romero JC, Vanhoutte PM (1986) Flow-induced release of endothelium-derived relaxing factor. Am J Physiol 250:H1145–H1149
38. Miller VM, Vanhoutte PM (1988) Enhanced release of endothelium-derived factor(s) by chronic increases in blood flow. Am J Physiol 255:H446–H451
39. Cooke JP, Stamler J, Andon N, Davies PF, McKinley G, Loscalzo J (1990) Flow stimulates endothelial cells to release a nitrovasodilator that is potentiated by reduced thiol. Am J Physiol 259:H804–H812
40. Korenaga R, Ando J, Tsuboi H, Yang W, Toyo-oka T, Kamiya A (1994) Laminar flow stimulates ATP- and shear stress-dependent nitric oxide production in cultured bovine endothelial cells. Biochem Biophys Res Commun 198:213–219
41. Yoshizumi M, Kurihara H, Sugiyama T, Takaku F, Yanagisawa M, Masaki T, Yazaki Y (1989) Hemodynamic shear stress stimulates endothelin production by cultured endothelial cells. Biochem Biophys Res Commun 161:859–864
42. Milner P, Bodin P, Loesch A, Burnstock G (1990) Rapid release of endothelin and ATP from isolated aortic endothelial cells exposed to increased flow. Biochem Biophys Res Commun 170(2):649–656
43. Sharefkin JB, Diamond SL, Eskin SG, McIntire LV, Dieffenbach CW (1991) Fluid flow decreases preproendothelin mRNA levels and suppresses endothelin-1 peptide release in cultured human endothelial cells. J Vasc Surg 14:1–9

44. Diamond SL, Eskin SG, McIntire LV (1989) Fluid flow stimulates tissue plasminogen activator secretion by cultured human endothelial cells. Science 243: 1483–1485
45. Diamond SL, Sharefkin JB, Dieffenbach C, Frasier-Scott K, McIntire LV, Eskin SG (1990) Tissue plasminogen activator messenger RNA levels increase in cultured human endothelial cells exposed to laminar shear stress. J Cell Physiol 143:364–371
46. Hsieh H-J, Li N-Q, Frangos JA (1991) Shear stress increases endothelial platelet-derived growth factor mRNA levels. Am J Physiol 260:H642–H646
47. Ono O, Ando J, Kamiya A, Kuboki Y, Yasuda H (1991) Flow effects on cultured vascular endothelial and smooth muscle cell functions. Cell Struct Funct 16:365–374
48. Gupte A, Frangos JA (1990) Effects of flow on the synthesis and release of fibronectin by endothelial cells. In Vitro Cell Dev Biol 26:57–60
49. Grimm J, Keller R, Groot PG (1988) Laminar flow induced cell polarity and leads to rearrangement of proteoglycan metabolism in endothelial cells. Thromb Haemostasis 60(3):437–441
50. Ohtsuka A, Ando J, Korenaga R, Kamiya A, Toyama-Sorimachi N, Miyasaka M (1993) The effect of flow on the expression of vascular adhesion molecule-1 by cultured mouse endothelial cells. Biophys Biochem Res Commun 193:303–310
51. Fry DL (1968) Acute vascular endothelial changes associated with increased blood velocity gradients. Circ Res 22:165–169
52. Jo H, Dull RO, Hollis TM, Tarbell JM (1991) Endothelial albumin permeability is shear dependent, time dependent, and reversible. Am J Physiol 260:H1992–H1996
53. Shibata M, Kamiya A (1992) Blood flow dependence of local capillary permeability of Cr-EDTA in the rabbit skeletal muscle. Jpn J Physiol 42:631–639
54. Davies PF, Dewey CF Jr, Bussolari SR, Gordon EJ, Gimbrone MA Jr (1984) Influence of hemodynamic forces on vascular endothelial function: in vitro studies of shear stress and pinocytosis in bovine aortic cells. J Clin Invest 73:1121–1129
55. Sprague EA, Steinbach BL, Nerem RM, Schwartz CJ (1987) Influence of a laminar steady-state fluid-imposed wall shear stress on the binding, internalization, and degradation of low-density lipoproteins by cultured arterial endothelium. Circulation 76:648–656
56. Worthen GS, Smedly LA, Tonnesen MG, Ellis D, Voelkel NF, Reeves JT, Henson PM (1987) Effects of shear stress on adhesive interaction between neutrophils and cultured endothelial cells. J Appl Physiol 63:2031–2041
57. Sato M, Levesque MJ, Nerem RM (1987) Micropipette aspiration of cultured bovine aortic endothelial cells exposed to shear stress. Arteriosclerosis 7:276–286
58. Ando J, Komatsuda T, Kamiya A (1988) Cytoplasmic calcium response to fluid shear stress in cultured vascular endothelial cells. In Vitro Cell Dev Biol 24: 821–877
59. Nollert MU, Eskin SG, McIntire LV (1990) Shear stress increases inositol trisphosphate levels in human endothelial cells. Biochem Biophys Res Commun 170(1): 281–287
60. Olesen S-P, Clapham DE, Davies PF (1988) Haemodynamic shear stress activates a K^+ current in vascular endothelial cells. Nature 331:168–331
61. Nakache M, Gaub HE (1988) Hydrodynamic hyperpolarization of endothelial cells. Proc Natl Acad Sci USA 85:1841–1843
62. Ando J, Ohtsuka A, Korenaga R, Kamiya A (1991) Effect of extracellular ATP level on flow-induced Ca^{++} response in cultured vascular endothelial cells. Biochem Biophys Res Commun 179:1192–1199

63. Nollert MU, Diamond SL, McIntire LV (1991) Hydrodynamic shear stress and mass transport modulation of endothelial cell metabolism. Biotechnol Bioeng 38: 588–602
64. Mo M, Eskin SG, Schilling WP (1991) Flow-induced changes in Ca^{2+} signalling of vascular endothelial cells: effect of shear stress and ATP. Am J Physiol 260: H1698–H1707
65. Dull RO, Davies PF (1991) Flow modulation of agonist (ATP)-response (Ca^{2+}) coupling in vascular endothelial cells. Am J Physiol 261:H149–H154
66. Luckhoff A, Busse R (1986) Increased free calcium in endothelial cells under stimulation with adenine nucleotides. J Cell Physiol 126:414–420
67. Caro CC, Nerem RM (1973) Transport of ^{14}C-4-cholesterol between serum and wall in the perfused dog common carotid artery. Circ Res 32:187–205
68. Ando J, Ohtsuka A, Korenaga R, Kawamura T, Kamiya A (1993) Wall shear stress rather than shear rate regulates cytoplasmic Ca^{++} responses to flow in vascular endothelial cells. Biochem Biophys Res Commun 190:716–723
69. Ando J, Ohtsuka A, Korenaga R, Kamiya A (1992) Intracellular calcium response to mechanical shearing force in cultured vascular endothelial cells. In Vitro Cell Dev Biol 28:152
70. Lansman JB, Hallam TJ, Rink TJ (1987) Single stretch-activated ion channels in vascular endothelial cells as mechanotransducers? Nature 325:811–813
71. Letsou GV, Rosales O, Maitz S, Vogt A, Sumpio BE (1990) Stimulation of adenylate cyclase activity in cultured endothelial cells subjects to cyclic stretch. J Cardiovasc Surg 31:634–639
72. Ingber D (1991) Integrins as mechanochemical transducers. Curr Opin Cell Biol 3:841–848

Part 8. Hand

Refined Three-Dimensional Finite-Element Models of the PIP Joint

Hiroyuki Hashizume[1], Toshiyuki Miyake[1], Qulin Shi[1], Hajime Inoue[1], and Yasuharu Sakamoto[2]

Summary. The initial phase of a program to develop a general method to measure patterns of stress concentration in hard tissue of the human body produced a modeling system that uses the three-dimensional (3-D) finite-element method (FEM) to precisely analyze the stress distribution in the proximal interphalangeal (PIP) joint. Accurate geometric data for the 3-D model were obtained with a noncontact 3-D digitizer (Orbis). A convoluted mesh was obtained by smoothly connecting the five sagittal plane configurations, also using b-spline curves. The relative positions of the two bones were determined in the 3-D computer-assisted design (CAD). A general FEM program (I-DEAS, SDRC) was used for computer simulation analysis. The validity of our simulation was tested by comparing the areas of stress concentration in three PIP joint positions with the fractured areas of 10 cadaveric specimens in an experimental fracture study. Stress analysis using FEM showed that the distribution of areas of concentrated stress changed according to the angle of the PIP joint. The analysis accurately predicted the location and type of fracture.

Key words: Finite-Element method—Computer simulation—Proximal interphalangeal joint

Introduction

We have been working on software to produce smooth and anatomically accurate shapes for finite-element models (FEMs) to more closely approximate real joint dynamics. The proximal interphalangeal (PIP) joint study is one of a

[1] Department of Orthopedic Surgery, Okayama University Medical School, 2-5-1 Shikata, Okayama, 700 Japan
[2] Western Industrial Technology Center, Hiroshima, Japan

series of studies that include analyses of stress in bones and joints. The purpose of these studies is to explore the applicability of finite-element methods to the study of the effects of stress on the human hard tissue. The focus is on the development of methods to record and project anatomically accurate shapes of bones, joints, and teeth that can be adjusted to reflect different population norms.

The FEM was selected because it was designed to precisely analyze the behavior of stress in objects of inorganic materials with complex shapes [1]. Our work as orthopedic surgeons deals primarily with shapes and the effects of stress. We hold the idea that shape reflects biological function [2]. If we can restore the natural physical shape, we can usually rely on normal biological processes to complete repair of the injury. Abnormal concentrations of stress can interfere with the long-term success of the repair. Sudden loading of such stress can result in fractures, and osteoarthritic changes can result when stress is lessened. Current diagnostic technology cannot provide adequate information on the effects of stress on shape and on the internal structure. Surgeons require more information about hidden damage, microscopic disturbances at the crystal level. If the FEM is adaptable to clinical biomaterials, we expect that analyses of the effects of stress will lead directly to more accurate diagnoses and more effective treatment.

Materials and Methods

The model for FEM analysis was developed in several steps with data from 10 PIP joints. A three-dimensional (3-D) coordinate system (X, Y, Z) of the surface of the bone of each middle and proximal phalanx was measured to form the PIP joint image with a noncontact 3-D digitizer (Orbis). The surface of each bone was marked by a parallel series of points 1 mm apart. The assembled profiles formed an angular 3-D image of the intact bone. Five b-spline curves containing 75 points crossed the group of profiles. Fifteen points on each profile were selected to constitute the 3-D surface with lines perpendicular to the surface of the five profiles, which gave a smooth surface connecting 75 points in a mesh (Fig. 1). The final step rounded or smoothed the 3-D surface by the projection of 75 additional points from the original data.

Two enlarged plastic replicas (×10) of the cross-sectional view of the articular surfaces were cut. (NC) data for the anatomic cutting process were derived from the smoothed surface of the computer image. A visual inspection of the articular contact surface was made of the enlarged smooth contact surface (Fig. 2).

An enlarged resin model (×10) of the computer image was made using the optical configuration system (SOUP) (Seemet) to verify the anatomical accuracy of the image (Fig. 3). Physical measurements of the resin model and bone specimens were made by hand and compared.

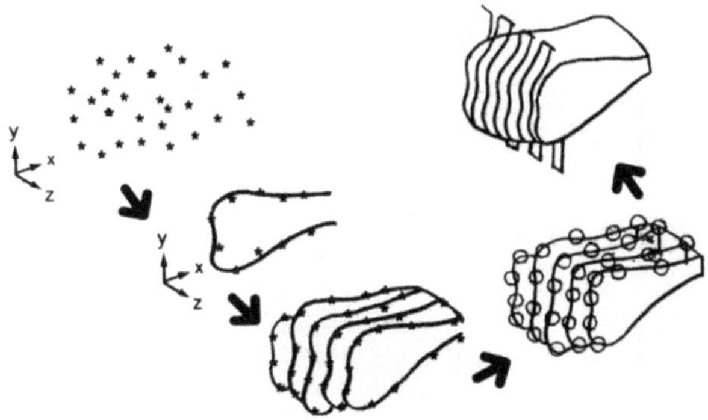

Fig. 1. Steps in the construction of the computer image

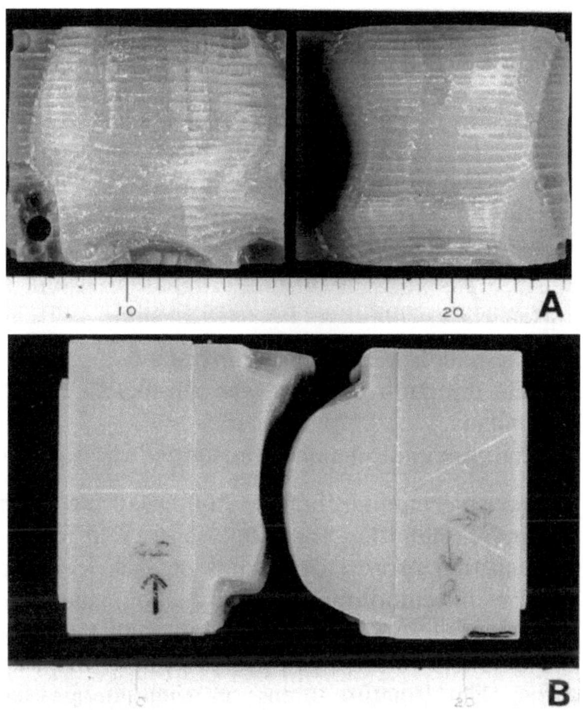

Fig. 2. A Enlarged plastic replicas. **B** Lateral view of **A** showing concave/convex contact surfaces

The middle and proximal phalangeal images were manipulated in such a way that smooth flexion and extension positions that conformed to the configuration data were achieved. After the smooth contact condition was achieved, the positions of the two bones were determined in the 3-D CAD.

The material properties were assumed to be linearly elastic and isotropic; Young's modulus 15 000 MPa and Poisson's ratio 3.0 [3]. The more appropriate dynamic, nonlinear, and anisotropic methods of analyses of biomaterials are unavailable at this date for FEM. (They are, in fact, a later objective of our

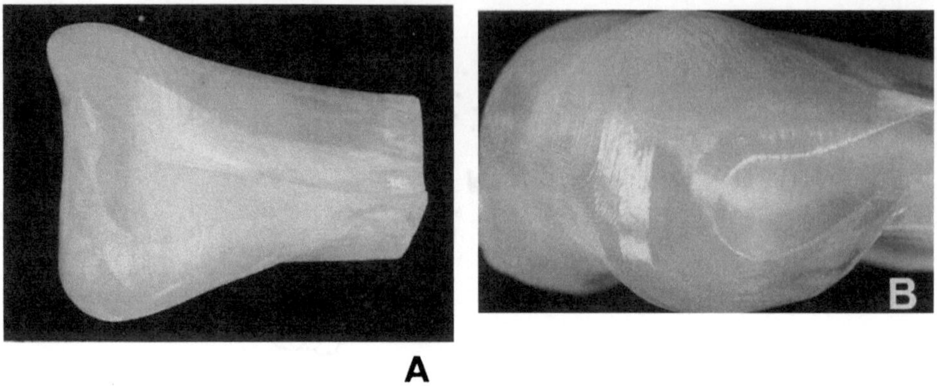

Fig. 3. Resin model. A Middle phalangeal base. B Proximal phalangeal head

program.) The middle phalanx mesh consisted of 372 elements and 722 nodal points. The proximal phalanx mesh consisted of 887 elements and 457 nodal points. The elements of the meshes were 2 mm wide. A general FEM program (I-DEAS, SDRC), computed on the "Work Station," was used for the simulation. The changes of the stress distribution in the PIP joint were analyzed in three positions: neutral, 20° hyperextension, and 20° flexion (Fig. 4).

Boundary conditions for each position were as follows:

1. The middle phalanx was restrained.
2. The direction of force was parallel to the longitudinal axis of the middle phalanx.
3. The proximal phalanx transmitted the force to the middle phalanx.

The experimental fracture study used specimens from cadavers that had been preserved with 10% formalin. Thirty PIP joints were affixed with dental resin in a custom fixator. Both ends of each joint were set at the same depth and angle as the computer analysis. The Shimazu Autograph S-2000 loading machine added axial loads statically and simultaneously recorded the deformation curve. The fracture event was defined by the abrupt reversal of the deformation curve. The fracture areas on roentgenographs taken before and after the experiment were compared with areas of stress concentration in the computer simulation.

Results

The configuration from the articular surface view and the lateral view of the joint were precisely represented in the resin model (see Fig. 3). The configuration of the joint obtained from the computer simulation was almost identical to the anatomical configuration.

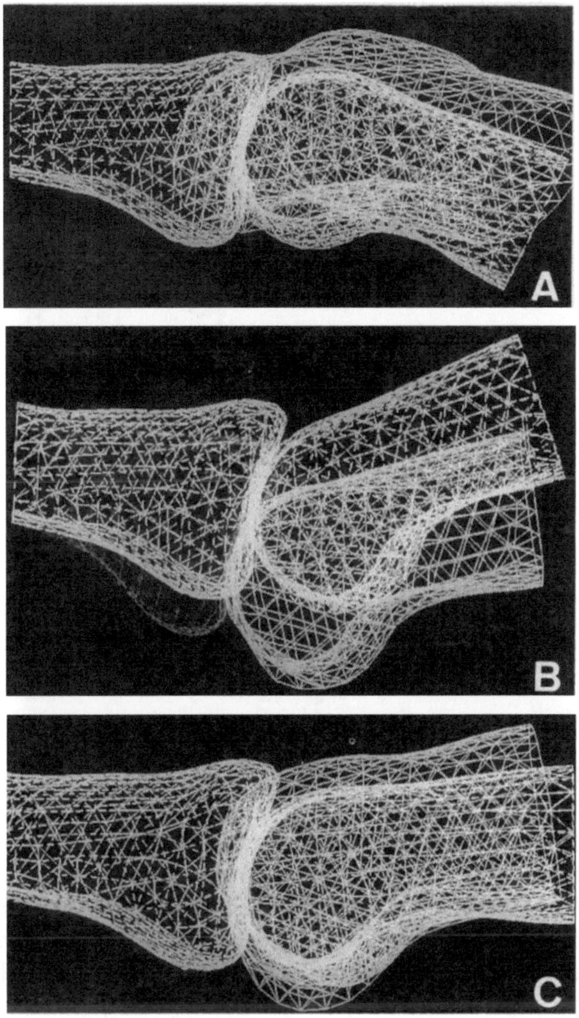

Fig. 4. Mesh image of the PIP joint. Middle phalangeal base is on the *left* and proximal phalangeal head is on the *right*. **A** Neutral position. **B** Hyperextension position. **C** Flexion position

In the von Mises stress analysis of the proximal phalangeal head, the stress concentration at the endmost point of the condyle was in the neutral position (Fig. 5). The areas of stress concentration shifted from intercondylar areas to the dorsal region in the hyperextension position. Stress was concentrated on one side of the protruded condyle in the flexion position. In the neutral position, the stress concentrated at the endmost point of the proximal phalangeal head. When the joint was in the hyperextension position, the area of stress was concentrated in the center of the articular surface [4].

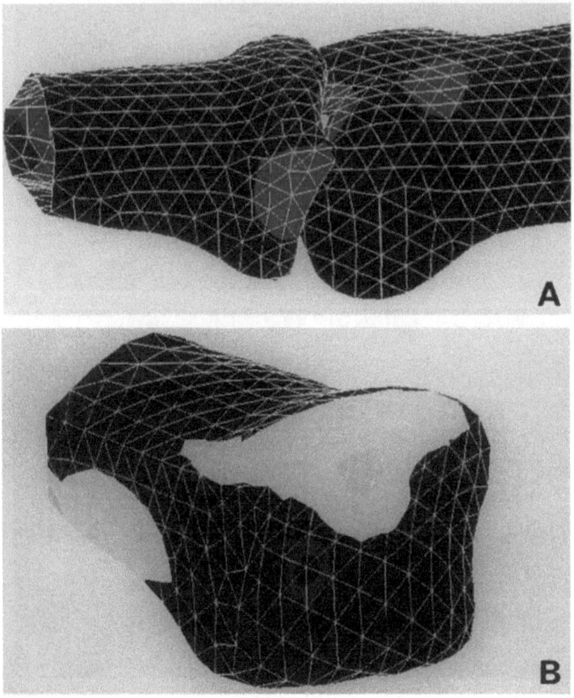

Fig. 5. Stress concentration appears as *light gray areas* on the surface mesh. **A** PIP joint in neutral position. Note stress concentration on both bones. **B** Middle phalangeal base in flexion position. Note the changed location of stress concentration

Discussion

Results from the study showed that the patterns of stress distribution in the bones of the joint changed according to the position of the joint during static loading. The locations of the fracture areas in the experimental fracture study coincided with the areas of stress concentration in the computer simulation under similar loading conditions. It can therefore be argued that fracture dislocation is largely determined by the position of the PIP joint at the time of injury.

The results of the preliminary phase of our investigations into the applicability of FEM to the analysis of stress in human hard tissues are promising. However, FEM was originally developed for materials with linear properties and linear interactions. It would be unwise to assume that such methods can be simply and directly transferred to nonlinear biological materials and interactions. The results of this and other experimental studies in the preliminary phase of the investigation indicate that FEM is accurate with simple structures and materials. Can FEM evolve far enough to permit analysis of compound structures composed of nonuniform materials? The next phase will require clarification and development in a number of areas. I list here a few ideas:

1. We need a method that will produce exact, digital, anatomical data to construct the computer image.

2. We need clarification of the differential effects of linear and nonlinear material properties and interactions in FEM. The range of established theoretical numerical values of the properties of biomaterials is limited.
3. Human hard tissues and skeletal structures present variables that FEM was not originally designed to accomodate [5,6]. The materials are not uniform. A single bone will include cortical bone, subchondral bone, cancellous bone, and cartilage, distributed in a nonuniform way [7]. A joint compounds the variables and raises the issue of continuity in a stress analysis. Methods must eventually be developed to test joints with intact intraarticular fluids if joint dynamics are to be accurately simulated for clinical purposes.
4. Much work needs to be done on the mathematics of contact models and their interpretation as computer images. Human skeletal articulation tends to be complex and convoluted. The computer image is compelling but it does not display enough data to allow the viewer to determine the validity of the model. Extensive testing for validity should mediate undesirable risk that simplistic theoretical contact models may carry into the clinical setting [8].
5. We need to establish indicators of appropriateness for the selection of the type of model to use, whether a shell (surface) model or a volume (bulk) model.
6. We have to consider the effects of not only axial stress but also traction stress induced by the tendons and ligaments.

References

1. Togawa H (1979) The guide to the finite element method (in Japanese). Science Publishing, Tokyo, pp 1–18
2. Wainwright SA (1988) Axis and circumference: the cylindrical shape of plants and animals. Harvard University Press, Cambridge, pp 1–37
3. Hayashi A, Tanaka S, Tsutsumi S (1991) Simulation analysis about development of osteoarthritis using finite element method (in Japanese). J Jpn Orthop Assoc [suppl] 65:1178
4. Hashizume H, Nagasawa F, Wake H, Hara S, Akahori O (1991) Study of middle phalangeal base fracture (in Japanese with English abstract). J Jpn Soc Surg Hand 8:709–713
5. Little RB, Wevers HW, Slu D, Cooke TDV (1986) A three-dimensional finite element analysis of the upper tibia. J Biomech Eng 108:111–119
6. Shirazi-ADL SA, Shrivastava SC, Ahmed AM (1984) Stress analysis of the lumbar disc-body unit in compression. A three-dimensional nonlinear finite element study. Spine 9:120–133
7. Okihara T, Kurosawa H, Yamagosi K, Yamanoi T (1990) Analysis of stress distribution human tibia using three-dimensional finite element method (in Japanese with English abstract). In: Proceedings of 1990 annual meeting of Japanese Society for Orthopaedic Biomechanics 12:61–64
8. Kawai T (1990) An application of the finite element method to the biomechanics (in Japanese). In: Proceedings of 1990 annual meeting of Japanese Society for Orthopaedic Biomechanics 12:21–31

Analysis of Thumb and Index Finger Joints During Pinching Motion and Writing a Cross, as Measured by Electrogoniometers

Akira Hara, Yasuo Yamauchi, and Koichi Kusunose

Summary. The purpose of this study is to determine the flexion angle of the thumb and the index finger joints while using a pinching motion to write a cross, observed with flexible goniometers. (1) The pinching motion: the flexion angles of the joints of the thumb and the index finger under 100-g and 1000-g pressure pinch were measured at 0, 1, 2, 3, and 5 cm distances of pulp pinch and tip pinch. The flexion angle of the index metacarpophalangeal (MP) joint was decreased as the distance of the pinch was increased, but the flexion angles of other joints were almost the same in any distance of the pinch. (2) Writing a cross: the size of the cross was set at either 4 or 6 cm. One cycle was started by writing a horizontal line from left to right and then a vertical line from up to down, returning to the starting point. The most mobile joint was the interphalangeal (IP) joint of the thumb, followed by the proximal interphalangeal (PIP) joint of the index finger. The less mobile were the MP joints of the both digits. This flexible goniometer will be useful in the motion analysis of the finger joints of both normal and diseased hands.

Key words: Hand—Fingers—Goniometer—Pinch—Biomechanics

Introduction

Numerous reports have appeared to date on motion analysis of the finger joints using goniometers [1,2]. The authors have been conducting finger joint motion analysis [3] using an electrogoniometer attached to a parallelogram linkage. However, this apparatus itself is bulky in size and has the danger of interfering with delicate motion analyses. Similarly, a three-dimensional video analysis

---

Department of Orthopaedic Surgery, Juntendo University School of Medicine, 2-1-1 Hongo, Bunkyo-ku, Tokyo, 113 Japan system [4] required time to use and was disadvantageous in that the analysis of the long and ring fingers was hindered by the overlapping adjacent fingers.

In this study, a set of flexible electrogoniometers ("goniometer") was used to measure the joint angles of the thumb and the index finger during pinching motion ("pinch"). To test the dexterity of the hand, moreover, the joint motion of the thumb and the index finger was examined when writing a cross. Compact in size, the goniometer used in our study was connected to the dorsum of the digits, enabling relatively easy observation of the angles during motion.

Materials and Method

The goniometer used in this study was a single axis flexible electrogoniometer for finger joints (type G35, produced by Penny & Giles) (Fig. 1). The flexion angle of two endblocks was measured using an installed strain gauge, with the distance between two endblocks being 35 mm. Accuracy of the goniometer calibrated beforehand was confirmed to be within 1° from 0 to 90° of flexion. The goniometer, placed on the dorsum of the fingers, was checked to see that it did not "buckle" from the axis of the underlying bone by the flexion and extension movements of the joints.

Joint Motion During Pinching

The dominant right hands of 12 normal men, ages ranging from 25 to 40 years (mean, 31) were tested. The goniometer endblocks were placed on the dorsum

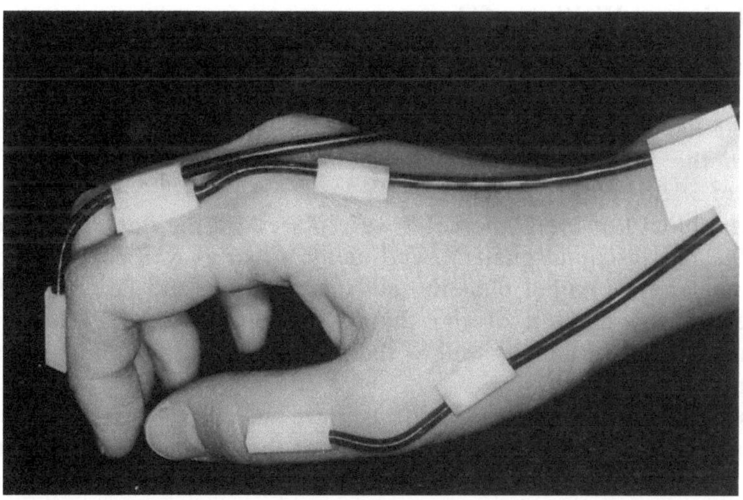

Fig. 1. Goniometer for the finger (Penny & Giles). The endblocks are placed on the dorsum of the thumb and the index finger

of the thumb distal phalanx, proximal phalanx, and metacarpus, as well as the index finger distal phalanx, middle phalanx, proximal phalanx, and metacarpus; the flexion angles at the dorsum of the fingers were measured.

Measurement of the angles was made during light pinch and power pinch with a pinch force of 100 and 1000 g, respectively. Before carrying out the pinch, subjects were taught the 100 g and 1000-g force by using an electropinch sensor (KYOWA, PSA-A). It was ascertained before measurement that in pinching a block 0 to 5 cm in thickness with the thumb and index finger, the distance between the first and second metacarpal heads for all the subjects remained unchanged, while the abducted position of the thumb was unaffected by changes in the pinch widths. The following tests were performed with two different forces, 100 g and 1000 g.

A. Pulp-Pinch
   1. Pinching between the pulps of the two digits without holding an object; the motion was repeated ten times.
   2. Pulp-pinching blocks of 1, 2, 3, 4, and 5 cm in thickness, respectively.

B. Tip-pinch
   1. Pinching between the tips of the two digits without holding an object; the motion was repeated ten times.
   2. Tip-pinching blocks of 1, 2, 3, 4, and 5 cm in thickness, respectively.

C. Motion analysis
   Pinching between the pulps of the two digits without holding an object was repeated ten times, with each stroke lasting 2 s each (ext.–pinch–ext.), and motion analysis was performed for both pulp-pinch and tip-pinch.

Joint Motion on Writing a Cross

The dominant right hands of eight normal subjects (seven men and one woman) aged between 20 and 24 years (mean, 22.9) were selected for this study. The size of the hand, expressed in terms of the distance from the distal volar wrist crease to the tip of the long finger, ranged from 165 to 192 mm, with an average of 182.6 mm (Table 1).

The end blocks of the goniometer were placed on the dorsum of the thumb distal phalanx, proximal phalanx, and metacarpus, as well as the index finger middle phalanx, proximal phalanx, and metacarpus, and flexion angles were measured at the dorsum of the thumb interphalangeal (IP) and metacarpophalangeal (MP) joints as well as the index finger proximal interphalangeal (PIP) and MP joints.

For simplification purposes, subjects were asked to write a cross of two varying sizes, 4 cm and 6 cm (Fig. 2a). To set the writing speed, a metronome was set to strike once for every second. With the forearm fixed on the table, the subjects were asked to hold the ballpoint pen as close to the tip as possible, then, synchronously with the metronome, were told to make the cross, first a horizontal line, then a vertical line, and to complete the sequence in 4 s.

Table 1. Data of the tested individuals

	Age	Sex	Distance from volar wrist crease to tip of long finger (mm)	Distance from proximal finger crease to tip of long finger (mm)
K.O.	22	F	165	72
Y.N.	24	M	168	74
T.O.	24	M	188	78
S.H.	21	M	190	79
H.S.	24	M	183	80
K.Y.	20	M	187	82
K.K.	24	M	192	82
H.I.	24	M	188	85
Mean ± SE	22.9		182.6 ± 3.6	79.0 ± 1.5

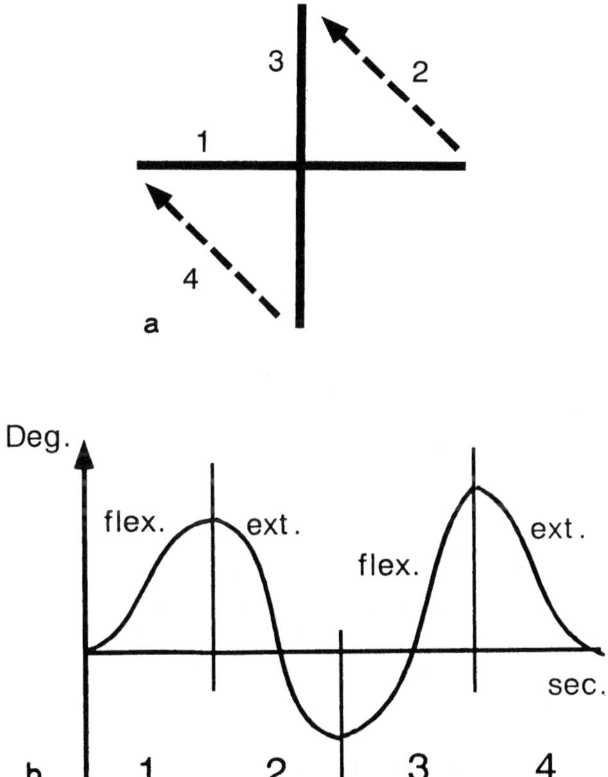

Fig. 2. **a** Writing order of a cross. **b** Example of the time–angle curve

The computer used in this study carried out measurements in two groups: first, a combination of three joints consisting of the thumb IP joint and the index PIP and MP joints, followed by a combination of the thumb MP joint and the index PIP and MP joints. Findings of the index finger PIP and MP joints were used as indicators to obtain the correlation of the four joints. The angles at the start of the motion were set at 0° of flexion. The horizontal axis showed the time, while the vertical axis showed the relative angle, and the motion of each angle was represented in a wave shape (Fig. 2b).

The relative flexion and extension angles were plotted every 0.01 s, and recorded using a personal computer. Each motion was measured for 10 s, with the time from the beginning of the motion to the end designated as one cycle. Roughly two cycles were measured for every 10 s. The motion was repeated five times, and the angles of the last cycle in each movement were averaged for use in the analysis.

Results

Joint Motion on Pinch

A. Pulp-pinch without holding an object
  1. Pulp-pinch
     a. The average joint angles with a pinch force of 100 g were 4.8° at the thumb IP joint, 18.0° at the thumb MP joint, 23.3° at the index distal interphalangeal (DIP) joint, 30.6° at the index PIP joint, and 58.9° at the index MP joint, indicating that the thumb joints moved less than the index joints.
     b. The average joint angles of the thumb with a pinch force of 1000 g were 2.5° at the IP joint and 20.5° at the MP joint. Compared with a pinch force of 100 g, the thumb IP joint was slightly more extended and the thumb MP joint slightly more flexed. The thumb IP joint was thought to have been extended by the index finger. While the index finger was made to extend by the thumb, the angle of the index PIP joint did not change significantly, although it tended to become slightly smaller.
  2. Pulp-pinch, with a pulp distance of 1–5 cm
     a. With the 100 g pinch force, the angles of the thumb IP and MP joints and the index DIP and PIP joints remained fairly constant, ranging from 4.1° to 6.9°, 13.8° to 18.0°, 21.2° to 24.8°, and 29.1° to 33.0°, respectively. In contrast, the angles of the index MP joint gradually extended from 58.9° to 33.8° as the pulp distance increased.
     b. With a 1000 g pinch force, the angles of the thumb IP and MP joints and the index DIP and PIP joints were fairly stable, as in the 100 g force, ranging from 2.0° to 3.2°, 16.0° to 20.5°, and 24.1° to 26.2°, and 27.2° to 29.3°, respectively, while the index MP joint angles

changed from 60.1° to 35.1° as the pulp distance increased, a tendency seen with the 100 g pinch test as well (Table 2A).

B. Tip-pinch
   1. Tip-pinch without holding an object
      a. With a pinch force of 100 g, the average joint angles of the thumb IP and MP joints and the index DIP, PIP, and MP joints were 20.6°, 12.5°, 40.3°, 41.8°, and 50.8°, respectively. Compared with pulp-pinch, the thumb IP and index DIP and PIP joints took a more flexed position, while the thumb MP joint was slightly more extended, with little or no change seen in the index MP joint.
      b. With a 1000 g force, the average angles of the thumb IP and MP joints and the index DIP, PIP, and MP joints were 26.9°, 8.7°, 43.0°, 39.6°, and 51.1°, respectively. Compared with a tip-pinch force of 100 g, the thumb IP joint was more flexed and the thumb MP joint more extended. This was in contrast to the pulp-pinch, where the thumb IP joint extended and the thumb MP joint flexed slightly as a result of increased force. The index DIP joint was flexed and the PIP joint extended.
   2. Tip-pinch, with a pulp distance of 1–5 cm
      a. With a 100 g force, the index MP joint likewise changed from 50.8° to 22.8°, with the index MP joint adjusting the pinch width. Compared with pulp-pinch, the index MP joint angle in tip-pinch was reduced by 8.1°–11.0°.
      b. Similarly, with a 1000 g force, the pinch width changed along with the index MP joint (Table 2B).

C. Motion analysis
   The time–angle curve for subjects with substantial learning experience showed a regular wave pattern. The angle showed a sharp rise at startup, indicating a lack of "play" in the apparatus itself. Flexion began at more or less the same time with index DIP, PIP, and MP joints. Depending on the stroke, the same subject showed a slight time lag. The thumb IP and MP joints tended to begin their motions between 0.1 and 0.25 s later than the index finger (Fig. 3).

Joint Motion on Writing a Cross

The wave pattern differed from subject to subject, while the same individual showed almost identical wave patterns when writing a cross 4 cm and 6 cm in size. All subjects showed clear bisferious patterns in their thumb IP joints, indicating that the joint began to flex as soon as the subjects began writing a horizontal line, extended markedly when they began writing a vertical line, then gradually flexed. Seven subjects showed bisferious patterns in their index PIP joints, while the thumb and index MP joints showed a smaller range of motion as compared with their respective IP and PIP joints. The thumb IP joints of all eight subjects and the seven who showed bisferious patterns in

Table 2. Flexion angles of the index MP, PIP, and DIP, and the thumb MP and IP joint in pulp pinch (**A**) and tip pinch (**B**)

A. Pulp pinch

	Distance (cm)		Index MP	Index PIP	Index DIP	Thumb MP	Thumb IP
100 g	0	Mean	58.9	30.6	23.3	18.0	4.8
		SD	3.9	10.5	9.5	6.6	2.0
	1	Mean	55.1	30.2	23.1	13.8	6.9
		SD	6.4	10.5	6.9	7.1	3.3
	2	Mean	51.9	33.0	21.2	16.4	6.1
		SD	5.5	12.0	6.6	5.0	5.3
	3	Mean	46.8	30.9	22.8	15.5	6.8
		SD	8.0	11.9	6.5	5.1	6.0
	5	Mean	33.8	29.1	24.8	15.2	4.1
		SD	5.2	8.9	6.8	11.1	5.9
1 kg	0	Mean	60.1	29.3	25.7	20.5	2.5
		SD	3.5	13.4	6.6	9.0	1.9
	1	Mean	56.5	28.7	24.1	17.4	2.6
		SD	8.2	5.8	4.9	4.9	1.7
	2	Mean	54.1	27.6	24.1	18.4	2.0
		SD	5.9	9.2	7.6	5.1	5.1
	3	Mean	47.9	27.3	24.8	16.0	2.0
		SD	7.4	12.4	3.3	9.5	3.7
	5	Mean	35.1	27.2	26.2	17.6	3.2
		SD	6.6	8.2	2.1	5.7	3.3

B. Tip pinch

	Distance (cm)		Index MP	Index PIP	Index DIP	Thumb MP	Thumb IP
100 g	0	Mean	50.8	41.8	40.3	12.5	20.6
		SD	9.8	11.1	13.1	8.3	8.8
	1	Mean	49.1	41.0	38.5	12.8	21.8
		SD	9.5	10.7	6.8	9.4	7.8
	2	Mean	42.8	41.0	37.7	14.2	21.9
		SD	12.0	12.3	5.0	8.0	9.1
	3	Mean	36.3	35.4	37.1	12.8	24.8
		SD	9.5	16.6	9.1	8.4	12.0
	5	Mean	22.8	42.4	39.0	14.1	20.9
		SD	8.1	13.7	11.9	9.2	11.9
1 kg	0	Mean	51.1	39.6	43.0	8.7	26.9
		SD	7.3	11.3	7.4	11.4	11.3
	1	Mean	49.3	36.4	42.9	7.8	28.4
		SD	4.8	12.7	11.0	8.3	12.7
	2	Mean	44.3	35.5	43.9	8.7	30.5
		SD	5.8	14.8	10.5	9.4	13.8
	3	Mean	37.0	33.2	40.1	9.1	31.5
		SD	1.3	12.5	11.5	13.4	11.5
	5	Mean	26.8	34.3	44.5	9.4	29.2
		SD	8.1	8.4	11.1	11.3	11.5

MP, metacarpophalangeal joint; PIP, proximal interphalangeal joint; DIP, distal interphalangeal joint; IP, interphalangeal joint.

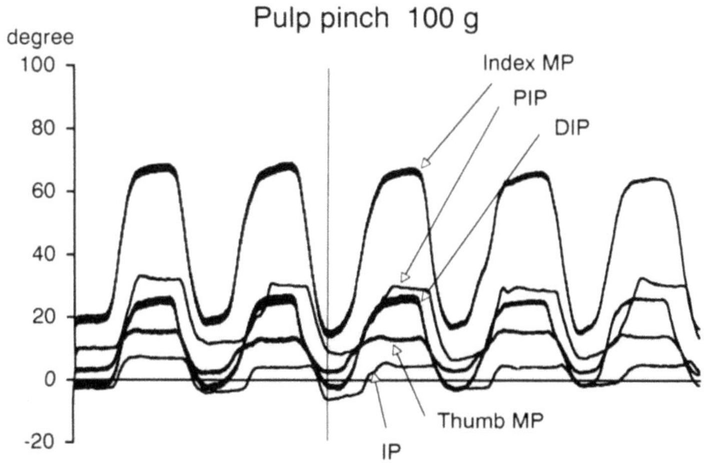

Fig. 3. Ordinary time–angle curves. The initiation of flexion of the thumb metacarpophalangeal (*MP*) and interphalangeal (*IP*) joints is slightly delayed compared with that of the index finger in many cases. *DIP*, distal interphalangeal joint

their index PIP joints were then compared with respect to their range of motion for writing horizontal and vertical lines. The range of motion tended to be larger when writing vertical, rather than horizontal, lines, and when writing a 6-cm rather than a 4-cm cross (Fig. 4).

In examining the range of motion throughout the sequence, the largest was seen with the thumb IP joint for all subjects, with an average angle of 57.5°, followed by the index PIP joint, with an average angle of 27.1°. In comparison, the motions of thumb and index MP joints were only 15° and 13.6°, respectively (Table 3). To determine the relationship between the range of motion and the size of the hand, a Student *t*-test was conducted. First, based on the length of the long finger, the distance between the proximal finger crease to the tip of the long finger was made the length of the long finger. The subjects were divided into two groups, those with the long finger shorter than 80 mm (four subjects) and those whose long finger was longer than 80 mm (four subjects), then underwent the test. A significant difference was seen in their thumb IP joints and index MP joints, with a larger range of motion seen in those with shorter long fingers. No significant differences were seen with the other two joints (Table 3).

Discussion

Because of its compact size, the goniometer used in the present study could be attached to the dorsum of the fingers of the hand, from the thumb to the little finger, enabling the relatively easy observation of angles in motion. However, a

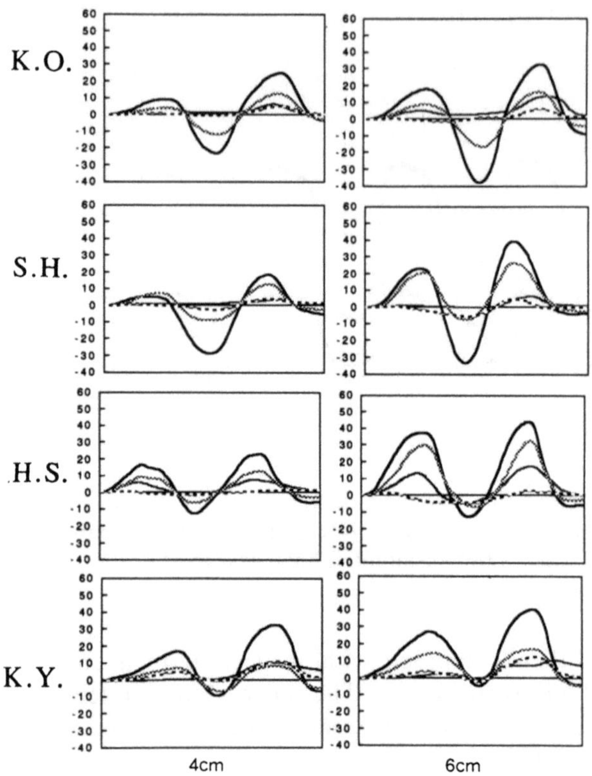

Fig. 4. Time–angle curves of the four tested hands

goniometer even this small was still considered slightly too bulky for measuring DIP joint angles, requiring extra care when attaching the apparatus to subjects with short fingers. Moreover, the angle in the dorsal surface did not always coincide with the angle at the midlateral line. The difference is considered substantial during flexion [5].

Adjustment of the pinch width was made by the index MP joint in all cases, with no changes in the DIP and PIP joints. As in previous reports [5,6], a delicate adjustment mechanism of the intrinsic muscles was deemed important in adjusting the flexion of the MP joint. In this case, the thumb was confirmed to be fixed at a certain angle. Increasing the pinch force increased the angle of the thumb MP joint in pulp-pinch, although only slightly, whereas with tip-pinch, the angle of the thumb IP joint increased. This was thought to reflect the force of the flexor pollicis brevis in pulp-pinch, and the flexor pollicis longus in tip-pinch. In the time-flexion angle curve, the start of the flexion of the thumb tended to lag slightly behind the start of the index flexion, indicating that in pinch, the thumb was thought to "accompany" the flexed index finger.

In studying the functional position of the "pinching" motion using these findings, it was shown that in trauma and RA, the conventional arthrodesis [7]

Table 3. Maximal flexion angle of the four joints in one cycle of writing a 6-cm cross[a]

	Thumb MP	Thumb IP	Index MP	Index PIP
K.O.	10.9	70.3	7.4	32.6
Y.N.	6.5	71	39.3	11.7
T.O.	29.2	49.3	8.7	27
S.H.	6.6	73	10.2	33.8
H.S.	22.3	56.5	7.3	39.5
K.Y.	10.9	44.9	14.4	19.2
K.K.	15.4	36.6	14	14.1
H.I.	18.1	58.4	7.8	39.2
Mean ± SE	15.0 ± 2.8	57.5 ± 4.7	13.6 ± 3.8	27.1 ± 3.9
Length of long finger < 80 mm (mean ± SE)	13.3 ± 5.4	65.9 ± 5.6	16.4 ± 7.7	26.3 ± 5.1
Length of long finger ⩾ 80 mm (mean ± SE)	16.7 ± 2.4	49.1 ± 5.1	10.9 ± 2.0	28.0 ± 6.6
t Value	0.572	2.220	2.129	0.206
P	NSD	<0.05	<0.05	NSD

NSD, no significant difference.
[a] The hands were divided into two groups according to the length of the long finger.

performed with the thumb IP and MP joints in extended positions as well as arthrodesis and tenodesis performed with the index DIP joint in slightly flexed positions were positions with relatively minor functional losses.

In writing a cross, the IP and PIP joints of both the thumb and index fingers had a larger range of motion than in pinch, while the MP joint moved less and was essentially fixed. This indicates the significance of the IP and PIP joints in writing a cross. At the same time, the MP joint seemed to move subtly to supplement the motions of both the IP and PIP joints.

Fixation of the MP joint was mainly made by the intrinsic muscle, with adductor pollicis, abductor pollicis brevis, and flexor pollicis brevis playing important roles [8] as far as the thumb was concerned. For index fingers, meanwhile, interossei were considered important, with both the first dorsal and first palmar interossei working to stabilize the MP joint [9]. With lumbricalis, in contrast, the MP joint reportedly has little or no flexion, and apparently is not involved in the fixation of the MP joint [9,10]. Other reports have indicated that lumbricalis pulls the flexor digitorum profundus to the periphery, and the contraction of the flexor digitorum profundus is transmitted to the extensor expansion through lumbricalis to extend the DIP and PIP joints. Paralysis of the flexor digitorum profundus is said to sharply weaken the lumbricalis ability to extend the IP joints [6,10,11].

During flexion, the lumbricalis works as the primary starter muscle to begin flexing the MP joint. It is said to have a delicate mechanism similar to a guiding apparatus that induces the actions of the flexor [12].

As can be seen, intrinsic as well as extrinsic muscles are involved in moving the IP and PIP joints by fixing and stabilizing the MP joint. In the motion of writing a cross, therefore, the interaction between the extrinsic and intrinsic muscles was deemed important.

In this study, the motion of writing a cross was used as the index to express hand dexterity. However, dexterity is a concept relying, to a certain degree, on one's subjective judgment, with no absolute evaluation criteria. The authors intend to increase the number of normal cases, and investigate these as well as other subjects with reduced manual function.

Conclusions

1. The angles of thumb and index finger joints during pinch were measured, using a single axis flexible electrogoniometer.
2. The goniometer used in this study was compact in size, making angle measurements relatively easy.
3. Adjustment of the pinch width was made by index MP joints.
4. The functional positions were confirmed by measuring the angles.
5. The angle of finger joints in writing a cross was measured with a goniometer.
6. When writing a cross, the MP joint was more or less fixed, while substantial motion was seen in the IP and PIP joints.
7. Actions of hand intrinsic as well as extrinsic muscles are involved in the motion of handwriting, and the collaborative motion between these two muscles was deemed especially important.

References

1. Akiyama K, Imamura K, Iwasaki K, Chiba G (1991) The effect of the speed for the joint motion of the index finger with one joint fixed. J Jpn Soc Surg Hand 8:104–108
2. Wise S, Gardner W, Sabelman E, Valainis E, Wong Y, Glass K, Drace J, Rosen JM (1990) Evaluation of a fiber optic glove for semi-automated goniometric measurements. J Rehabil Res Dev 27:411–424
3. Yanagihara Y (1983) An electrogoniometric study of human finger joint movement. J Jpn Orthop Assoc (Nippon Seikeigeka Gakkai Zasshi) 57:293–308
4. Morifusa S, Tamai K, Okuda Y, Takai S, Azuma T, Hirasawa Y (1991) Three dimensional motion analysis of thumb-index pinch. Jpn Soc Orthop Biomech 13:23–27
5. Okada Y (1989) Study of a detecting device for finger motions. Bull Nippon Bunri Univ 17:59–62
6. Citron N, Foster A (1988) Some aspects of lumbrical function. J Hand Surg [Br] 13B:54–59

7. Carroll RE, Hill NA (1969) Small joint arthrodesis in hand reconstruction. J Bone Joint Surg [Am] 51A:1219–1221
8. Imaeda T, An KN, Cooney III WP (1992) Functional anatomy and biomechanics of the thumb. Hand Clin 8:9–15
9. Sasao M (1974) Functional analysis of the intrinsic muscles at the finger motions. J Jpn Orthop Assoc (Nippon Seikeigeka Gakkai Zasshi) 48:265–279
10. Long C (1968) Intrinsic-extrinsic muscle control of the fingers. J Bone Joint Surg [Am] 50A:973–984
11. Stack HG (1962) Muscle function in the fingers. J Bone Joint Surg [Br] 44B:899–909
12. Ueba Y (1986) The hand—its function and anatomy. 2nd edn. Kinpodo, Kyoto, p 132

Biomechanical Analysis of Three-Dimensional Osteotomy at the Distal End of the Radius for Kienböck's Disease with Flexion Deformity of the Lunate

Hiromasa Miura, Hiroshi Tsumura, Kazuhiko Ihara, Seiichi Kawamura, and Yoichi Sugioka

Summary. Failure of radial closing wedge osteotomy for advanced Kienböck's disease frequently reveals a flexion deformity of the lunate and the collapse and fragmentation of its volar half. We performed biomechanical analyses including a rigid-body spring model and the finite-element method and demonstrated that the flexion deformity of the lunate increases contact pressure at the volar half of the radiolunate and capitolunate joints. The increased contact pressure could therefore lead to collapse and fragmentation of the volar half of the lunate. A new three-dimensional osteotomy, consisting of a radial closing wedge osteotomy and an extension osteotomy, was developed to achieve better alignment of the capitolunate joints. The three-dimensional osteotomy may therefore be an effective procedure for Kienböck's disease having a flexion deformity of the lunate.

Key words: Kienböck's disease—Finite-Element method—Rigid-body spring model—Biomechanics—Lunate

Introduction

Late-stage Kienböck's disease can produce collapse and fragmentation of the lunate. However, most advanced cases demonstrate flexion deformity of the lunate and collapse and fragmentation of the volar half of the lunate (Fig. 1). We have used a radial closing wedge osteotomy for Kienböck's disease on the basis of results of a rigid-body spring model analysis by Tsumura et al. [1]. Approximately 70% of the patients who underwent the osteotomy showed excellent or good results according to the Nakamura scoring system [2]. The

Department of Orthopedic Surgery, Faculty of Medicine, Kyushu University, 3-1-1 Maidashi, Higashi-ku, Fukuoka, 812 Japan

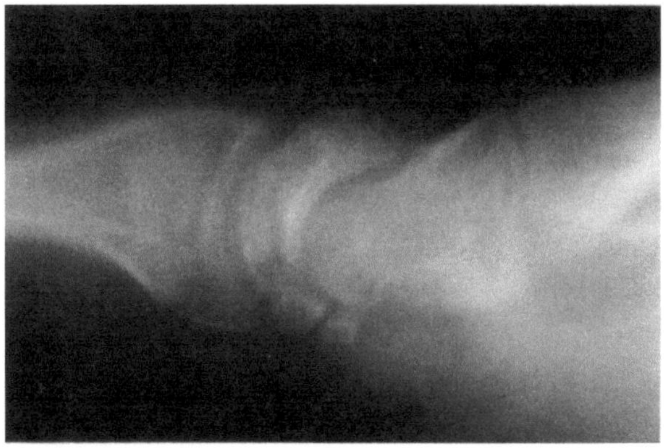

Fig. 1. Flexion deformity of the lunate and collapse and fragmentation of the volar half of the lunate

majority of patients with poor results demonstrated flexion deformities of the lunate and collapse and fragmentation at the volar half of the lunate [3]. We therefore postulated that flexion deformity of the lunate is one of the most important factors affecting the collapse and fragmentation of the lunate in advanced Kienböck's disease. Biomechanical analyses, using a rigid-body spring model and the finite-element method, were performed to evaluate the effect of a flexion deformity of the lunate on stress distribution around the lunate itself.

A new three-dimensional osteotomy for Kienböck's disease, developed to correct the incongruity of the capitolunate joint caused by flexion deformity of the lunate, combines a radial closing wedge osteotomy with an extension osteotomy. Biomechanical analysis using a two-dimensional rigid-body spring model was also performed to evaluate the effect of the three-dimensional osteotomy on stress distribution around the lunate.

Technique of Three-Dimensional Osteotomy

In Kienböck's cases with flexion deformity of the lunate, the capitate articulates with collapsed or fragmented volar parts of the lunate. Flexion of the wrist, however, would make better congruity between the lunate and capitate (Fig. 2). The three-dimensional osteotomy combines a radial closing wedge osteotomy with an extension osteotomy of the radius to achieve better alignment of the capitolunate joint in preoperative flexion position of the wrist (Fig. 3).

The operative technique of the three-dimensional osteotomy is similar to that of the radial closing wedge osteotomy. A longitudinal incision is made along the radial side of the flexor carpi radialis tendon at the distal forearm. The flexor pollicis longus is displaced radially, and the flexor digitorum superficialis is displaced ulnarly to expose the pronator quadratus muscle. The pronator quadratus is incised at its radial insertion, and the distal third of the radius is exposed. Fifteen degrees of a radial closing wedge osteotomy is made

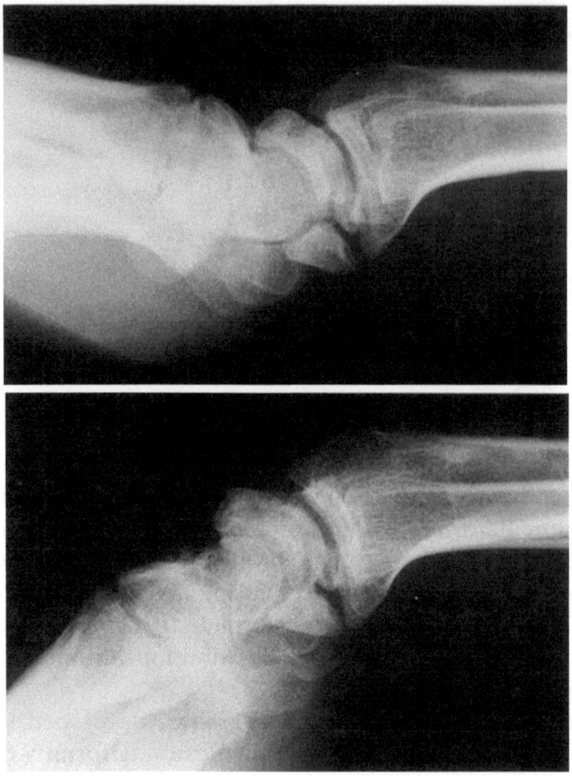

a

b

Fig. 2. Wrist flexion improves congruity of the capitolunate joint: **a** neutral; **b** flexion

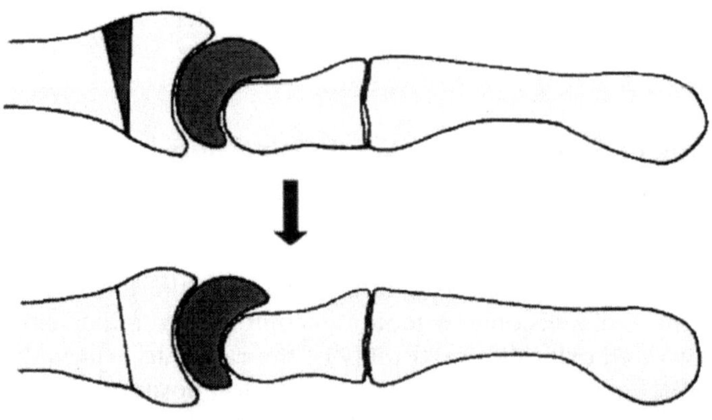

Fig. 3. Schema of the extension osteotomy

first and then an extension osteotomy is performed to achieve good alignment of the capitolunate joint. The osteotomies should be performed 3 to 4 cm distal to the tip of the radial styloid process to minimize the postoperative deformity of the forearm. Internal fixation is achieved using a four-hole plate. The pronator quadratus is sutured to the original position and the wound is then closed. A long arm splint is applied with the elbow flexed 90° and the wrist at neutral position for 6 weeks.

This procedure has the combined effects of a decrease in the contact pressure around the lunate by the radial closing wedge osteotomy and improvement of congruity of the capitolunate joint by the extension osteotomy.

Methods

Analysis of Flexion Deformity of the Lunate

Two-dimensional rigid-body spring models, which consisted of four elements including the radius, lunate, capitate, and third metacarpal bone, were constructed from 10 normal subjects and 9 patients with Kienböck's disease. Normalization of size was done using the length of third metacarpal bone. Compressive springs with stiffnesses of $10\,\mathrm{N\,mm^{-1}\,mm^{-2}}$ were applied to the articular surface, and intraarticular shear force was neglected. Six tensile springs equivalent to the intercarpal ligaments or capsule were considered. The proximal end of the radius was fixed, and a compressive load of 100 N was applied along the longitudinal axis of the third metacarpus (Fig. 4). The average pressure on the radiolunate and capitolunate joints was compared between the two groups. Further, displacement of the lunate under a compressive load was investigated in the group with Kienböck's disease.

Intraosseous stress of the lunate was also analyzed by the two-dimensional finite-element method. Lunate models with 80 and 94 elements, respectively, were constructed from a normal individual and a typical case of Kienböck's disease (Fig. 5). Isotropic and homogenous models with mechanical properties of cancellous bone were considered. Pressure distribution at the capitolunate joints obtained from the rigid-body spring model analysis was applied to the corresponding elements. The radiolunate joint was assumed to have zero deformation. Von Mises' equivalent stress was compared between the normal person and the Kienböck's case.

Analysis of Three-Dimensional Osteotomy

Two-dimensional rigid-body spring models, which consisted of the radius, lunate, capitate, and third metacarpal bone, were constructed from nine

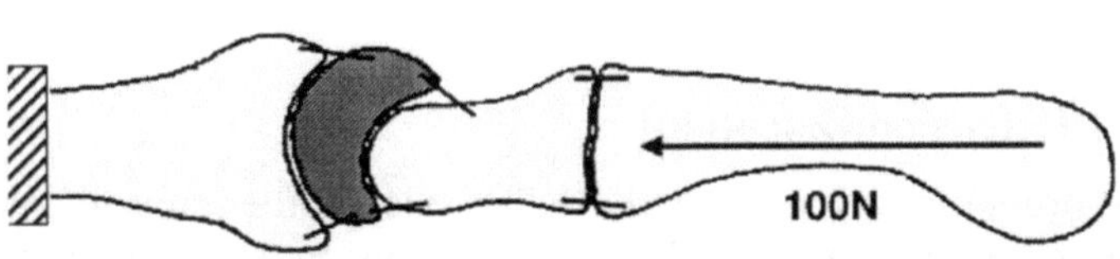

Fig. 4. Rigid-body spring model

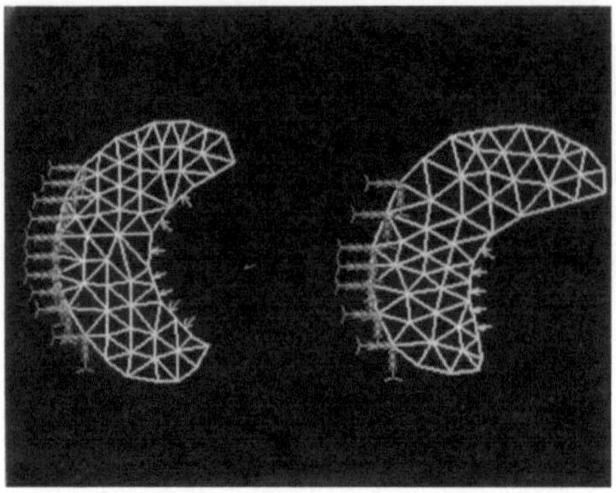

Fig. 5. Model of the finite-element method: *left*, normal; *right*, flexion deformity of the lunate

patients with Kienböck's disease who had undergone the radial closing wedge osteotomy. Normalization of size used the length of the third metacarpal bone. Compressive springs with stiffnesses of $10\,\mathrm{N\,mm^{-1}\,mm^{-2}}$ were applied to the articular surface. Intraarticular shear force was neglected. Six tensile springs equivalent to the intercarpal ligaments or capsule were considered. The proximal end of the radius was fixed, and a compressive load of 100 N was applied along the longitudinal axis of the third metacarpus. The extension osteotomy was simulated, and the changes in average pressure at the radiolunate and capitolunate joints were compared.

Results

Analysis of Flexion Deformity of the Lunate

Radiolunate angles in the normal group and Kienböck's group were 3.3° and 19.8°, respectively. In the normal group, even pressure distribution was seen around the lunate. In the group with Kienböck's disease, however, high peak pressure was seen at the volar side of the capitolunate and radiolunate joints. The mean contact pressure at the capitolunate and radiolunate joints is significantly higher in the group with Kienböck's disease (Fig. 6). Displacement of the lunate in Kienböck's cases shows the increased volar flexion of the lunate under compressive loading (Fig. 7). The finite-element method showed that there was severe stress concentration at the volar half of the lunate and moment, increasing a flexion deformity of the lunate in Kienböck's disease (Fig. 8).

Analysis of Extension Osteotomy

High peak pressure was seen at the volar side of the capitolunate and radiolunate joints preoperatively. After the extension osteotomy, the pressure distribution around the lunate was even. Seven of nine cases showed decreased

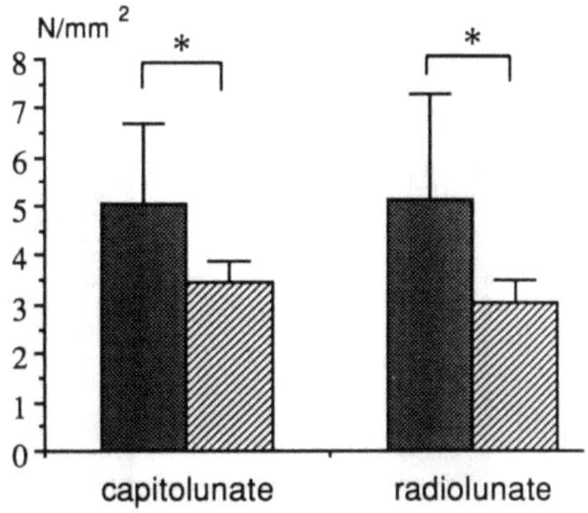

Fig. 6. Mean contact pressure is significantly higher in the group with Kienböck's disease (*stippled bars*) than in the normal group (*shaded bars*); *, $P < 0.01$

7 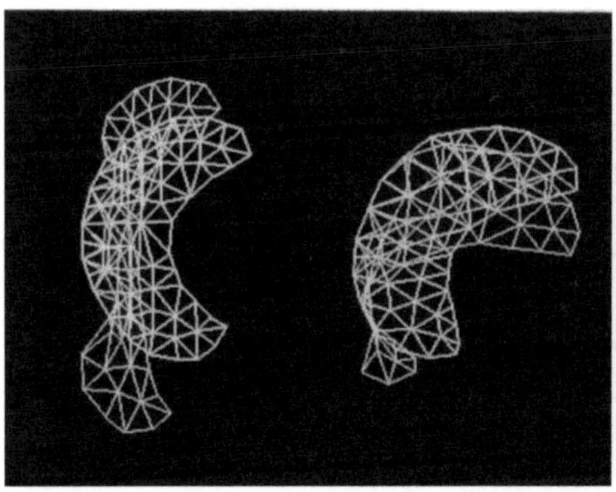

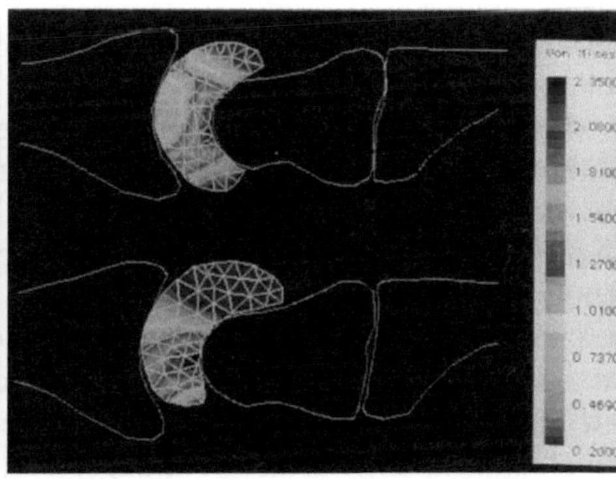 8

Fig. 7. Displacement of the lunate in Kienböck's cases shows an increased volar flexion of the lunate under compressive loading: *left*, normal; *right*, flexion deformity of the lunate

Fig. 8. Stress concentration at the volar half of the lunate in Kienböck's disease: *upper*, normal; *lower*, flexion deformity of the lunate

average pressure after the osteotomy. The mean contact pressures at the capitolunate and radiolunate joints were significantly lower in the postoperative cases (Fig. 9).

Discussion

Many theories exist concerning the cause of Kienböck's disease. Although repetitive trauma, acute trauma, and the minus variant theory by Hulten [4] are dominant, little is known about the mechanism of the collapse and fragmentation of the lunate. In general, a decrease in the elastic modules of

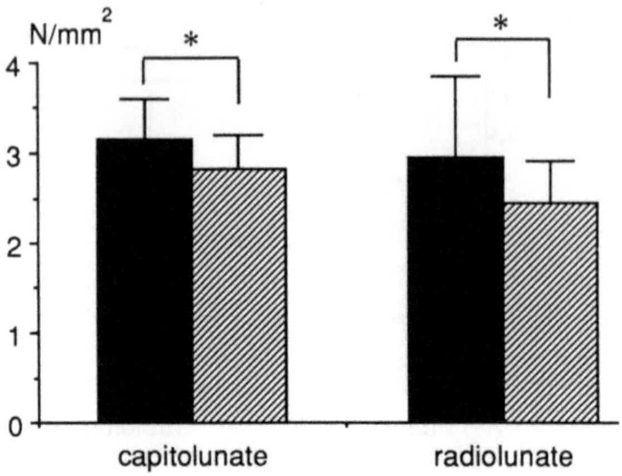

Fig. 9. The mean contact pressures at the capitolunate and radiolunate joints are significantly lower in postoperative cases (*shaded bars*) of the extension osteotomy than preoperatively (*solid bars*); *, $P < 0.05$

necrotic cancellous bone and the nonunion of the lunate are thought to lead to collapse.

Most advanced cases of Kienböck's disease demonstrate the flexion deformity of the lunate and collapse and fragmentation of its volar half. Yoshida et al. [5] have characterized the flexion deformity of the lunate in their clinical study of Kienböck's disease. Some highly advanced cases show laceration of extensor tendon because of dorsal displacement of the fragment. The typical clinical course of lunate collapse in Kienböck's disease includes an early stage, when an affected side has a flexion deformity compared with the normal side, and a late stage, when the collapse and fragmentation of the volar side of the lunate develops with osteoarthrotic changes on the volar articular surface of the radius (Fig. 10). We therefore hypothesized that the flexion deformity of the lunate affected the biomechanical condition around the lunate itself, as supported by our results of biomechanical analysis using the rigid-body spring model and the finite-element method. Although Armistead et al. [6] have reported that excessive tensile force in lunotriquetral and radiolunate ligament in wrist extension caused the fragmentation of the lunate at its volar half, the Armistead work can not entirely explain the volar flexion of the lunate.

After evaluation of the clinical course in Kienböck's disease and the biomechanical analysis presented here, we propose a mechanism for the collapse of the lunate as follows: When the elastic module of cancellous bone decreases, the normally slight volar flexion of the lunate increases because of a weaker dorsal intercarpal ligament and moment. The flexion deformity of the lunate increases contact pressure on the volar half of the radiolunate and capitolunate joints and increases the stress concentration in the lunate itself. These changes could lead to collapse and fragmentation of the volar half of the lunate and induce osteoarthritic changes at radiocarpal joint.

The radial closing wedge osteotomy was developed in 1980 by Tsumura et al. [1] on the basis of the rigid-body spring model analysis. Its biomechanical significance stems from the radial deviation of loading vectors on the lunate.

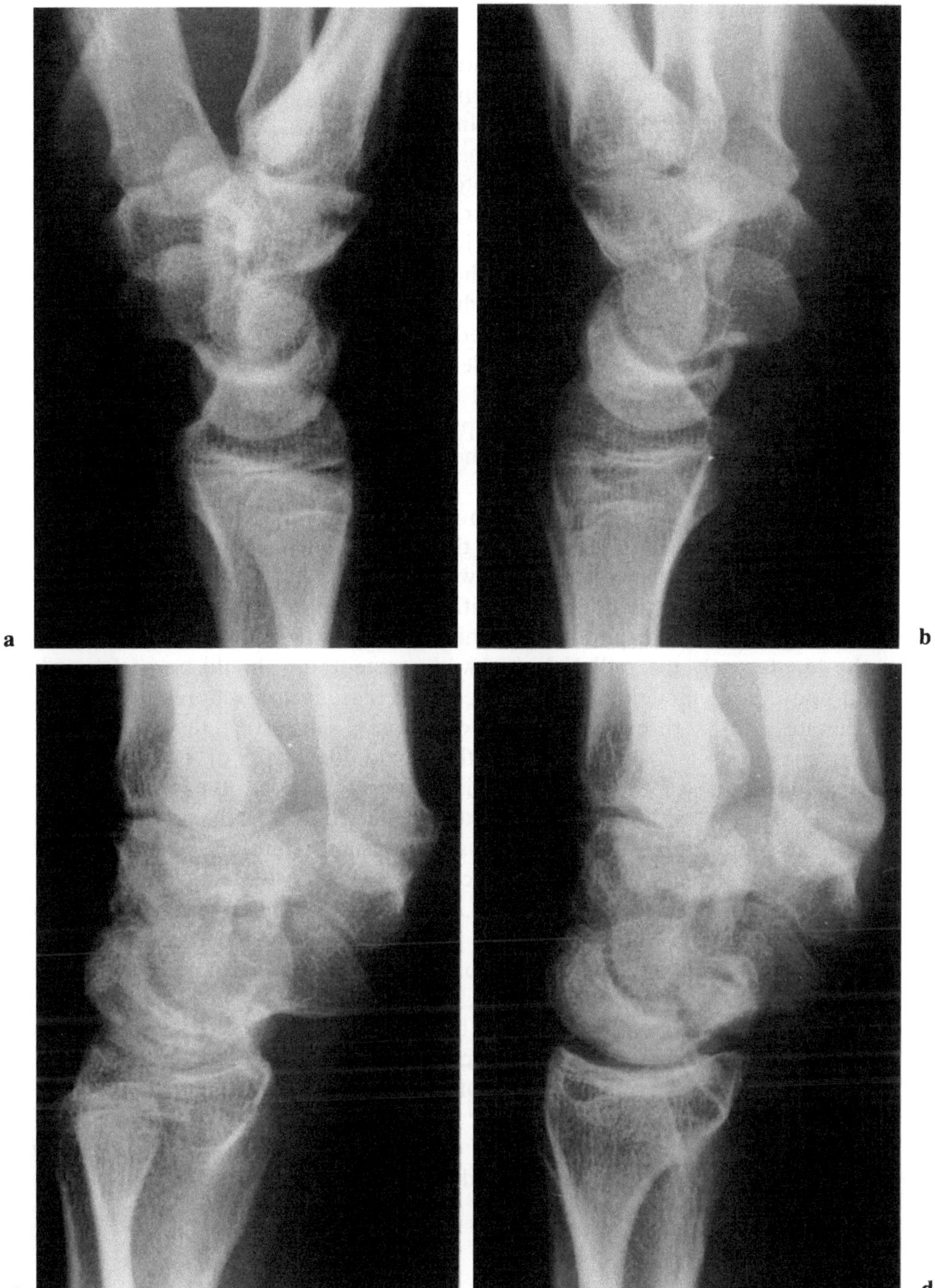

Fig. 10. Typical course of lunate collapse in Kienböck's disease: **a** normal side; **b** affected side; **c** after 1 year; **d** after 2 years and 8 months

According to the rigid-body spring model analysis, a decrease in the radial inclination without radial deviation of the loading vector does not decrease the contact pressure around the lunate. Palmer and Werner [7] and Mani et al. [8] have reported experimental studies in cadavers that demonstrated the ineffectiveness of the radial closing osteotomy, although radial deviation of the loading vector was not included in their methods.

The radial closing wedge osteotomy is an effective treatment for Kienböck's disease. Of 35 patients undergoing the procedure, 95% were content with the results and approximately 70% showed improvement according to the Nakamura scoring system [2]. Advanced cases, however, revealed poor results partly because of the presence of a flexion deformity of the lunate. From a biomechanical perspective, mechanical correction of both the anteroposterior view as well as the lateral view, are important in improving the clinical results. A procedure to prevent or correct the underlying flexion deformity of the lunate would be necessary.

Combining extension osteotomy with radial closing wedge osteotomy obtains better alignment of wrist flexion in the preoperative condition, producing an improvement in the congruity between the capitate and the lunate. The procedure is recommended for cases with severe flexion deformity of the lunate and incongruity of the capitolunate joint. For cases without a flexion deformity of the lunate, only the radial closing wedge osteotomy should be performed. Rigid-body spring model analysis showed a decrease in contact pressure around the lunate after a three-dimensional osteotomy, suggesting the procedure's effectiveness for Kienböck's disease with a flexion deformity of the lunate. The surgical technique of this procedure is similar to that of the radial closing wedge osteotomy, although the osteotomy site would ideally be as distal as possible to minimize cosmetic problems and ensure extension. The rigid-body spring model analysis of the radial closing wedge osteotomy showed that the optimal osteotomy angle was 15°. The optimal angle for the extension osteotomy should be determined by preoperative dynamic roentgenogram. Better congruity of the capitolunate joint can be usually obtained with 20° of maximal extension osteotomy.

A possible complication of this osteotomy is an increase in the flexion deformity of the lunate. However, because motion occurs mostly at the capitolunate joint during small degrees of wrist flexion, movement at the radiolunate joint should be negligible after 20° of osteotomy.

We conclude that the addition of extension osteotomy to radial closing wedge osteotomy is an effective procedure for Kienböck's disease with flexion deformity of the lunate.

References

1. Tsumura H, Himeno S, An KN, Cooney WP, Chao EYS (1987) Biomechanical analysis of Kienböck's disease. Orthop Trans 11:327

2. Nakamura R, Tsuge S, Watanabe K, Tsunoda K (1991) Radial wedge osteotomy for Kienböck disease. J Bone Joint Surg 73A:1391–1396
3. Miura H, Uchida Y, Arima J, Kojima T, Sugioka Y (1992) Radial closing wedge osteotomy for Kienböck's disease. Annual meeting of the Japanese Society for Surgery of the Hand, Kyoto, Japan
4. Hulten O (1928) Über anatomische Variationen der Handgenkknochen. Acta Radiol 9:155–168
5. Yoshida K, Yamanaka K, Shigeri T, Inoue A (1990) Follow-up study of Kienböck's disease treated with radius shortening osteotomy. J Jpn Soc Surg Hand 7:738–742
6. Armistead RB, Linscheid RL, Dobyns JH, Beckenbaugh RD (1982) Ulnar lengthening in the treatment of Kienböck's disease. J Bone Joint Surg 64A:170–178
7. Palmer AK, Werner FW (1988) Biomechanical evaluation of operative procedures performed for the treatment of Kienböck's disease. Annual meeting of the AAOS, Baltimore, Maryland
8. Mani T, Nakabe N, Hara T, Sasagawa K, Hatano Y, Shibata M (1994) Contact pressure distribution in the wrist joint: the effect of radial wedge osteotomy. In: Proceedings, 3rd bioengineering symposium, p 144

Part 9. Spine

Biomechanical Sequelae of Short Lumbar Fusion

Daniel H.K. Chow[1], John H. Evans[1], Keith D.K. Luk[2], and John C.Y. Leong[3]

Summary. The biomechanical effects of "floating" L4–L5 anterior interbody fusion on the adjacent, unfused segments were investigated experimentally using cadaveric specimens. The mechanical characteristics of isolated lumbar spine specimens were measured before and after in vitro simulated L4–L5 anterior interbody fusion. It was found that there was a significant increase in segmental flexion deformation at L5–S1 and a significant decrease in segmental extension deformation at L3–L4 and L5–S1. These findings suggest that the functional range of motion of the adjacent, unfused segments, especially the level below the fusion, i.e., L5–S1, was biased toward flexion after a floating L4–L5 fusion. This result also suggests that the unfused segments more frequently have to operate toward the extremes of their normal ranges of motion to partially compensate for the loss of mobility at the fused level. However, the largest increase in segmental movement was found at the L5–S1 segment, which exhibited the largest deformation in the flexion range. Thus, this segment has the highest possibility of working beyond its physiological limit in flexion.

Key words: Anterior lumbar interbody fusion—Segmental mobility

Introduction

Short segmental fusions in the lumbar spine, involving one or more levels, are indicated in various pathological conditions including degenerative disc disease and those infective and malignant conditions that also induce spinal instability. A number of structurally different surgical procedures have been developed.

[1] Jockey Club Rehabilitation Engineering Centre, Hong Kong Polytechnic, Hunghom, Kowloon, Hong Kong
[2] The Duchess of Kent Children's Hospital, 12 Sandy Bay Road, Sandy Bay, Hong Kong
[3] Department of Orthopaedic Surgery, University of Hong Kong, Hong Kong

These may be categorized according to the site of fusion and to the surgical approach. Grafts are commonly sited in the interbody space of the anterior column and between the lamina or transverse/spinous processes of the posterior column. Fusion of the posterior elements has been demonstrated experimentally to be incapable of preventing all motion between the vertebrae [1] because of the flexible nature of the neural arch. Fusion of the principal weight-bearing elements of the spine, the vertebral bodies, produces a structure with mechanical characteristics that are intrinsically very sound. The wide base of the three points of contact (the graft and the two facet joints between the vertebral bodies) should produce a very stable segment. This concept has been validated experimentally by Lee and Langrana [2] using cadaveric lumbosacral spines. They found that the stiffness of the fusion mass was maximum after interbody fusion, being less so following posterior column fusions.

With the development of the techniques of spinal fusion and consequent improvements in outcome, more thought was given to the long-term consequences of fusion. There is a common belief that fusion of part of the spine throws extra load onto the adjacent, unfused segments and that the longer and stiffer the fusion mass, the greater would be the resulting stress. An analytical study by Lee and Langrana [2] indicated that there would be a redistribution of mobility in the lumber spine following short lumbosacral fusion (L5–S1) with the level immediately proximal to the fusion (L4–L5) accommodating most of the extra percentage deformation. These findings correlate well with the clinical observation that degeneration is usually found at this level subsequent to fusion [3]. Hypermobility of the adjacent, unfused segment following fusion was therefore proposed as a cause of disc degeneration in these segments. In a long-term study of patients (mean follow-up, 12.7 years) who had short anterior interbody fusions, Leong et al. [4] found that disc degeneration adjacent to fusion was present in 52.5% of the patients. They also found that the degenerative changes in the adjacent segments were worse following single-level fusions than double-level fusions. No satisfactory explanation for this was given, but it was concluded that anterior interbody fusion was unlikely to have accelerated the degenerative process. Obviously, any relationship between the local rigidity created by the fusion mass and the accelerated degeneration reported at the adjacent, unfused intervertebral discs is still not clear. In the current study, the biomechanical effects of "floating" L4–L5 anterior interbody fusion on the adjacent, unfused segments were investigated experimentally in cadaveric specimens.

Materials and Methods

Six fresh cadaveric lumbar spine specimens comprising L1 to S3, with the adjacent ilia, were harvested at autopsy. Each specimen was prepared by totally excising the paraspinal muscles while preserving all the ligamentous components and was then frozen at −25°C in sealed bags until it could be tested. Anteroposterior and lateral radiographs of the specimens were examined to ensure that the specimen was free of any bony or structural abnormality that might affect its mechanical properties. The mean age of the subjects was 66 years (Table 1).

Before testing, a specimen was thawed at room temperature (23°C) for 24 h and, during this period, caps were affixed at both ends of the specimen with the aid of a mounting jig. This procedure ensured that before loading the specimen was mounted in the upright neutral position with the L3–L4 disc in the horizontal plane [5,6]. Secure mounting of the end vertebrae was achieved while preserving the full range of movement of the L5–S1 segment.

After thawing, the mounted specimen was attached to a loading frame. The lower end of the specimen was rigidly attached to the platform of the loading frame, and two cables were fastened to the cap on the L1 vertebra. A constant vertical force of 30% body weight was then applied as preload through the cap to the top of the specimen. The tension of the two cables, which were used to simulate the action of erector spinae muscles in vivo, could be varied to balance the bending moment resulting from the preload, and the equilibrium of the specimen was maintained by adjusting the tension in the cables. The loading frame was specially designed to ensure free three-dimensional movement of the specimen under the three applied forces without any additional constraints. The neutral position of the specimen was defined as that adopted when the specimen was in equilibrium under the three applied forces and the upper loading cap was in the horizontal plane. In the neutral position, the preload was positioned at 50 mm anterior to the geometric center of the L3–L4 disc while the pair of cables was situated at 68 mm posterior to the center of the L3–L4 disc and symmetrically, 48 mm apart. The location of the applied preload and of the two cables was based on data obtained in vivo by other investigators [7,8]. Flexion and extension were simulated with reference to the neutral position by varying the lengths of the two cables. Wedged metal blocks with known angulation were located on the loading cap to indicate when the desired posture had been achieved. In this study, 20° of flexion and 15° of extension were simulated and studied. Reflective markers were attached to each vertebral body so that intersegmental mobility could be measured at each level in the sagittal plane with a three-dimensional motion analysis system

Table 1. General information of cadaveric specimens

No.	Sex/Age	Cause of death	Body weight (kg)
1	M/68	Lymphoma	51
2	F/79	Per rectum bleeding	37
3	M/72	Ventricular tachycardia	54
4	F/50	Chronic renal failture	31
5	M/48	Hydrocephalus	57
6	F/81	Respiratory arrest	64

(Elite, Bioengineering Technology & Systems, Milan, Italy). The standard errors of the system for linear and angular measurements were ±0.2 mm and ±0.5°, respectively.

Before the experiment proper, the specimen was put through the extremes of the motion to be simulated three times as a preconditioning procedure as it had been demonstrated in the pilot study that the repeatability of the measurements improved after this procedure. The test was then performed with the specimen intact and repeated after L4–L5 anterior interbody fusion simulated with bone cement and screws (Fig. 1).

Results

The results derived from the normal intact specimens demonstrated that the relative contributions of each motion segment to the total mobility of the lumbar spine in the flexion and extension ranges were different. Segmental movements of each level in the flexion and extension ranges were therefore not combined as a single quantity but considered separately. The measurements from the intact specimens and the same specimens with simulated L4–L5 anterior interbody fusion were compared using the one-tailed paired t test.

At the simulated "fused" level, the mean segmental mobility in both flexion and extension was not significantly different from zero with $P > 0.05$. In the flexion range, there was an increase in segmental mobility at L1–L2, L3–L4, and L5–S1 after fusion. The increase in segmental flexion mobility at the L5–S1, i.e., the level below the fusion, was statistically significant with $P < 0.05$ (Fig. 2).

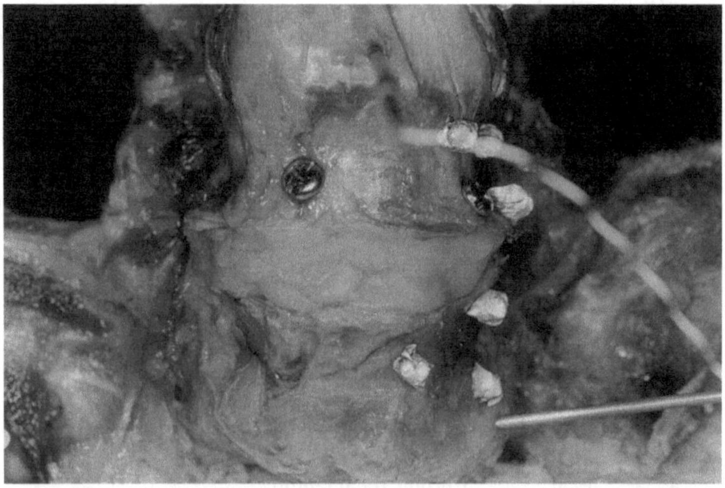

Fig. 1. A "floating" L4–L5 anterior interbody fusion was simulated using bone cement and screws

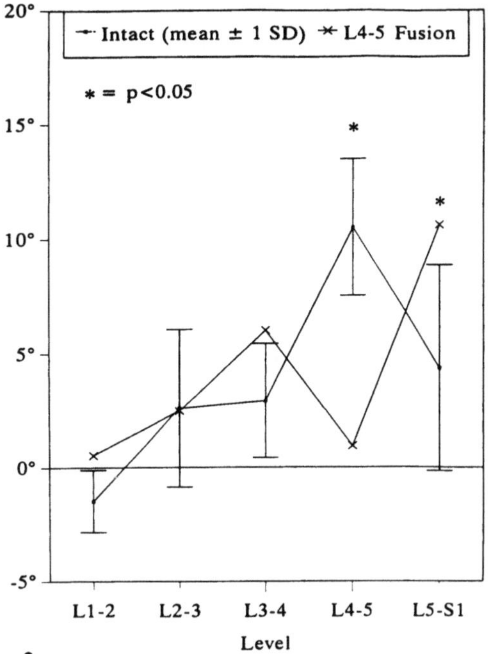

2

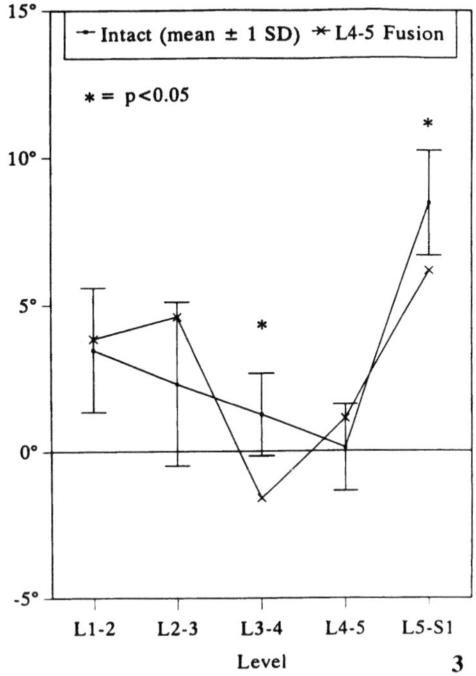

3

Fig. 2. Mean segmental flexion deformation of the lumbar spine (neutral to 20° of flexion) of intact specimens and specimens with simulated L4–L5 anterior interbody fusion

Fig. 3. Mean segmental extension deformation of the lumbar spine (neutral to 15° of extension) of intact specimens and specimens with simulated L4–L5 anterior interbody fusion

In the extension range, there was a decrease in segmental mobility at both L3–L4 and L5–S1 after fusion. The decrease in segmental extension mobility at the L3–L4 level was statistically significant with $P < 0.05$ (Fig. 3).

Discussion

It is believed that the actions of the muscles, abdominal pressure, and other structural components of the trunk, such as the thoracolumbar fascia, are important for maintaining stability of the lumbar spine. If all actions were included in the simulation of physiological movements of the lumbar spine, an indeterminate situation would result. In this experimental study, only the actions of the two erector spinae muscles spanning the whole lumbar spine specimen were simulated.

To preserve the normal natural lordosis of the lumbar spine, the fusion procedures were performed with the spine being fixed in a rigid transfer jig to ensure that fusion was accomplished in the neutral position.

In analyzing the results, segmental movements of each motion segment in flexion and extension were studied separately because the results derived from the normal intact specimens demonstrated that the relative contributions of each motion segment to the total mobility of the lumbar spine in the flexion and extension ranges were different.

Following a simulated "floating" L4–L5 anterior interbody fusion, there was a significant increase in segmental flexion deformation at the level below, i.e., L5–S1. This finding agreed well with those observed in vitro by other investigators [2,10]. However, there was a decrease in segmental movement at both L3–L4 and L5–S1 levels when the specimens were subjected to extension. This indicated that the functional range of motion of the level below the fusion, i.e., L5–S1, had been biased toward flexion. This implies that the L5–S1 has to operate deeper into its functional range of motion toward the limit of flexion than do the segments more distant from the fusion. These results also suggest that the unfused segments more frequently must operate toward the extremes of their normal ranges of motion to partially compensate for the loss of mobility at the fused level. However, the largest increase in segmental movement was found at the L5–S1 segment, which exhibited the largest deformation in the flexion range. Thus, the increase of segmental movement at this level has the highest possibility of working beyond its physiological limit, in flexion.

References

1. Rolander SD (1966) Motion of the lumbar spine with special reference to the stabilizing effect of posterior fusion. An experimental study on autopsy specimens. Department of Orthopaedic Surgery, University of Göteborg, Sweden
2. Lee CK, Langrana NA (1984) Lumbosacral spinal fusion. A biomechanical study. Spine 9:574–581
3. Lee CK (1988) Accelerated degeneration of the segment adjacent to a lumbar fusion. Spine 13:375–377
4. Leong JCY, Chun SY, Grange WJ, Fang D (1983) Long-term results of lumbar intervertebral disc prolapse. Spine 8:793–799
5. Goel VK, Voo LM, Weinstein JN, Liu YK, Okuma T, Njus GO (1988) Response of the ligamentous lumbar spine to cyclic bending loads. Spine 13:294–300
6. Tracy MF, Gibson MJ, Szypryt EP, Rutherford A, Corlett EN (1989) The geometry of the muscles of the lumbar spine determined by magnetic resonance imaging. Spine 14:186–193
7. Wirhed R (1982) Athetic ability and the antomy of motion. Wolfe Medical, Orebro, Sweden, pp 63–77
8. Németh G, Ohlsén H (1986) Moment arm lengths of trunk muscles to the lumbosacral joint obtained in vivo with computed tomography. Spine 11:158–160

9. Pearcy MJ, Burrough S (1982) Assessment of bony union after interbody fusion of the lumbar spine using a biplanar radiographic technique. J Bone Joint Surg 64B:228–232
10. Quinnell RC, Stockdale HR (1981) Some experimental observation of the influence of a single lumbar floating fusion on the remaining lumbar spine. Spine 6:263–267

Distribution of Axial Compressive Elastic Modulus in Degenerated Intervertebral Discs

S. Umehara[1], S. Tadano[2], K. Kaneda[1], K. Abumi[1], S. Asano[1], and T. Ukai[2]

Summary. The objectives of this study were to measure the distribution of axial compressive elastic modulus in the lumbar intervertebral disc and to clarify the differences in distribution between normal and degenerated discs. The authors have proposed a method to measure the local compressive elastic modulus of the intervertebral disc by an indentation test. To estimate the distribution of elastic modulus, sheet specimens were sliced from cadaveric lumbar discs. Two normal and two degenerated discs were utilized to clarify the effect of disc degeneration on the distribution of elastic modulus. The degree of disc degeneration was evaluated macroscopically and radiographically. The distribution of elastic modulus in normal intervertebral discs was symmetric to the midsagittal plane. The elastic modulus in the region of the nucleus pulposus was lower than that of the annulus fibrosus, but was nearly the same as that in the posterolateral regions of the annulus fibrosus. The elastic moduli in the anterior region of the annulus fibrosus were the highest in any area of a disc. In comparison with normal discs, degenerated discs showed asymmetric and irregular distribution of elastic modulus. The elastic moduli of the degenerated nucleus pulposus and annulus fibrosus were much higher than those in the same areas of normal discs.

Key words: Biomechanics–Lumbar spine–Intervertebral disc–Indentation–Compressive elastic modulus–Disc degeneration

Introduction

The spine has at least three fundamental biomechanical functions. First, it transfers force and moment. Second, it allows sufficient physiological motion.

Departments of [1]Orthopaedic Surgery and [2]Mechanical Engineering II, Hokkaido University, Kita 15, Nishi 7, Kitaku, Sapporo, Hokkaido, 060 Japan

Finally, it protects the delicate spinal cord and nerve roots from potentially damaging load or motion. In particular, the lumbar spine is subject to larger loads and more complex motions than the other levels. The lumbar intervertebral disc, therefore, is the most important component for maintaining the stability and mobility of the lumbar spine. The intervertebral disc has a complex structure that consists of two distinct parts, the nucleus pulposus and the annulus fibrosus. The nucleus pulposus is located in the central portion of the disc and is composed of very loose and translucent networks of fine fibrosus strands that lie in the mucoprotein gel. The annulus fibrosus is composed of fibrosus tissue including concentric laminated bands. Many diseases of the human lumbar spine are caused by the defects in mechanical function of the intervertebral disc, such as disc degeneration or disc herniation. Therefore, basic research on the biomechanical characterization of the intervertebral disc is essential to provide useful therapy for lumbar diseases.

Many experimental [1–6] or mathematical studies [7–13] on the biomechanical properties of the lumbar intervertebral disc have been reported. Stress state in the disc calculated by a mathematical model depends greatly on the elastic modulus. The elastic modulus of the annulus fibrosus or the nucleus pulposus is, of course, not uniform as regards location. In most previous mathematical studies, however, each elastic modulus was set as a unique value, which left some unresolved problems relative to analytical accuracy. The method of determining the material constant in the intervertebral disc has not been discussed well previously.

This chapter presents a method to estimate the compressive elastic modulus in a local region of the intervertebral disc and to evaluate the distribution of the elastic modulus on the axial plane in the disc. Further, the relationship between the distribution of elastic modulus and disc degeneration is studied by use of disc specimens with varying degrees of degeneration.

Materials and Methods

Indentation Test Setup

The experimental setup designed for the indentation test is shown in Fig. 1. A digital dial gauge was equipped vertically on the usual X–Y stage. When a weight is put on the tray mounted on the top of the axis rod of the gage, the compressive displacement can be read directly. The weight of the axis rod is 5 g, which is reduced by a balancing weight. Figure 2 shows the shape of the tip of the indenter which is a circular plate 5 mm in diameter.

Estimation of Elastic Modulus by the Model Test

To estimate the compressive elastic modulus in a local region of the intervertebral disc by indentation test, the relationship between the elastic modulus and the indentation behavior must be investigated in advance. In this work, a

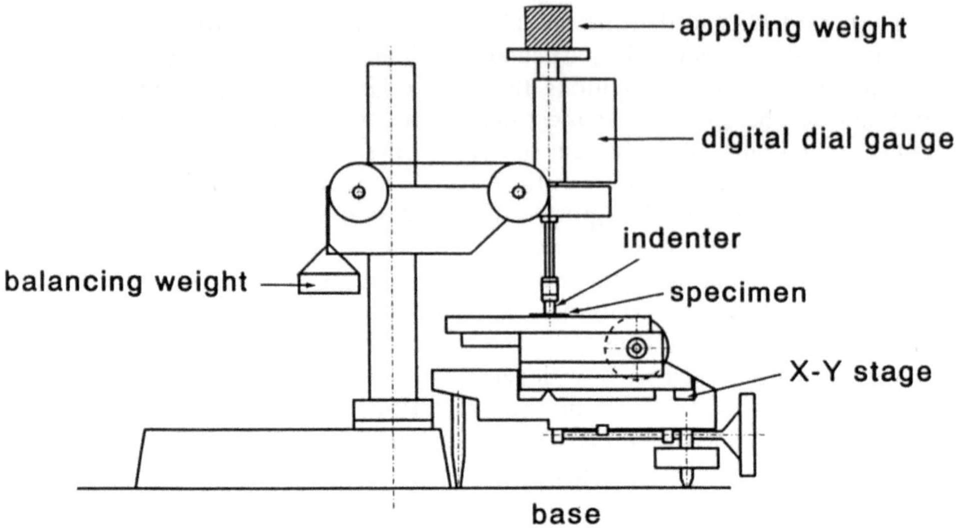

Fig. 1. Experimental setup for the indentation test

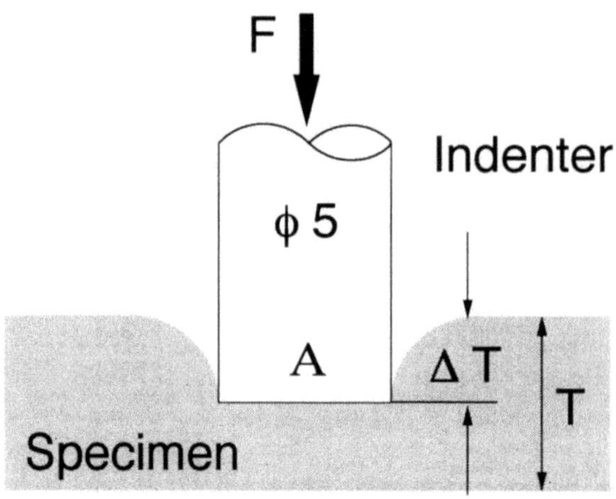

Fig. 2. Schematic of indentation of soft tissue

model indentation test was carried out using sheet specimens of polyurethane elastomer [14].

Eight kinds of polyurethane elastomer with different hardness (PU01–PU08) were utilized in this model test. These materials were specially made for this experiment by Tokai Rubber Industries (Komaki, Japan). Before the indentation test, the compressive elastic modulus of each model material was confirmed by an axial compression test using cylindrical specimens 30 mm in diameter and 30 mm in height. A desk-type tensile-compressive testing machine (SV-301-E, Imada Industrial, Tokyo, Japan) was utilized, and the loading velocity was constant at 0.83 mm/s. Figure 3 shows the compressive stress–strain relationship obtained by this compression test. Until the point of 40% strain is reached in each specimen, the relationship can be approximated by a linear equation, and each compressive modulus was calculated from the gradient of the curve. The elastic moduli of each specimen are listed in Table 1.

Because the elastic modulus of each polyurethane elastomer could be determined as just described, the indentation test was carried out using sheet specimens (30 × 30 mm) of polyurethane. To confirm the effect of the thickness of specimen on the indentation behavior, five different thicknesses of the specimen were selected: 1, 3, 5, 7, and 10 mm. The location where indented was the geometric center of each specimen. The stress σ and the strain ε under indentation were calculated by $\sigma = F/A$ and $\varepsilon = dT/T$, where A and F were the tip area of the indenter and the applying load, respectively, and T and dT were the initial thickness and the indentation displacement, respectively, as shown in Fig. 2. The applied load was constant at F = 50 *g*, which was determined from various preliminary tests.

Because polyurethane elastomer is soft and viscoelastic, the creep behavior of the material under constant loading must be confirmed. Figure 4 shows the creep behavior under indentation obtained from the 3-mm-thick specimen under loading of F = 50 *g*. As a result, the strain in each specimen became almost constant by 10 s after the start of loading. Therefore, the strain ε at 50 s after loading was used in the following numerical process for the experimental results.

Figure 5 shows the relationship between the compressive elastic modulus and the indentation strain, taking account of initial thickness. The experimental results were plotted with different symbols for each strain value (see Figure). Figure 5 shows that the strain depends greatly on the initial thickness. The strain under constant loading increases with decrease of the initial thickness. This relationship could be approximated by a linear equation in log-log scale, as equation 1:

$$\ln \varepsilon = A_0 + A_1 \ln T \tag{1}$$

where A_0 and A_1 could be expressed only by the elastic modulus E (kPa), as follows:

$$A_0 = a + b \ln E \tag{2}$$

$$A_1 = c + d \ln E \tag{3}$$

From equations 1 to 3, elastic modulus would be expressed as equation 4:

$$\ln E = \frac{\ln \varepsilon - a - c \ln T}{b + d \ln T} \tag{4}$$

The constants of this equation were $a = -1.6$, $b = -2.9 \times 10^{-1}$, $c = -8.7 \times 10^{-2}$, and $d = -9.4 \times 10^{-1}$. Equation 4, drawn with dotted curves in Fig. 5,

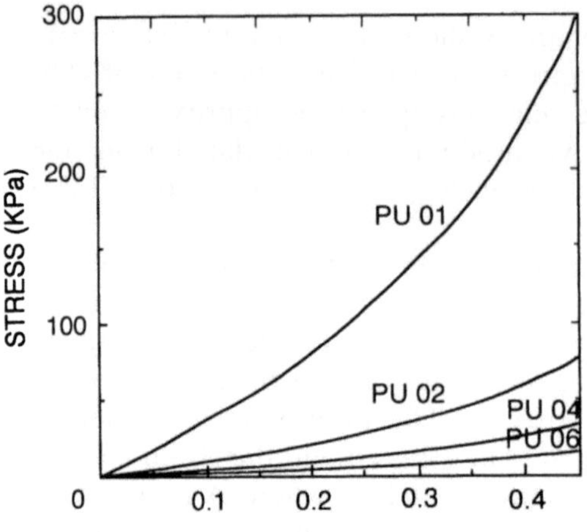

3

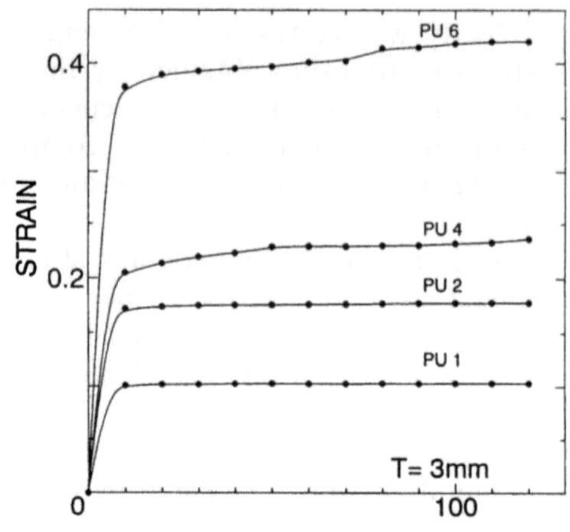

4

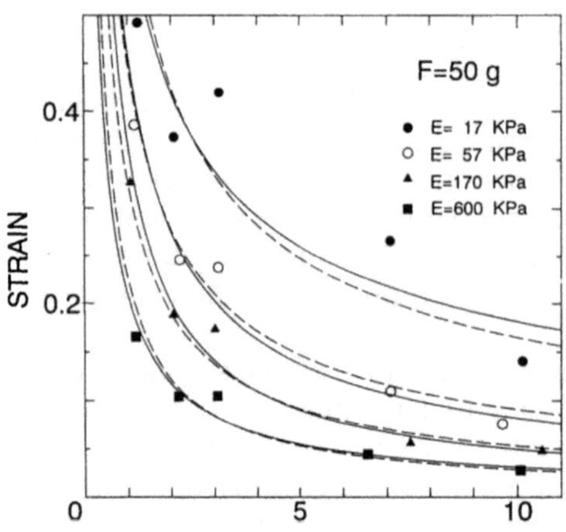

5

Fig. 3. Compressive stress–strain relationship of the cylinder specimen of polyurethane

Fig. 4. Creep behavior under indentation

Fig. 5. Elastic modulus: strain–initial thickness relationship

Table 1. Compressive elastic modulus of polyurethane specimens

Specimen No.	PU01	PU02	PU03	PU04	PU05	PU06	PU07	PU08
E (KPa)	600	170	96	57	45	17	12	3

approximated well the experimental results. Therefore, the elastic modulus at a local region in the intervertebral disc could be estimated from equation 4 by substitution of the values of initial thickness and strain measured.

Distribution of Elastic Modulus in the Intervertebral Disc

Four fresh cadaveric intervertebral discs were obtained from routine autopsies within 6h after death. These specimens were double wrapped in plastic bags and stored in a deep freezer at −26°C. Three specimens were L4/L5, one was L3/L4. For the indentation test, sheet specimens were sliced from the frozen intervertebral discs and thawed at room temperature; local thickness of thawed sheet specimens varied from 2 to 5 mm. An example of the distribution of initial thickness is shown in Fig. 6.

The degree of disc degeneration of the specimens was confirmed macroscopically using Galante's grading [15] and radiographically by Kellgren's classification [16]. Axial views of the specimens are shown in Fig. 7. One specimen showed almost no macroscopic degeneration, one was graded as slight degeneration, one showed moderate degeneration, and one was classified as severe degeneration. The macroscopic classification, radiographic grading, and other specimen information are summarized in Table 2.

Using these specimens, the indentation test was carried out using the experimental setup shown in Fig. 1. The nodal points of a 10×10 mm network on a specimen were loaded by the indenter (Fig. 8). The displacement was measured at 50 s after under constant weight of $F = 50$ g.

Results

The compressive elastic modulus in a local region of the intervertebral disc was estimated from equation 4 and the data measured. An example of the initial thickness of the specimen and the location tested is shown in Fig. 6, and the strain measured is shown in Fig. 9. The distributions of elastic modulus are shown in Figs. 10 and 11. Each figure shows the value of elastic modulus calculated and the location tested.

Specimens 1 and 2 in Table 2 were classified as minimal or slight degeneration. They demonstrated similar pattern of the distribution of elastic modulus,

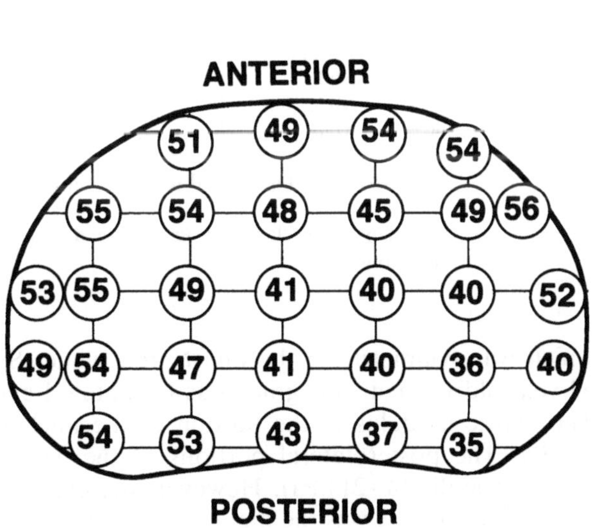

Fig. 6. Distribution of initial thickness ($\times 10^{-1}$ mm) in a disc

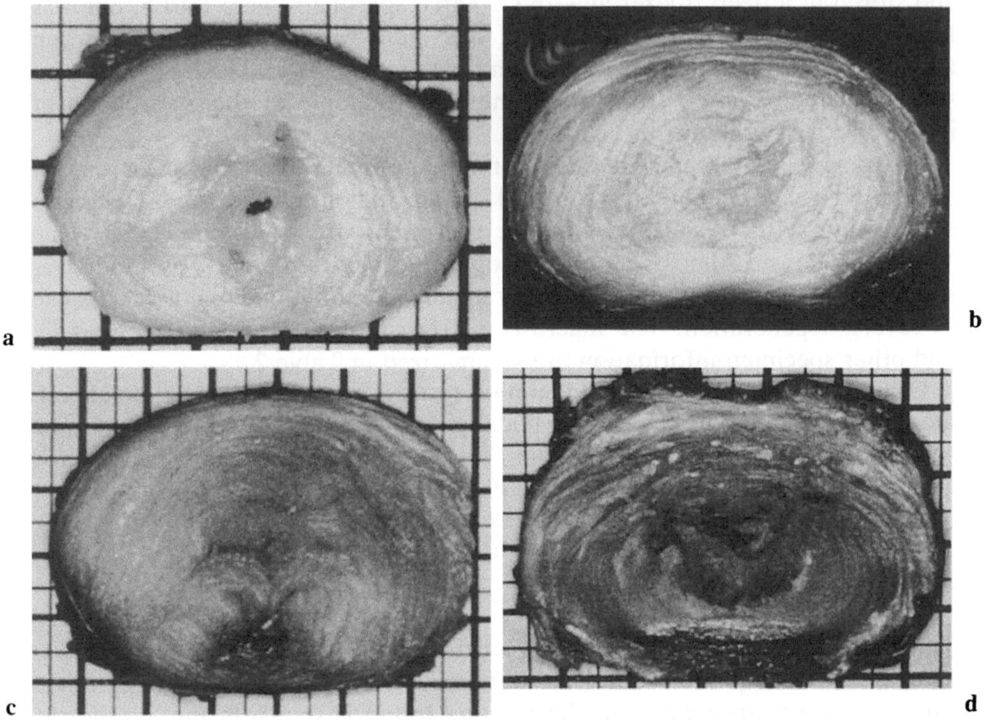

Fig. 7. Specimens of the cadaveric intervertebral disc: **a**, Specimen 1; **b**, specimen 2; **c**, specimen 3; **d**, specimen 4

Table 2. Characteristics of test specimens

No.	Sex	Age	Level	Galante[a]	Kellgren[b]
1	Female	39	L4/5	I	I
2	Male	48	L3/4	II	I
3	Male	48	L4/5	III	I
4	Male	74	L4/5	IV	III

[a] Macroscopic grading of disc degeneration presented by Galante [15].
[b] Radiographic classification of disc degeneration presented by Kellgren [16].

which was symmetric to the midsagittal plane. Specimen 1, from L4/L5 of a 39-year-old woman, was graded as Galante's I and Kellgren's I (see Table 2). The elastic modulus in the portion of the nucleus pulposus varied from 4 to 12 kPa. The posterolateral portions of the annulus fibrosus showed minimal elastic moduli (14–21 kPa). However, the elastic modulus in the anterior portion of the annulus was the highest (30–62 kPa). The value in the middle of the

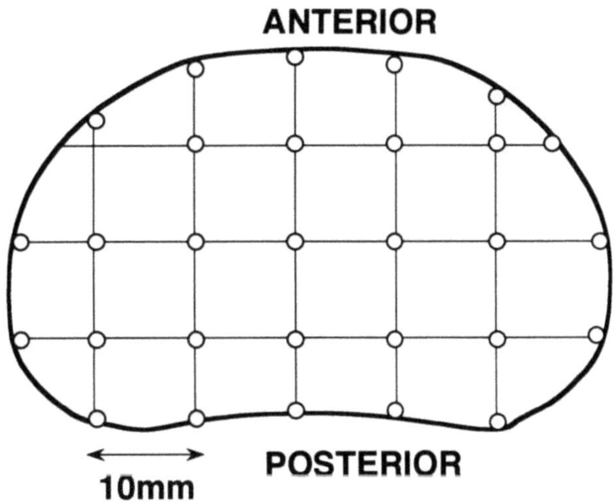

Fig. 8. Measuring points

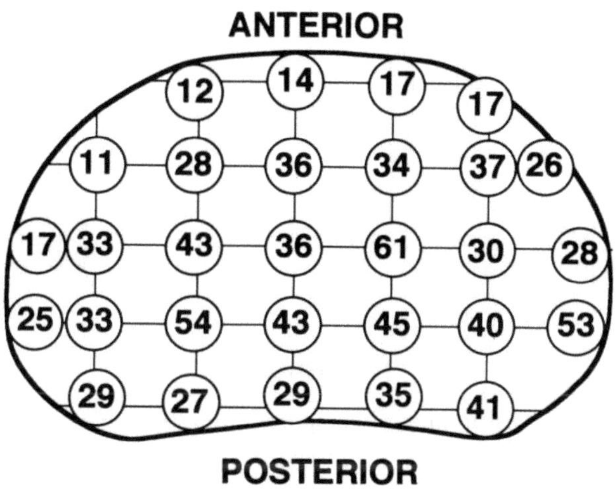

Fig. 9. Distribution of indentation strain (%) in a disc

posterior annulus fibrosus was 22.5 kPa and in the lateral portions of the annulus about 26 kPa (Fig. 10a). Specimen 2 was from L3/L4 of a 48-year-old man and was classified as Galante's II and Kellgren's I (see Table 2). The elastic moduli in the nucleus pulposus were 5–10 kPa. The posterolateral portions of the annulus fibrosus showed minimal elastic moduli (5–13 kPa). However, the elastic modulus in the anterior portion of the annulus was the highest (30–80 kPa). The values in the middle portions of the posterior annulus fibrosus were from 13 to 17 kPa and in the lateral portions from 13 to 36 kPa (Fig. 10b).

In comparison with specimens 1 and 2, specimens 3 and 4 were graded as moderate or severe degeneration. The distribution of elastic modulus in these

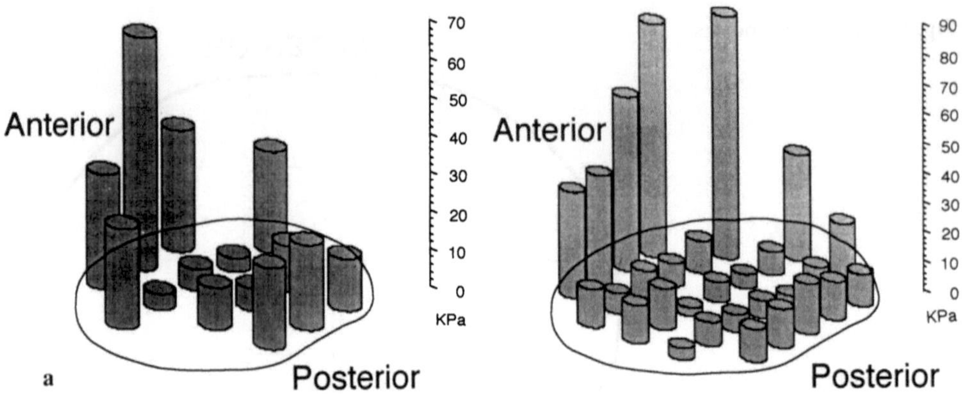

Fig. 10. Distribution of compressive elastic modulus in normal discs: **a**, specimen 1; **b**, specimen 2

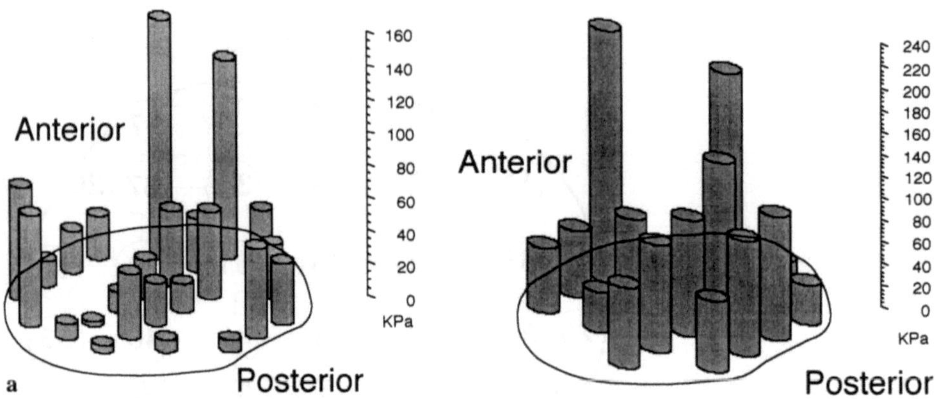

Fig. 11. Distribution of compressive elastic modulus in degenerated discs: **a**, specimen 3; **b**, specimen 4

specimens with degeneration was asymmetric and irregular. Specimen 3, from L4/L5 of a 48-year-old man, was graded as Galante's III and Kellgren's I (see Table 2). The elastic moduli in the nucleus pulposus were from 13 to 53 kPa and in the annulus fibrosus from 25 to 146 kPa. The values in several regions of the nucleus pulposus exceeded those of the annulus fibrosus (Fig. 11a). Specimen 4 was from L4/L5 of a 74-year-old man and was classified as Galante's IV and Kellgren's III (see Table 2). The elastic moduli in the nucleus pulposus were 50–145 kPa and in the annulus fibuosus 21–230 kPa. The values in the interface region of the nucleus and the annulus were relatively less (23–39 kPa, Fig. 11b).

The values of elastic modulus in the portion of the nucleus pulposus and the annulus fibrosus of specimens 3 and 4 were much higher than those of specimens 1 and 2: 6- to 10-fold more in the portion of the nucleus pulposus and 3- to 5-fold more in the portion of the annulus fibrosus.

Discussion

Previous mechanical studies on the functional spinal unit or the intervertebral disc have been reported [1–6]. Brown et al. [2] measured the distribution of elastic modulus by the tensile test using bone–disc–bone specimens cut from the intervertebral disc. In the present study, the local compressive elastic modulus was estimated by the indentation test using sliced sheet specimens of the lumbar intervertebral disc. Further, the differences of the distribution of elastic modulus were confirmed among specimens with various amounts of degeneration.

In this study, the distribution of elastic modulus of normal discs was confirmed to be symmetric to the midsagittal plane. The elastic moduli of the nucleus pulposus were, of course, lower than those of the annulus fibrosus. In most of the previous analysis for the intervertebral disc, the nucleus pulposus and the annulus fibrosus have been treated as apparently different materials in the numerical modeling [7–13]. However, these results of the distribution of elastic modulus showed that it is difficult to clearly find the interface between them. Even the elastic modulus of the annulus fibrosus appeared not to be constant and showed various values along its circumference. In particular, the elastic modulus of the normal annulus fibrosus showed the lowest value in the posterolateral portions and the highest in the anterior portions. The posterolateral portion of the annulus fibrosus is where disc herniation is likely to occur and cause clinical symptoms. This distribution of elastic modulus would be inherent in the normal intervertebral disc and relate to the pathophysiology of disc herniation.

Disc degeneration has been evaluated clinically by radiograms including functional lateral views or computed tomograms. However, the alteration of mechanical properties of the intervertebral disc following degeneration has been poorly understood. Our study clarified that the "radiographic degeneration" did not indicate the "macroscopic degeneration," as shown in Table 2. Further, this study presented "mechanical degeneration," which was detected as the alteration of the distribution of elastic modulus, and this mechanical degeneration was well correlated to the macroscopic degeneration but not to the radiographic degeneration. This result implied that it is difficult to diagnose radiographically the earlier stage of degeneration, which can however be detected macroscopically and mechanically.

The local elastic moduli measured by this indentation method can be used as essential data to construct more precise analytical models of the lumbar intervertebral disc. Use of these models may allow the progression and mechanism of disc degeneration to be well understood.

Conclusions

This article describes a method to estimate the compressive elastic modulus in a local region of the lumbar intervertebral disc. The differences of the distribution of elastic modulus among discs with various degrees of degeneration were evaluated. The following conclusions were obtained for the internal mechanical characterization of the lumbar intervertebral disc.

1. The distribution of elastic modulus in discs with minimal or slight degeneration was symmetric to the midsagittal plane. The values of elastic modulus in the portion of the nucleus pulposus were lower than those in the portion of the annulus fibrosus. The interface between the nucleus and the annulus became unclear with increase of disc degeneration, determined by macroscopic observation and the disbribution of elastic modulus.
2. Elastic moduli of the annulus fibrosus did not show a unique value along its circumference in minimally or slightly degenerated discs. The highest value was obtained in the anterior portion of the annulus fibrosus and the lowest value in its posterolateral portions.
3. The distribution of elastic modulus in discs with moderate or severe degeneration was asymmetric and irregular. The values of elastic modulus in degenerated discs were much higher than those in minimally or slightly degenerated discs.
4. It is difficult to detect radiographically the earlier stage of degeneration that can be detected macroscopically and mechanically.

Acknowledgment. This study has been supported by a Grant-in-Aid for Scientific Research from the Ministry of Education, Science and Culture in Japan. Special thanks to Kazuaki Katagiri, who assisted in performing mechanical tests.

References

1. Asano S, Kaneda K, Umehara S, Tadano S (1992) The mechanical properties of the human L4–5 functional spinal unit during cyclic loading—The structural effects of the posterior elements. Spine 17:1343–1352
2. Brown T, Hanson R, Yorra A (1957) Some mechanical test on the lumbosacral spine with particular reference to the intervertebral discs. J Bone Joint Surg 39A:1135–1164
3. Goel VK, Nishiyama K, Weinstein JN, Liu YK (1986) Mechanical properties of lumbar spinal motion segments as affected by partial disc removal. Spine 11:1008–1012
4. Markolf KL (1972) Deformation of the thoracolumbar intervertebral joints in response to external load. J Bone Joint Surg 54A:511–533
5. Panjabi MM, Krag MH, Chung TQ (1984) Effects of disc injury on mechanical behavior of the human spine. Spine 9:707–713

6. Schultz AB, Warwick DN, Berkson MH, Nachemson AL (1979) Mechanical properties of human lumbar spine motion segments: Part 1. Response in flexion, extension, lateral bending, and torsion. J Biomech Eng 101:46–52
7. Kulak RF, Belytschko TB, Schultz AB, Galante JO (1976) Nonlinear behavior of the human intervertebral disc under axial load. J Biomech 9:377–386
8. Laible JP, Pflaster DS, Krag MH, Simon BR, Haugh LD (1993) A poroelastic-swelling finite element model with application to the intervertebral disc. Spine 18:659–670
9. Shirazi-Adl SA, Shrivastava SC, Ahmed AM (1984) Stress analysis of the lumbar disc-body unit in compression—a three dimensional nonlinear finite element study. Spine 9:120–134
10. Shirazi-Adl A, Ahmed AM, Shrivastava SC (1986) A finite element study of a lumbar motion segment subjected to pure sagittal plane moments. J Biomech 19:331–350
11. Simon BR, Wu JSS, Carlton MW, Evans JH, Kazarian LE (1985) Structural models for human spinal motion segments based on a poroelastic view of the intervertebral disc. J Biomech Eng 107:327–335
12. Spilker RL, Jakobs DM, Schultz AB (1986) Material constants for a finite element model of the intervertebral disc with a fiber composite annulus. J Biomech Eng 108:1–11
13. Ueno K, Liu YK (1987) A three-dimensional nonlinear finite element model of lumbar intervertebral joint in torsion. J Biomech Eng 109:200–209
14. Tadano S, Ukai T, Umehara S, Kaneda K (1993) Distribution of compressive elastic modulus in a human intervertebral disc. In: 1993 Advances in bioengineering, vol 26. American Society of Mechanical Engineers, Washington, DC, pp 427–430
15. Galante JO (1967) Tensile properties of the human lumbar annulus fibrosus. Acta Orthop Scand [suppl 100]
16. Kellgren JH (1963) The epidemiology of chronic rheumatism, vol 2. Blackwell, Oxford, p 63

Biomechanical Properties of the Rabbit Disc Body Unit Under Cyclic Torsional Loading

Takenori Oda[1], Sohei Ebara[1], Masao Tanaka[2], Tadashi Kuraya[2], and Keiro Ono[1]

Summary. To understand the biomechanical effect of repetitive torsional loading to the spine, fresh cadaveric lumbar disc body units from Japanese white rabbits were tested. The specified angular displacement was loaded to the specimen in a cyclic manner. The amplitude of axial rotation ranged from 10° to 19°, and the average rate of loading was 24°/s. The peak values of torque were monitored as a time series. The torque angle curves were recorded intermittently during cyclic loading, which was continued until failure of the specimen occurred. As the number of torsional cycles increased, the peak torque decreased. It was represented as two portions of a linear segment plotted against the logarithm of the number of torsional cycle. The decrease of peak torque showed the two phases, which were, first, the gradual decrease phase and second, the sudden decrease phase. The torque angle curves showed that the stiffness decreased and the initial toe part of the curve increased after repetition of the load. Histological examination after testing showed that the failure site was a growth plate at the upper vertebral body. However, distortion of stratification and rupture of the laminated bands were also observed at the intervertebral disc. These were considered as the accumulation of damage resulting in the failure caused by the cyclic load. These findings suggest that the response to the repetitive loading, in other words, the fatigue phenomenon, is strongly associated with the etiology of disc degeneration and spondylosis.

Key words: Spine—Spondylosis—Degeneration—Fatigue—Cyclic Loading—Torsion—Biomechanics

---

[1] Department of Orthopedic Surgery, Osaka University Medical School, 2-2 Yamada-Oka, Suita, Osaka, 565 Japan
[2] Department of Mechanical Engineering, Osaka University, Faculty of Engineering Science, Machikaneyama, Tokonaka, Osaka, 560 Japan

Introduction

Spondylosis is defined as a condition of progressive degeneration that begins in an intervertebral disc and leads to changes in the surrounding bones and soft tissues of the spine. This disease is usually recognized as a phenomenon associated with aging [1], but mechanical environment is considered to be one of the important etiologic factors, because cervical spondylosis occurs at a younger age and its progression is rapid in patients with athetoid cerebral palsy [2,3], whose necks move repetitively [4]. Experimentally, Wada et al. [5] created a spondylosis in the rabbit cervical spine by repetitive extension-flexion movement that was generated by electric stimulation of the trapezius muscle. They suggested that overuse from repetitive movement is an important factor in the pathogenesis of the spondylosis.

There have been several studies concerning fatigue tests of the spine [6–10]. However, the mechanical properties due to repetitive loading are still unclear. The purpose of this study was to demonstrate the biomechanical effect of repetitive torsional loading to the spine in vitro. Based on their results, we would like to discuss the degradation of the biomechanical properties of the spine.

Materials and Methods

Fresh cadaveric lumbar spinal functional units of L2–L3 and L4–L5 levels were harvested from Japanese white rabbits (ages, 8–9 weeks; weight, 1.9 kg). Specimens were frozen at −40°C until testing. After being thawed to room temperature, the disc body units (vertebral body–intervertebral disc–vertebral body units), after removing muscles and the posterior elements, were used for the biomechanical test.

The specified angular displacement was loaded to the specimen in a cyclic manner. The loading apparatus is shown in Fig. 1. This apparatus converts the translational movement of the piston head of the electrohydraulic universal testing machine (Hydropuls system PSA-01, Tokyokoki-Schenck, Tokyo, Japan) into the angular movement to the specimen by using the parallelogram mechanism. The center of angular movement of the specimen was a geometric center of the intervertebral disc. Upper and lower vertebral bodies were firmly held by special designed clamps. A pair of strain gauges were bonded to the driving rods to measure the torque acting on the specimen.

We determined the condition of cyclic loading on the basis of torque angle behavior of the specimens that was obtained by preliminary one-time failure tests. Figure 2 shows the typical torque angle curves of the L2–L3 disc body unit when the specimens were twisted until failure at the two different constant loading rates. These curves start with the nonlinear toe part followed by the linear part with constant stiffness and the nonlinear part showing moderate decrease of the stiffness, and then arrived at the point with the maximum

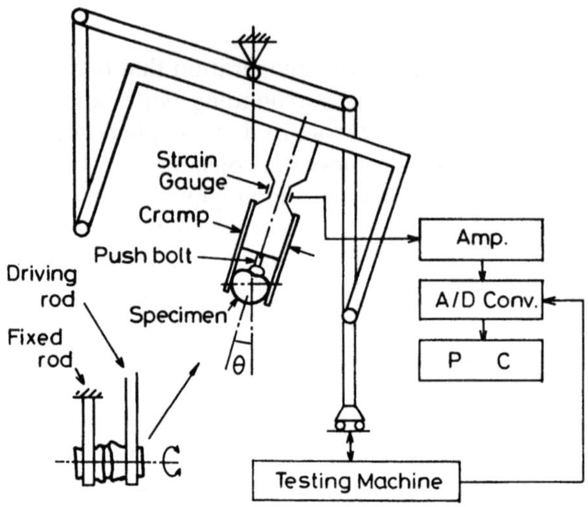

Fig. 1. The loading apparatus

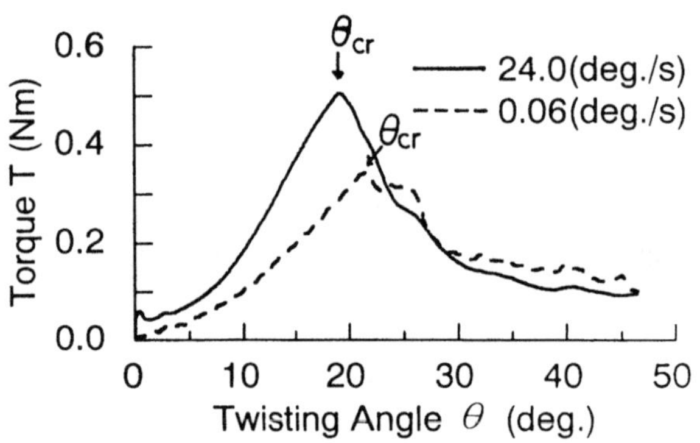

Fig. 2. Typical torque angle behavior of the L2–L3 disc body unit

torque. The torque angle curves differed with the loading rate. Although the angle at the maximum torque (critical angle: Θ_{cr}) depended on the level of the specimen, it was not dependent on the loading rate. To keep the specimen free from failure from one-time twisting, the amplitude of axial rotation (Θ_a) of cyclic testing was below the critical angle (Θ_{cr}) obtained by preliminary tests. It was less than 20° for the L2–L3 level and 16° for the L4–L5 level. The sinusoidal rotation to one side was used. To accelerate the repetitive test, the average rate of loading was fixed at the preliminary faster velocity of 24°/s. Therefore, the loading frequency was dependent on the amplitude of axial rotation (Θ_a). For example, when the amplitude of axial rotation (Θ_a) was 12°, the frequency of sinusoidal loading was 1 Hz. The cyclic loading was continued until failure of the specimen occurred.

The peak values of torque was monitored as a time series during cyclic loading. The torque angle curves were recorded intermittently. During the experiment, the specimen was placed in a bath with physiologic saline at the constant temperature of 20°C. In this situation, mechanical properties of the specimens were preserved even after 96h (Fig. 3).

Histological examination was performed after completion of the biomechanical test. Failure sites of the specimen and the condition of intervertebral discs were evaluated.

Results

Mechanical Properties

The values of peak torque (T) were plotted in terms of the number of torsional cycle (N). A typical curve (N–T curve) is shown in Fig. 4. The peak torque decreased as the number of torsional cycles increased. It was represented as two portions of a linear segment plotted against the logarithm of the number of

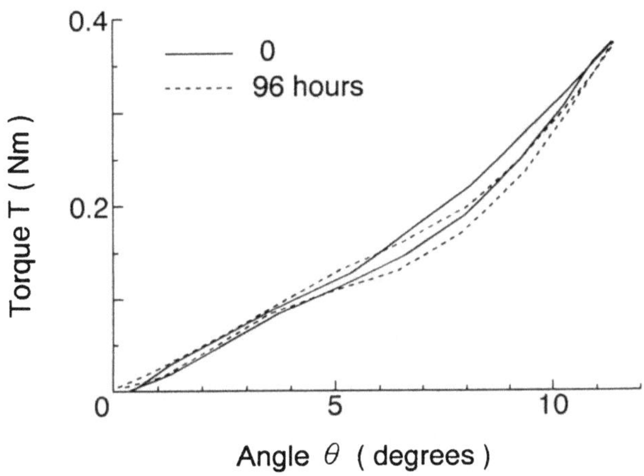

Fig. 3. The effect of preservation on the mechanical properties of the disc body unit

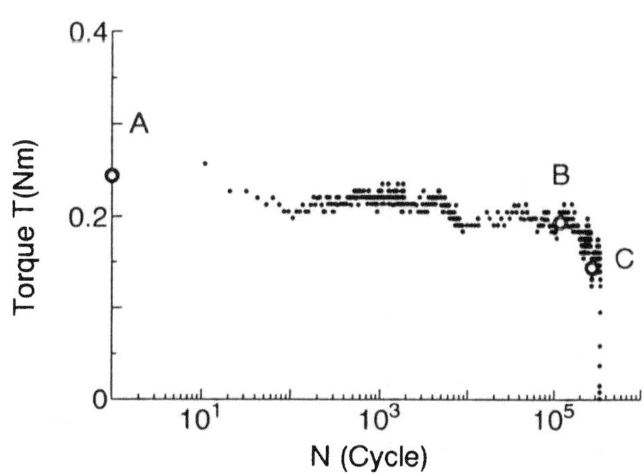

Fig. 4. Typical peak torque decrease under cyclic torsional loading

torsional cycle. That is, the decrease of peak torque demonstrated the two phases, which were the first, gradual decrease phase and the second, sudden decrease phase. After sudden decrease of the peak torque, macroscopic failure of the specimen occurred. These features were commonly observed in every specimen regardless of the specimen level.

The torque angle curves for the three different cycles are shown in Fig. 5. It represented the increase of the initial toe part of the torque angle curve after repetition of the load. In other words, the range of angle at which torque was less readily detected increased as the number of torsional cycles increased. Figure 6 shows the change of the tangential stiffness at the maximum twisting angle. The stiffness decreased as the number of torsional cycles increased, although its decrease was small.

The number of cycles to failure (N_f) was defined as the number at the crosspoint of zero torque on the second linear segment of the N–T curve. There was a logarithmic correlation between the number of cycles to failure (N_f) and the amplitude of rotation (Θ_a) (Fig. 7).

$$\Theta_a = a + b \log_{10} N_f$$

Coefficients a and b are shown in Table 1. The critical angle (Θ_{cr}) obtained by preliminary one-time failure tests (see Fig. 2) is also shown. Significant differences occurred between the values of critical angle (Θ_{cr}) and the values of coefficient a ($\Theta_a = a$ when $N_f = 1$). The value of coefficient a was larger than the value of the critical angle (Θ_{cr}) at both L2–L3 and L4–L5 levels.

Histological Examination

Histological examination showed that the failure site of the specimen was a growth plate at the upper vertebral body. Separation at the growth plate was observed (Fig. 8). However, distortion of the stratification and rupture of the laminated bands were also detected at the intervertebral disc (Fig. 9). The delamination was prominent in the posterolateral part of the annulus.

Discussion

There have been a few studies of torsional fatigue tests to the spine. Farfan et al. [6] first described the effects of repeated torsional loading to the spine. They used a human cadaveric lumbar spine, and repeated torque was applied to the specimen for a maximum of 48 h (144 000 cycles). They demonstrated the initial torque loss; however, further torque loss and rupture of the disc were not detected. The effects of cyclic torsional loads on the behavior of intact lumbar intervertebral joints to a maximum of 10 000 cycles were investigated by Liu et al. [9]. Their results showed that failure occurred when the initial angular displacement exceeded the critical threshold value. Several types of failure were observed in their study, one of which was the tear in the

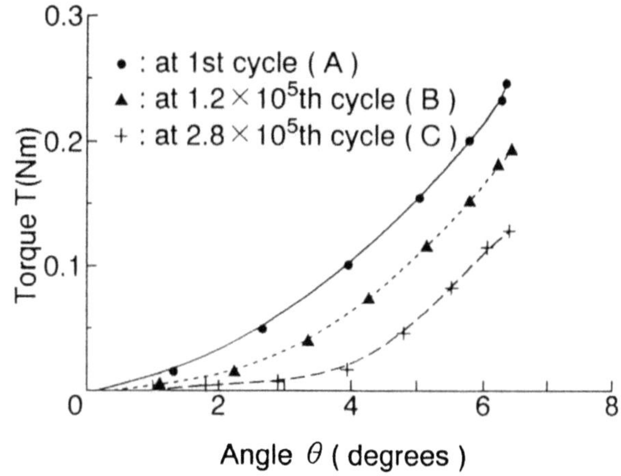

Fig. 5. Torque angle behavior during cyclic torsional loading

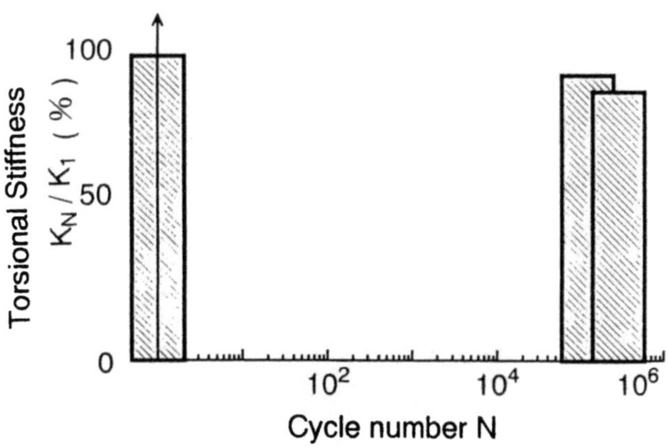

Fig. 6. Decrease of stiffness under cyclic torsional loading K_1, stiffness at the first cycle; K_N, stiffness at the cycle number N

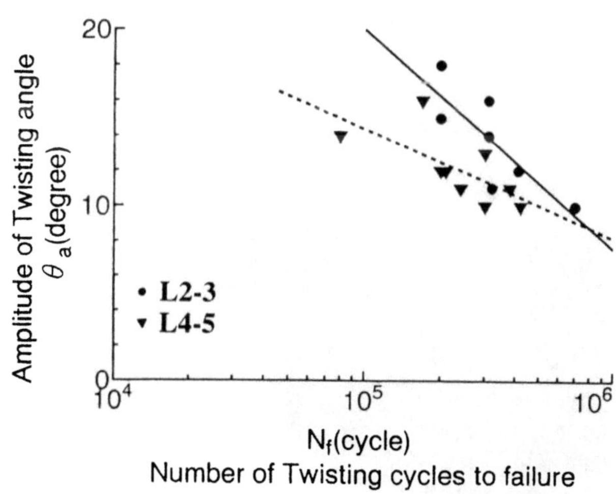

Fig. 7. Correlation between the number of cycles to failure (N_f) and the amplitude of rotation (Θ_a)

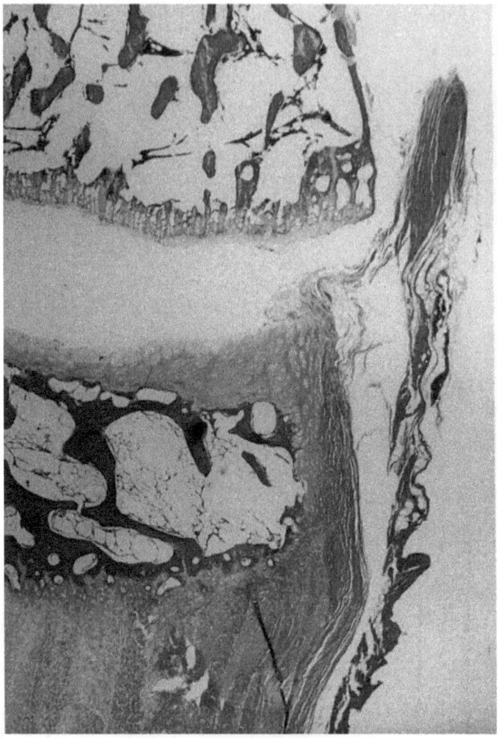

8

Fig. 8. Separation at the growth plate

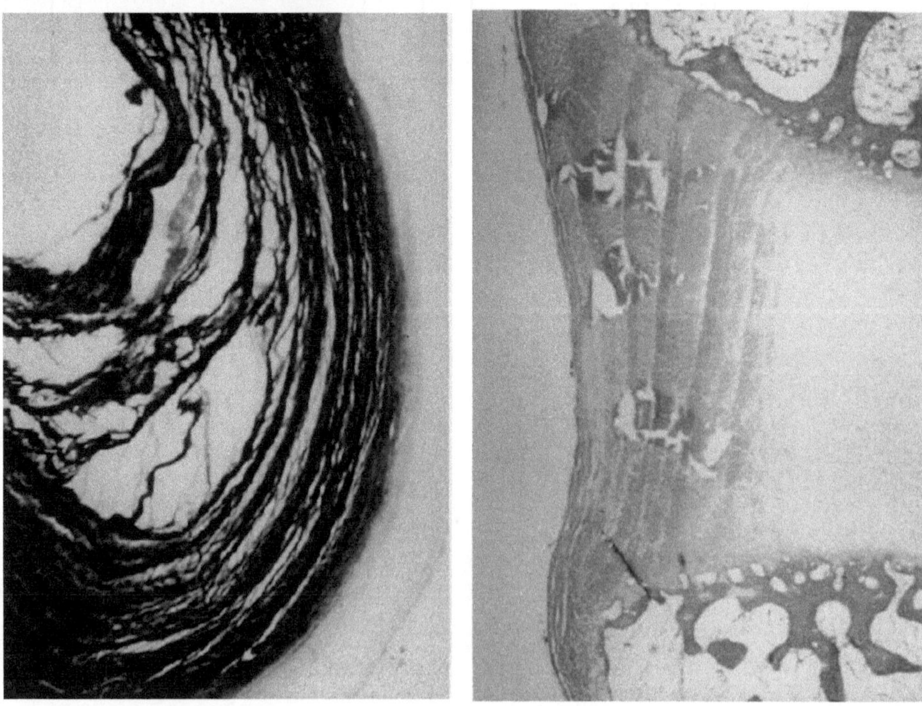

9

Fig. 9. Histological findings at the intervertebral disc

Table 1. The coefficients a and b and the critical angle Θ_{cr}

Level	a	b	Θ_{cr}
L2–L3	82	−12	22.1 ± 2.4
L4–L5	45	−6	18.3 ± 2.6

annulus. They speculated that cumulative damage occurred in both the soft and hard tissues and failures caused by fatigue lead to weakening and improper functioning of apophyseal joints and disc. In this study, repetitive loading to a maximum of 1 000 000 cycles was applied and the fatigue failure was observed as well.

Biomechanical properties from cyclic loading were evaluated in more detail in this study. The decrease of stiffness caused by repetitive loading suggested increase of instability and weakening for the motion unit. The increase of the initial toe part of the torque angle curve meant approximately the increase of the neutral zone, which was defined by Panjabi [11] and is a good parameter of instability. These findings may be attributed to the cumulative damages in the disc body unit, one of which was distortion of the stratification in the disc.

We speculate that the two different phases of the peak torque decrease are related to the histological findings. Because the failure of the specimen occurred finally at the growth plate, the second sudden decrease phase in the N–T curve is considered to be related to the weakening of this lesion. On the other hand, the first gradual decrease phase is supposed to be attributed to the distortion and rupture of the laminated bands in the intervertebral disc. To clarify this speculation, histological examination of the specimen at the first gradual decrease phase is necessary.

Our study also demonstrated the difference between the repetitive loading and one-time loading with regard to the mechanical parameter at the failure (see Table 1). The negative value of the coefficient b means the number of cycles to failure (N_f) decrease exponentially as to the increase of the amplitude of rotation (Θ_a). This kind of exponential decrease is a typical feature found in the fatigue phenomenon, and is not found in the one-time failure phenomenon.

Several limitations were existent in this study. First, biological response could not be evaluated because cadaveric specimens were used. Second, biochemical changes, which were not evaluated in this study, may exist during the fatigue test. Third, excessive nonphysiological axial rotation was applied because the posterior elements of the spine were removed. Fourth, the specimens possessed the growth plate, which seemed to be the weakest structure for the motion unit.

Although these limitations existed in this study, mechanical and histological results suggest that prolonged exposure to cyclic torsional loads enhances the vulnerability of the spine. As a result, distortion of the stratification and rupture of the laminated bands occur in the disc. These minor failures may lead to and accelerate the degeneration of the disc. In fact, Osti et al. [12] reported that the peripheral tears of the annulus lead to secondary degenerative changes in other disc components. Putting these findings together, the fatigue phenomenon may be one of the important etiologic factors of disc degeneration and spondylosis.

Acknowledgments. Support was provided by a grant from the Ministry of Education, Science and Culture, Grant-in-Aid for Scientific Research on Priority Areas 04237106, Japan.

References

1. Lestini WF, Wiesel SW (1989) The pathogenesis of cervical spondylosis. Clin Orthop 239:69–93
2. Harada T, Ebara S, Hiroshima K, Yonenobu K, Fujiwara K, Ono K, Kajiura I, Ohkawa A (1990) The natural history of cervical disc degeneration in athetoid cerebral palsy (in Japanese). Cent Jap J Orthop Traumat 33:80–82
3. Ebara S, Harada T, Yamazaki Y, Hosono N, Yonenobu K, Hiroshima K, Ono K (1989) Unstable cervical spine in athetoid cerebral palsy. Spine 14:1154–1159
4. Ebara S, Yamazaki Y, Harada T, Hosono N, Morimoto Y, Tang L, Seguchi Y, Ono K (1990) Motion analysis of the cervical spine in athetoid cerebral palsy. Extension-flexion motion. Spine 15:1097–1103
5. Wada E, Ebara S, Saito S, Ono K (1992) Experimental spondylosis in the rabbit spine. Overuse could accelerate the spondylosis. Spine 17:S1–S6
6. Farfan HF, Cossette JW, Robertson GH, Wells RV, Kraus H (1970) The effects of torsion on the lumbar intervertebral joints: The role of torsion in the production of disc degeneration. J Bone Joint Surg [Am] 52:468–497
7. Liu YK, Njus G, Buckwalter J, Wakano K (1983) Fatigue response of lumbar intervertebral joints under axial cyclic loading. Spine 8:857–865
8. Adams MA, Hutton WC (1983) The effect of fatigue on the lumbar intervertebral disc. J Bone Joint Surg [Br] 65:199–203
9. Liu YK, Goel VK, Dejong A, Njus G, Nishiyama K, Buckwalter J (1985) Torsional fatigue of the lumbar intervertebral joints. Spine 10:894–900
10. Goel VK, Voo L-M, Weinstein JN, Liu YK, Okuma T, Njus GO (1988) Response of the ligamentous lumbar spine to cyclic bending loads. Spine 13:294–300
11. Panjabi MM (1988) Biomechanical evaluation of spinal fixation devices: I. A conceptual framework. Spine 13:1129–1134
12. Osti OL, Vernon-Roberts B, Fraser RD (1990) Anulus tears and intervertebral disc degeneration. An experimental study using an animal model. Spine 15:762–767

Part 10. Transplantation

Biomechanical Properties of Autoclaved and γ-Irradiated Bony Autografts After Replantation and Reintegration in Dogs

Paul Böhm[1], Michael Scherer[2], Hans Joachim Früh[2], and Rudolf Ascherl[3]

Summary. The reconstruction of bony defects after resection of bone tumors by extracorporeal treatment and replantation of the "sterilized" tumorous bone segments might represent an interesting alternative to common methods. Diaphyseal segments from the left tibia of 24 adult beagle dogs were resected. After extracorporeal treatment by autoclaving or γ-irradiation, the segments were replanted and fixed by plate osteosynthesis or intramedullary solid interlocking nails. During the 36 weeks, anteroposterior (AP) and lateral X-rays and scintigraphic images were done in 4-week intervals. After 36 weeks the animals were killed by means of an overdose barbiturate injection. The tibiae were resected, and the specimens were tested biomechanically after removing the osteosynthesis material. For comparison, the tests were made at the nonoperated side in the same way. Each measuring point was tested in the AP, mediolateral, and lateromedial direction because of the triangular cross-sectional shape. The X-rays and microradiographs showed bony union in 46 of the 48 osteotomy lines of the 24 transplants. A pseudarthrosis was seen only in two distal osteotomy lines of the autoclaved and nailed group. Considering the biomechanical properties, there is a reduction of stiffness to 32%–65% after replantation and reintegration of extracorporeally devitalized tumorous bone segments. In combination with plate osteosynthesis or intramedullary nailing, stability of bone seems to be sufficient because there was no fracture either during the animal study or in our six patients treated by this method to date. The experimental combination of γ-irradiation and plate osteosynthesis provides the best biomechanical results.

[1] Orthopädische Universitätsklinik Tübingen, Hoppe-Seyler-Str. 3, 72076 Tübingen, Germany
[2] Institut für Experimentelle Chirurgie der Technischen Universität München, Ismaninger Str. 22, 81675 München, Germany
[3] Klinik für Orthopädie der Medizinischen Universität Lübeck, Ratzeburger Allee 160, 23562 Lübeck, Germany

Key words: Bone tumors—Bony autografts—Extracorporeal devitalization—Autoclaving—γ-irradiation—Biomechanical properties

Introduction

Malignant primary bone tumors mostly affect children and young persons. Since introduction of the adjuvant chemotherapy in the therapy regimen of high-grade sarcomas, the disease-free and long-term survival of these patients has improved dramatically [1–4]. By means of computed tomography (CT) and magnetic resonance imaging (MRI), it is possible to determine the exact extent of the tumor preoperatively with high precision [5–7]. In consequence, there have been an increasing number of limb salvage operations in the therapy of malignant primary bone tumors [1,2,4].

When reconstructing bony defects by means of endoprostheses following resection of the tumors, we have to expect considerable subsequent problems. The most important following complications are aseptic loosening, infection, and fatigue fracture. At 80 months postoperatively, between 35% [8] and 60% [9] of tumor prostheses must be removed. Thus, we must expect an increasing number of secondary amputations in the following years. Taking into account the increasing number of young patients who will survive disease-free, there is an intensified demand for long-lasting biological reconstruction methods.

Apart from the autogenous transplants, which are available only in limited size and number, some centers have extensive experience with allografts. These bone banks require a lot of time, energy, and money [10]. There are medical problems, also, such as a high risk for local infection [11,12] or the risk of transferring viral infections [13].

When the tumorous bone shows no considerable loss of mechanical stability, replantation of the tumorous bone segment after extracorporeal devitalization by means of autoclaving [14–20] or γ-irradiation [21–23] may be an interesting alternative reconstruction method. We tested the possibilities of this reconstruction technique in an experimental study. The results of one essential aspect, the biomechanics of the replanted bone after healing, are presented in this article. The influence of the type of devitalization and the kind of osteosynthesis technique on healing and biomechanics were investigated in particular.

Materials and Methods

Following the approval of the Bavarian Laender-state government, 25-mm-long diaphyseal segments of the left tibia of 24 adult beagle dogs were resected under general anesthesia. After extracorporeal treatment by autoclaving (134°C, 3 min) or γ-irradiation (5000 Gy of ^{60}Co), the bony segments were replanted orthotopically. Stabilization was carried out by plate osteosynthesis or solid interlocking nails.

The animals were grouped according to the type of extracorporeal treatment and the type of osteosynthesis:

AP, autoclaving and plate osteosynthesis
AN, autoclaving and nailing
IP, γ-irradiation and plate osteosynthesis
IN, γ-irradiation and nailing

During the 36 weeks of observation anteroposterior (AP) and lateral X-rays and three-phase scintigraphies using the ROI technique were done at 4-week intervals. After 36 weeks, the animals were killed by means of an overdose pentobarbital injection and the tibiae of each dog were resected on both sides respectively. The osteosynthesis material was then removed.

The biomechanical tests were carried out in a nondestructive manner so as not to diminish the meaningfulness of the following histologic examination. Stiffness and the extent of bending between 5 and 25 newton (N) were tested by means of a three-point bending-test in the middle of the transplant and at the proximal and distal osteotomy lines, respectively. The speed of the testing stamp of the TZZ-707 Wolpert universal testing machine was adjusted to 5 mm/min. The maximum load was limited to 25 N. Because of the almost triangular cross-sectional shape, the position of the bone was stable without additional fixation (Fig. 1). To be able to define stability as accurately as possible, each measuring point was tested in each of the three possible supporting surfaces, that is, in the anteroposterior, mediolateral, and lateromedial direction. Each time, three endurance tests were made and the third curve was evaluated. For comparison, the tibiae of the nonoperated right side were tested in the same manner. Thus, a total of 1296 (24 × 2 × 3 × 3 × 3) measuring curves were drawn.

Fig. 1. Three-point-bending test of a tibia after removal of the AO 2.7-mm DC plate. The specimen is tested in the anteroposterior direction. Manufacturer: Zwick, Ulm, Germany

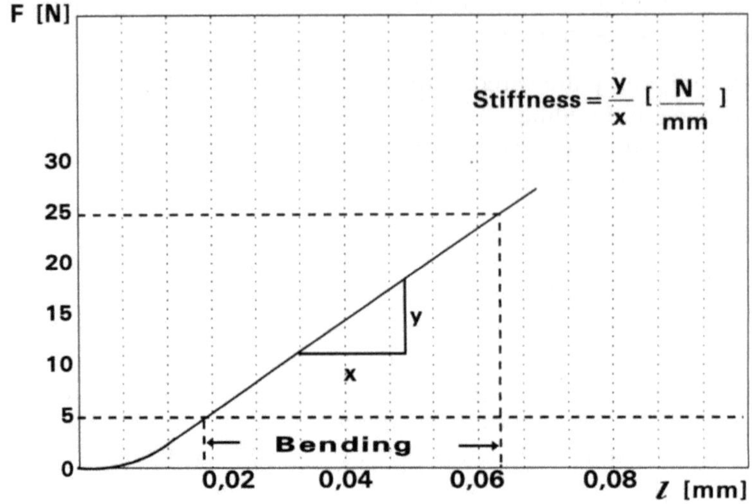

Fig. 2. Evaluation biomechanics: bending and stiffness are determined graphically. To ascertain the extent of bending, a reference line at 5 N parallel to the x-axis is drawn. Bending is then defined as the distance between the point of intersection of the measuring curve with the reference line and the point of intersection of the perpendicular dropped from the measuring curve at 25 N with the reference line. Stiffness is defined as gradient in the force-way diagram as y/x (N/mm). The values are also determined graphically by means of the measure curves

To ascertain the extent of bending, a reference line was drawn at 5 N parallel to the x-axis. Bending was then defined as the distance between the point of intersection of the force diagram with the reference line and the point of intersection of the perpendicular dropped from the diagram at 25 N with the reference line (Fig. 2).

Stiffness was ascertained graphically by means of the gradient in the force diagram according to the formula: stiffness = y/x (N/mm).

Results

There was no local infection. One animal of group IN died during the 8-week control scintigraphy period following a narcosis accident and had to be replaced by another dog. Forty-six of the 48 osteotomies of the 24 transplants showed bony union. After plating there was a direct bone healing according to the definition of Perren [24] in all cases. This could be confirmed both radiologically and histologically. After nailing regularly, we saw indirect bone healing. Two pseudarthroses developed in the group AN. Revascularization and revitalization of the transplants were documented by microangiographies and sequential fluorescent stained histological slides.

Analysis of the results of the biomechanical tests had to take into account the high extent of the SD. The extreme values of the SD, which amount to

111% of the mean bending and to 89% of the mean stiffness, however, are found in the group AN with the two pseudarthroses only. No tendency to weakening in a particular direction of loading could be ascertained either in the proximal osteotomy or in the transplant or in the distal osteotomy.

The results of bending and stiffness are indicated in Tables 1 and 2, respectively. Because of greater accuracy there are no absolute results but relative results are indicated in percent compared to the nonoperated right control side.

Table 1. Bending[a]

		Anteroposterior	Mediolateral	Lateromedial
AN	p	312 ± 228	295 ± 154	263 ± 64
	t	453 ± 505	291 ± 244	590 ± 656
	d	236 ± 103	257 ± 145	219 ± 129
AP	p	376 ± 337	449 ± 230	228 ± 60
	t	278 ± 205	185 ± 94	371 ± 235
	d	368 ± 163	259 ± 208	229 ± 143
IN	p	201 ± 43	283 ± 153	176 ± 76
	t	229 ± 109	132 ± 40	243 ± 90
	d	197 ± 86	241 ± 151	198 ± 157
IP	p	323 ± 215	395 ± 155	353 ± 186
	t	157 ± 98	138 ± 100	266 ± 142
	d	253 ± 165	293 ± 273	141 ± 65

AN, autoclaving and nailing; AP, autoclaving and plate osteosynthesis; IN, γ-irradiation and nailing; IP, γ-irradiation and plate osteosynthesis; p, proximal osteotomy; t, transplant; d, distal osteotomy.

[a] Relative values in percent compared to the nonoperated control side. Indicated is the mean value ± SD:

$$\frac{\text{Bending operated side}}{\text{Bending control side}} \times 100 \pm \text{standard deviation } [\%]$$

Table 2. Stiffness[a]

		Anteroposterior	Mediolateral	Lateromedial
AN	p	57 ± 51	32 ± 11	34 ± 13
	t	35 ± 26	39 ± 34	22 ± 16
	d	41 ± 12	35 ± 14	45 ± 22
AP	p	47 ± 20	31 ± 12	47 ± 14
	t	42 ± 16	55 ± 31	36 ± 16
	d	34 ± 12	62 ± 51	49 ± 21
IN	p	49 ± 12	31 ± 11	51 ± 20
	t	45 ± 13	57 ± 11	38 ± 11
	d	46 ± 16	47 ± 26	56 ± 39
IP	p	50 ± 15	34 ± 18	34 ± 12
	t	86 ± 55	67 ± 31	43 ± 24
	d	58 ± 33	48 ± 18	76 ± 52

[a] Relative values in percent compared to the nonoperated control side. Indicated is the mean value ± standard deviation:

$$\frac{\text{Stiffness operated side}}{\text{Stiffness control side}} \times 100 \pm \text{standard deviation } [\%]$$

The order of magnitude of the absolute extreme values of the transplant region of the right reference side are indicated by these data:

Bending: Minimum, 0.0370 mm (IN/AP)
Maximum, 0.0852 mm (IP/mediolateral)
Stiffness: Minimum, 440 N/mm (AP/mediolateral)
Maximum, 722 N/mm (IN/lateromedial)

The means of the three directions of loading are shown graphically for bending (Fig. 3) and for stiffness (Fig. 4). The figures show that the biomechanical tests resulted in an increase of bending two- to fourfold that of the nonoperated side. Stiffness was reduced to 32%–65% of the control side.

With regard to biomechanics, the irradiation-plating (IP) group, with an average relative bending in the transplant region of 187% and a relative stiffness of 65%, is the best group. In order of ranking, it is followed by the IN, AP, and AN groups. The corresponding biomechanical data for the poorest group (AN) amount to 445% for bending and 32% for stiffness.

Discussion

The extracorporeal heat treatment and high-dose γ-irradiation in combination with the following replantation of bone has, especially in the treatment of bone tumors, some decisive advantages in comparison with other reconstruction methods. In this biological reconstruction, the main problems of tumor endoprostheses, aseptic loosening, and late infection, are not to be feared in the long term. In comparison to the transplantation of allografts, there is no

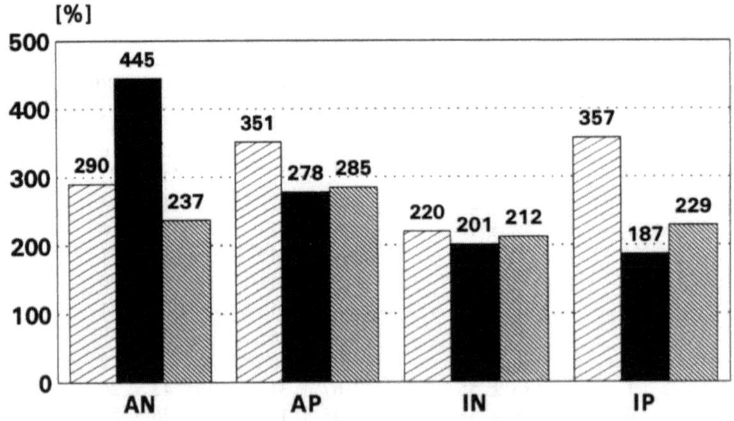

Fig. 3. Bending: the means of the three directions of loading are presented graphically. The values are indicated as percent of the nonoperated control side. *Light hatching*, proximal osteotomy; *heavy stippling*, transplant; *medium hatching*, distal osteotomy

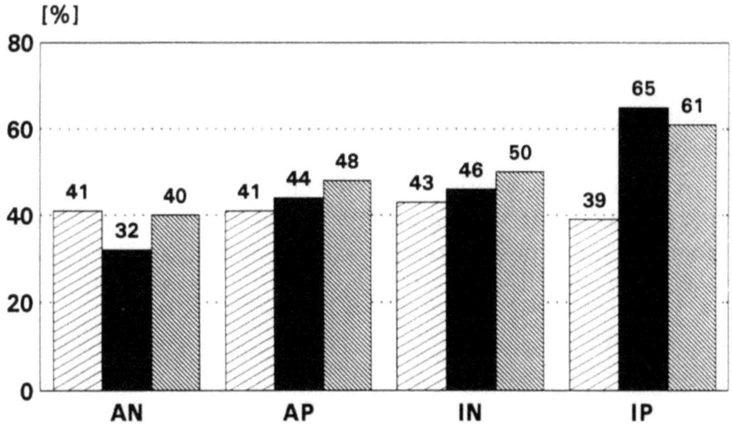

Fig. 4. Stiffness: the means of the 3 directions of loading are presented graphically. The values are indicated in per cent to the nonoperated control side. *Light hatching*, proximal osteotomy; *heavy stippling*, transplant; *medium hatching*, distal osteotomy

danger of transferred bacterial and viral infections. A prerequisite of the replantation of extracorporeal devitalized bone is a sufficient primary stability of the bone. The main problems of the method are to achieve a bony healing and a sufficient long-term stability. Concerning the load-bearing capacity of bone, we have to assume that there will be a considerable reduction in relation to the other side even if there is a good healing of the transplant. Taking into account the bending and stiffness in the middle of the transplant, irradiation is superior to autoclaving and plate osteosynthesis is superior to nailing.

The measured values are a result of the overlapping of several factors. These factors are the primary weakening of the transplant by devitalization, the strength of the transplant-bed connection, the influence of rigid internal fixation [25,26], and the changes of the biomechanics of the transplant and bed by means of the remodeling, inasmuch as the results cannot be compared with biomechanical data that only take into consideration the method of devitalization and not the biologic reactions [27,28]. With regard to long resections in humans, we may have to reckon with an even greater reduction of the biomechanical parameters of stability.

Regarding the dynamics of bone remodeling, we expect in the starting phase an initial reduction of bone and later a reconstruction. In this first phase it is inevitable that a further weakening of the transplant will result. The chronological course of the changing mechanical stability of bone transplants is yet to be investigated in animal experiments. In these experiments the initial reduction of the stability parameters and the following reincrease could be confirmed. Apart from radiological observation [29,30], there are also biomechanical data. According to these results, we can expect the minimum stability between 1.5 and 6 months after transplantation of nontreated autogenous transplants. After

1 year, the primary stability is nearly achieved [31]. Ku and coworkers found 50% of stiffness relative to the control side in autoclaved transplants of sheep after 12 weeks. One year postoperatively, there was no difference between the operated and the nonoperated side [32].

The hypothesis of a loss of mechanical stability that is only temporary is also confirmed by clinical observations. Berrey and coworkers found fractures in 16% of massive allografts. The majority of the fractures took place between the second and the third postoperative year. By more than 4 years after surgery there was no longer any fracture [33].

Conclusions

As a result of our study, we can conclude:

1. The replantation of extracorporeal devitalized tumorous bone segments is a practicable reconstruction method in limb salvage surgery.
2. Bony union can be achieved by adequate osteosynthesis and will be followed by slow revitalization.
3. The combination of γ-irradiation and plate osteosynthesis provides the best biomechanical results in our experiments.
4. The reduced biomechanical properties after replantation and reintegration of the extracorporeal treated bone seem to ensure enough stability in combination with adequate osteosynthesis, because there were no fractures during the animal study nor in our six patients treated by this method thus far.

References

1. Eilber FR, Eckhardt J, Morton DL (1990) Advances in the treatment of sarcomas of the extremity. Current status of limb salvage. Cancer 54:2695–2701
2. Winkler K (1990) Diagnose und Therapie maligner Knochen- und Weichteiltumoren. Dtsch Ärztebl 87:224–225
3. Goorin AM, Andersen JW (1991) Experience with multiagent chemotherapy for osteosarcoma. Clin Orthop 270:22–28
4. Glasser DB, Lane JM (1991) Stage IIB osteosarcoma. Clin Orthop 270:29–39
5. Sundaram M, McGuire MH, Herbold DR, Wolverson MK, Heiberg E (1986) Magnetic resonance imaging in planning limb-salvage surgery for primary malignant tumors of bone. J Bone Joint Surg 68A:809–819
6. Böhm P, Yousry T, Kraus B (1991) Role of magnetic resonance imaging in the radiological management of musculoskeletal tumors. Arch Orthop Trauma Surg 110:127–131
7. Szendröi M, Antal I, Liszka G, Monostori Z (1993) Examination and evaluation of surgical margins in bone tumours. Int Orthop 17:93–97
8. Sim FH, Beauchamp CP, Chao EYS (1987) Joint arthrodesis. In: Coombs R, Friedlaender G (eds) Bone tumour management. Butterworths, London, pp 102–113

9. Salzer M, Knahr K, Kotz R, Ramach W (1987) Cementless prosthetic implants for limb salvage of malignant bone tumors of the knee joint: clinical results of three modes of anchorage. In: Enneking WF (ed) Limb salvage in musculoskeletal oncology. Churchill Livingstone, New York, pp 26–29
10. Wagner M, Pesch HJ (1989) Autoklavierte Knochenspäne beim Prothesenwechsel an der Hüfte. Orthopäde 18:463–467
11. Mankin HJ (1984) Allograft transplantation in the management of bone tumors. In: Uhthoff HK, Stahl E (eds) Current concepts of diagnosis and treatment of bone and soft tissue tumors. Springer, Berlin Heidelberg New York Tokyo, pp 147–162
12. Delloye C, de Nayer P, Allington N, Munting E, Coutelier L, Vincent A (1988) Massive bone allografts in large skeletal defects after tumor surgery: a clinical and microradiographic evaluation. Arch Orthop Trauma Surg 107:31–41
13. Simonds RJ, Holmberg SD, Hurwitz RL, Coleman TR, Bottenfield S, Conley LJ, Kohlenberg SH, Castro KG, Dahan BA, Schable CA, Rayfield MA, Rogers MF (1992) Transmission of human immunodeficiency virus type 1 from a seronegative organ and tissue donor. N Engl J Med 326:726–732
14. Orell S (1937) Surgical bone grafting with "os purum," "os novum," and "boiled bone." J Bone Joint Surg 19:873–885
15. Thompson VP, Steggall CT (1956) Chondrosarcoma of the proximal portion of the femur treated by resection and bone replacement. J Bone Joint Surg 38A:357–367
16. Harding RL (1957) Replantation of the mandible in cancer surgery. Plast Reconstr Surg 19:373–383
17. Harrington KD, Johnston JO, Kaufer HN, Luck JV, Moore TM (1986) Limb salvage and prosthetic joint reconstruction for low-grade and selected high-grade sarcomas of bone after wide resection and replacement by autoclaved autogeneic grafts. Clin Orthop 211:180–214
18. Sijbrandi S (1986) Resektion von Knochentumoren im Kniegebiet und Rekonstruktion mit gekochtem Autotransplantat. Spätergebnisse. In: Blauth, Ulrich (Hrsg) Spätergebnisse in der Orthopädie. Springer, Berlin Heidelberg New York
19. Lauritzen C, Alberius P, Santanelli F, Vällfors B, Lilja J, Stephensen H (1991) Repositioning of craniofacial tumorous bone after autoclaving. Scand J Plast Reconstr Hand Surg 25:161–165
20. Böhm P (1991) Kniegelenksarthrodese nach Autoklavierung und Replantation des tumortragenden Knochens. In: Böhm P, Küsswetter W (Hrsg) Maligne primäre Knochentumoren. Thieme, Stuttgart, pp 55–61
21. Spira E, Brenner HT, Lubin E (1970) Extracorporeal irradiation of malignant bone tumors. In: Chapchal G (ed) Operative treatment of bone tumors. Thieme, Stuttgart, pp 136–140
22. Uyttendaele D, de Schryver A, Claessens H, Roels H, Berkvens P, Mondelaers W (1988) Limb conservation in primary bone tumours by resection, extracorporeal irradiation and reimplantation. J Bone Joint Surg 70B:348–353
23. Hamaker RC (1981) Irradiated autogenous mandibular grafts in primary reconstructions. Laryngoscope 91:1031–1051
24. Perren SM (1992) Biomechanische Grundlagen der Frakturheilung. Orthopäde 21:3–10
25. Laftman P, Sigurdsson F, Strömberg L (1980) Recovery of diaphyseal bone strength after rigid internal plate fixation. Acta Orthop Scand 51:215–222
26. Strömberg L (1975) Diaphyseal bone in rigid internal plate fixation. Acta Chir Scand [suppl 456]:1–34

27. Kreicbergs A, Köhler P (1989) Bone exposed to heat. In: Aebi M, Regazzoni P (eds): Bone transplantation. Springer, Berlin Heidelberg, pp 155–161
28. Früh HJ, Voggenreiter G, Ascherl R, Scherer MA, Siebels W (1990) Zur Biomechanik autoklavierter, bestrahlter und kältekonservierter Corticalis. Hefte Unfallheilkd 212:681–682
29. Schweiberer L (1971) Der heutige Stand der Knochentransplantation. Chirurg 42:252–257
30. Jahn K (1980) Die Einheilung allogener Compacta nach Vorbehandlung des Empfängers mit Knochengewebe (im Tierversuch). Nova Acta Leopold [suppl 12]:1–240
31. Enneking WF, Burchardt H, Puhl JJ, Piotrowski G (1975) Physical and biological aspects of repair in dog cortical-bone transplants. J Bone Joint Surg 57A:237–252
32. Ku JL, Smith PA, Goldstein SA, Matthews LS (1985) An experimental investigation of the fate of autogenous autoclaved bone grafts. Orthop Trans 9:340
33. Berrey BH, Lord F, Gebhardt MC, Mankin HJ (1990) Fractures of allografts. J Bone Joint Surg 72A:825–833

Effects of Constant Load on Autograft Healing Compared with Those of Cyclic Load

TAKESHI MUNETA[1], SHUNICH MURAKAMI[1], YOUICHI EZURA[1], SHINTARO ASAHINA[1], KAZUO TAKAKUDA[2], HARUYASU YAMAMOTO[1], and KOHTARO FURUYA[1]

Summary. The rabbit bone-to-bone (BTB) model has been developed to investigate the effects of controlled load on the graft healing process. The previous rabbit BTB model was modified using a NiTi spring to investigate the effects of controlled constant load on autograft healing in this study. The results were evaluated biomechanically 6 weeks after implantation. The modulus of the constant-loading group was significantly higher than that of the unloaded control, and there was a significant difference in the linear stiffness between the constant-loading and the cyclic-loading group. The linear stiffness of the cyclic-loading group was higher than that of the constant-loading group. In conclusion, constant loading is effective to prevent graft tissue weakening in the regeneration process. However, the effectiveness of constant tension was inferior to that of the cyclic tension.

Key words: ACL reconstruction—Rabbit model—Controlled load—Mechanical test—Cyclic load—Constant load

Introduction

Autogenous tissue for anterior cruciate ligament (ACL) reconstruction, regardless of type, has been observed to follow a similar healing process of regeneration and maturation to reach the natural ACL-like structure. Animal studies, however, have shown that the mechanical strength of a reconstructed ACL decreases within 3 months of grafting and never regains normal strength. The mechanical strength becomes as much as 50% of the normal control value

[1]Department of Orthopedic Surgery and [2]Institute for Medical and Dental Engineering, Tokyo Medical and Dental University, 1-5-45 Yushima, Bunkyo-ku, Tokyo, 113 Japan

[1–3]. If a means to prevent graft weakening could be developed, it would be valuable for improvements of ACL reconstruction. If graft weakening could be prevented in the natural healing process by making mechanical modifications to the graft, application of such a method should be clinically considered in terms of modified postoperative management [4,5].

We have developed a rabbit model using a patellar tendon bone-to-bone (BTB) specimen to investigate the effects of controlled load on graft healing in the subcutaneous space. A preliminary report suggested that a manually applied cyclic load could significantly reduce graft weakening [6].

The objective of this study is to develop a rabbit BTB model to investigate the effects of a constant load on graft healing by modifying the rabbit model previously reported, and to compare the effects of constant load with those of cyclic load in terms of mechanical properties.

Materials and Methods

Six mature male New Zealand White (NZW) rabbits (average weight, 3.5 kg) were used for the study of constant load. The whole BTB specimen was removed from the right knee using an anterolateral straight incision. The patellar tendon was cut in the middle along the entire length including the tibial attachment into the same halves, and each side of the tendon was trimmed to a 3-mm width. The center of the patella and the medial tubercle were drilled using a 1.5-mm-diameter Kirschner wire. The BTB specimen was installed in an aluminum open rectangular frame with inner dimensions of 3 cm × 7 cm × 3 cm. The specimen was fixed to the frame using #0 multifilament nylon sutures through the drill holes (Fig. 1).

A 2 N constant tensile force was applied to the medial portion of the patellar tendon using a TiNi spring (Sentalloy, Super Elastic Nickel Titanium Alloy, Tomy International, Tokyo, Japan) that was positioned between the medial tubercle and the frame using sutures. The NiTi spring could exert a constant 2 N tension over an elongation of 5–15 mm from the original length. The frame with the BTB specimen was located in a subcutaneous pouch though a 4-cm straight incision on the back. The lateral half of the patellar tendon was used as an unloaded control.

In five NZW rabbits used for cyclic loading, a randomly selected site in the BTB was stretched manually using a calibrated spring scale with a peak force of 10 N at 60 cycles for 5 min three times per week. Six weeks after implantation, all the rabbits were killed with excess pentobarbital given intravenously. The BTB specimen including surrounding tissue was removed and the surrounding tissue was then carefully removed before the mechanical test.

In the constant-loading group, the patella as well as the medial and lateral sites of patellar tendon was potted with dental methylemethacrylate into a 9-mm-diameter cylindrical shape, then installed on the Instron 1185 materials testing machine (Instron Corp., Canton, MA, USA). Before the tensile test,

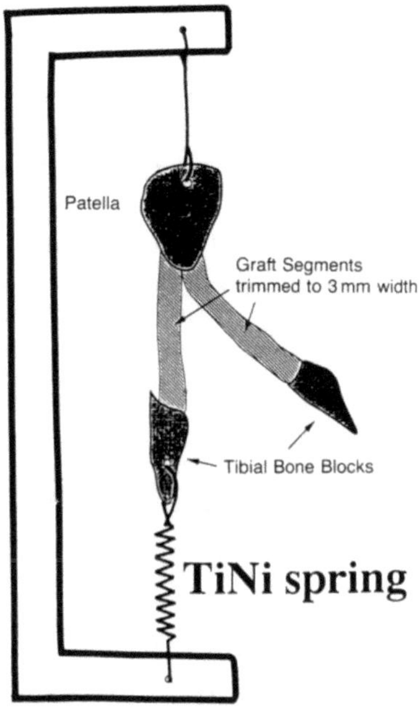

Fig. 1. Constant tensioning method. The whole patellar tendon bone-to-bone (BTB) specimen was removed from the right leg. The patellar tendon and the tibial tubercle were cut in the middle to form identical halves. The medial and lateral portions of the patellar tendon were trimmed to a 3-mm width. The medial portion of the BTB specimen was fixed in an aluminum frame with sutures through each hole in the patella and the medial tibial tubercle. A NiTi spring was positioned between the medial tibial tubercle and the frame to apply a 2 N constant load onto the medial side of the patellar tendon

the central cross-sectional area of the patellar tendon was measured with a specially made cross-section measurement device according to Butler et al. [8].

The distance between the fixation sites of the patella and tibia was measured at the position of minimum tension. After stretching the tissue five times from 0 to 20 N at a speed of 5 mm/min, the BTB specimen then reached failure at a speed of 1 mm/s and the failure mode was recorded.

The following parameters were statistically analyzed using a nonpaired *t*-test between the loaded and unloaded group: tendon cross-sectional area, modulus at 20 N from the repeated stretching curve, modulus in the linear portion, linear stiffness, and maximum load from the load-displacement curve. A P value of <0.05 was considered significant. In the cyclic-loaded group, the BTB specimen was stretched to failure on a MTS material testing machine (Model 858 Bionix Test System, MTS Systems, Minneapolis, MN, USA) with the linear stiffness and maximum load recorded from the load displacement curve. Statistical analysis was carried out using the *t*-test for linear stiffness and maximum load between the constant-loaded group and the cyclic-loaded group ($P < 0.05$).

Results

All specimens failed in midsection in the tensile test. The cross-sectional area and modulus at 20 N in the constant-loaded group and control are shown in

Table 1. The modulus in the linear portion of the load-displacement curve, linear stiffness, and maximum load are also shown in Table 1.

Statistical analysis indicated that the modulus at 20 N in the constant-loaded group was significantly higher than in the unloaded control. The modulus in the linear portion for the constant-loaded group showed a significant difference with a P value of 0.07.

A comparison between the constant-loaded and cyclic-loaded groups showed a significant difference in linear stiffness (Fig. 2) but not in maximum load (Fig. 3). The linear stiffness in the cyclic-loaded group was higher than that in the constant-loaded group.

Table 1. Biomechanical data of the constant tension group

Parameter	Constant tension group	Untensioned control
Cross-sectional area (mm^2)	3.2 ± 0.8	2.9 ± 0.8
Modulus at 20 N (MPa)	361.2 ± 76.9[a]	205.6 ± 86.7
Modulus at linear portion (MPa)	437.3 ± 129.3	275.4 ± 149.1
Linear stiffness (N/mm)	55.6 ± 14.1	39.3 ± 25.4
Maximum load (N)	92.1 ± 32.5	66.5 ± 48.0

[a] There was a statistically significant difference between the loaded group and the control ($P < 0.05$).

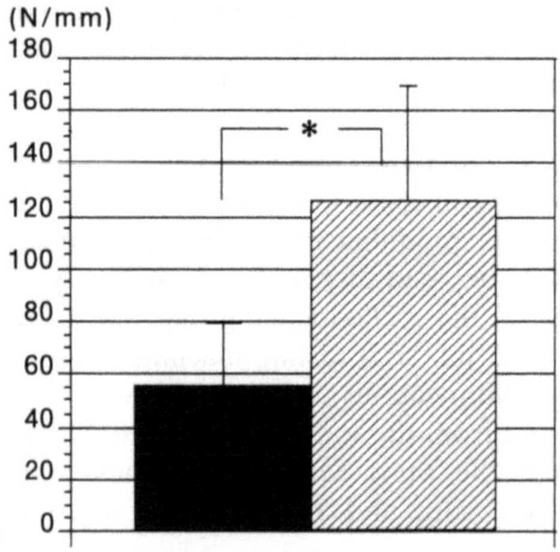

2

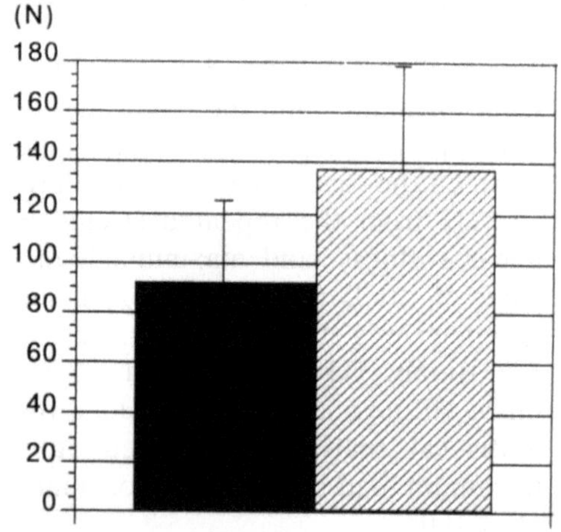

3

Fig. 2. The unpaired t-test showed that there was a significant difference in the linear stiffness between the constant loading group (*solid bar*) and the cyclic loading group (*hatched bar*)

Fig. 3. There was no statistical difference in the maximum load between the constant loading group (*solid bar*) and the cyclic loading group (*hatched bar*)

Discussion

This experimental model has some limitations concerning the state of graft healing and graft fixation if the results of the current study are directly applied to ACL graft remodeling. Graft remodeling was investigated subcutaneously using the model when an ACL graft was present in the synovial joint. The graft in this study was not fixed to any structure in the subcutaneous space, while an ACL graft is fixed in the bone tunnel. However, the main objective of this study was to investigate the effects of controlled load not only on ACL but on general soft tissue remodeling. In terms of realizing "controlled" load, the model is thought to be unique and of experimental value.

The results of this study suggested that a constant load shows some effect in preventing strength reduction of soft tissue during the healing process. The results also suggest that grafted tissue that bears some tension at initial fixation will result in a stronger structure during healing. There are some differences in the effects of constant load on the remodeling process between bone and soft tissue. Rubin et al. [9] reported that constant load had no effect on grafted bone healing in terms of mechanical strength compared to that of cyclic load. However, the study left some questions for future investigation, such as: what is the optimum initial load to preserve the best graft strength during the remodeling process? How does a variety of loads affect the remodeling process?

The model in this study has some advantages in terms of effects of load on graft remodeling over a long period. In most studies that have investigated the effects of load on soft tissue healing, the load was controlled only in the initial stage or during fixation of the tissue [10–14]. In those models, it would have been impossible to control the load on tissue as time went on. The model used in this study should be a good model because it actually exerts stress on tissues. However, the problem might be that the loading pattern is nonphysiological.

It is assumed that cyclic loading is more effective in preventing strength reduction of the graft during the healing process than constant loading in terms of loading history. If the cross-sectional area of the natural ACL is 60 mm^2 when it is 30 mm long, and the cross-sectional area of the rabbit BTB that was used in this study was 3 mm^2 with a length of 15 mm, the volume of the human ACL is 40-fold greater than that of the rabbit BTB. A 10 N force for the rabbit BTB used in the cyclic-loading model would be equal to the force experienced during running in the natural model ACL. A 2 N force for the rabbit BTB used in the constant-loading model indicates a slightly lower force than that exerted on the natural ACL during level walking. Both loads used in this study were considered to be within the physiological range.

Conclusion

A constant load can prevent weakening of soft tissue during healing; however, the effectiveness is less than that observed in cyclic loading.

References

1. Alm A, Gillquist J (1974) The anterior cruciate ligament—a clinical and experimental study on tensile strength, morphology and replacement by patellar ligament. Acta Chir Scand 445(suppl):1–49
2. Ballock RT, Woo SL-Y, Lyon RM, Hollis JM, Akeson WH (1989) Use of patellar tendon autograft for anterior cruciate ligament reconstruction in the rabbits—a long-term histologic and biomechanical study. J Orthop Res 7:474–485
3. Butler DL, Grood ES, Noyes FR, Olmstead ML, Hohn RB, Arnoczky ST, Siegel MG (1989) Mechanical properties of primate vascularized vs. nonvascularized patellar tendon graft—changes over time. J Orthop Res 7:68–79
4. Clancy WG, Narechania R, Rosenberg T, Gmeiner J, Wisnefske D, Lange T (1981) Anterior and posterior cruciate ligament reconstruction in rhesus monkeys—a histological, microangiographic, and biomechanical analysis. J Bone Joint Surg 63A:1270–1284
5. Muneta T, Yamamoto H, Takakuda K, Sakai H, Furuya K (1993) Effects of postoperative immobilization on reconstructed anterior cruciate ligament in rabbits. Am J Sports Med 21:305–313
6. Woo SL-Y, Inoue M, McGurk-Burleson E, Gomez MA (1987) Treatment of the medial collateral ligament injury 2: Structure and function of canine knees in response to different treatment regimens. Am J Sports Med 15:22–29
7. Muneta T, Lewis JL, Stewart NJ (1992) A rabbit model for assessing effects of controlled load on graft healing. Trans ORS 17:672
8. Butler DL, Matthew DK, Stouffer DC (1986) Comparison of material properties in fascicle-bone units from human patellar tendon and knee ligaments. J Biomech 19:425–432
9. Rubin CT, Lanyon LE (1987) Osteoregulatory nature of mechanical stimulifunction as a determinant for adaptive remodeling in bone. J Orthop Res 5:300–310
10. Bush-Joseph CA, Buseck MS, Cumming JF, Grood ES, Noyes FR, Bylski-Austrow DI, VanGinkel LA (1990) The healing response and structural properties of the ACL following freezing and repositioning of the tibial insertion. Trans JOR 15:59
11. King G, Edwards P, Frank C (1991) Rabbit medial collateral ligament laxity normalizes during early healing of loose or tight autograft. Transactions of combined meeting of the ORS of USA, Japan, and Canada, p 35
12. Majima T, Yasuda K, Yamamoto N, Hayashi K, Kaneda K (1992) Effects of stress shielding on the mechanical properties of in situ frozen patellar tendon augmented with polyester artificial ligament. Trans ORS 17:674
13. Ohno K, Yasuda K, Yamamoto N, Kaneda K, Hayashi K (1993) Effects of complete stress-shielding on the mechanical properties and histology of in situ frozen patellar tendon. J Orthop Res 11:592–602
14. Yoshiya S, Andrish JT, Manley MT, Bauer TW (1987) Graft tension in anterior cruciate ligament reconstruction—an in vivo study in dogs. Am J Sports Med 15:464–470

Biomechanical Evaluation of Extraarticularly Transplanted Patellar Tendon Grafts in the Rat: Fresh Versus Frozen Grafts

Akira Maeda[1], Konsei Shino[2], Shuji Horibe[1], Ken Nakata[1], Norimasa Nakamura[1], Yukiyoshi Toritsuka[1], and Yoshiki Shiozaki[1]

Summary. The purpose of this study was to evaluate the effect of freezing on the mechanical properties of rat patellar tendon autografts after transplantation. The medial half of the patellar tendon with the tibial bone block was harvested from each knee of 28 sixteen-week-old male Lewis rats. Two experimental models were designed: in the fresh autograft model, grafts harvested from the rats were immediately transplanted to other Lewis rats (28 knees of 14 rats) in an extraarticular location; in the frozen autograft model, the grafts were frozen, then thawed and transplanted to other Lewis rats (28 knees of 14 rats). Normal patellar tendons of 16-week-old Lewis rats were used as a control ($n = 16$). The recipient rats were killed at 4, 8, 12, and 24 weeks after transplantation for tensile-to-failure tests. Only 59 of 72 specimens that failed at their midsubstance were used for data analysis. At 4 weeks, there was no significant difference in the mechanical properties between the fresh and the frozen grafts. The fresh grafts showed higher tensile strength than the frozen grafts at 12 and 24 weeks after transplantation. At 24 weeks, the mean tensile strength of the fresh and the frozen groups were 44.2 MPa (80% of the control) and 30.5 MPa (55% of the control), respectively. These results suggest that freezing had an adverse effect on mechanical changes of the extraarticularly transplanted grafts.

Key words: Autograft—Deep-freezing—Ligament reconstruction—Mechanical properties—Patellar tendon

Introduction

Ligament reconstruction has recently been performed by using fresh-frozen allografts as well as fresh autografts [1,2]. The allografts are frozen and then

[1] Department of Orthopaedic Surgery, Osaka University Medical School, 2-2, Yamada-oka, Suita, Osaka, 565 Japan
[2] Department of Orthopaedic Sports Medicine, Osaka Rosai Hospital, 1179-3, Nagasone, Sakai, Osaka, 591 Japan

thawed before transplantation to reduce their immunogenicity. The differences in the remodeling process between fresh autografts and frozen allografts have been reported previously. Shino et al. [3] compared fresh patellar tendon autografts with fresh-frozen patellar tendon allografts in a canine anterior cruciate ligament (ACL) reconstruction model. They showed that there were no obvious differences between the two types of grafts in terms of mechanical properties. Sabiston et al. [4,5] showed the differences between frozen allografts and fresh autografts in a rabbit medial collateral ligament (MCL) model. Frozen allografts were inferior to fresh autografts in terms of mechanical properties up to 48 weeks posttransplantation. Jackson et al. [6] showed that fresh autografts had a larger maximum force to failure than did frozen allografts at 6 months after transplantation in a goat ACL model. However, the question remains unresolved as to whether these differences resulted from different materials (e.g., autografts or allografts) or different preservation methods (e.g., fresh or frozen) or both factors.

To clarify the effect of freezing, in this study we compared the mechanical properties of frozen-thawed autografts to those of fresh autografts using the rat patellar tendon transplantation model.

Materials and Methods

Experimental Design

Sixteen-week-old male inbred Lewis rats (LEW/crj, from Charles River, Kanagawa, Japan) were used. Two experimental groups were designed (Fig. 1): in the fresh model, grafts from Lewis rats were transplanted into other Lewis rats immediately after graft harvesting; in the frozen model, grafts were harvested from Lewis rats and stored at −80°C for 2–5 weeks. The grafts were then thawed and transplanted into other Lewis rats.

Graft Harvesting

The medial halves of the patellar tendons with tibial bone strips were harvested from 16-week-old male Lewis rats as donor tendons. The bone blocks were 4 mm wide and 8 mm long, and the tendon grafts were 2.5 mm wide and 8 mm long. The grafts were harvested from each knee of 28 Lewis rats for a total of 56 grafts. Before transplantation, the cross-sectional area of 4 grafts of each group for each posttransplantation period was measured.

Transplantation Procedures

Transplantation was done under general anesthesia. All the transplantation procedures were performed aseptically. In each recipient rat, defects matching the grafts were made in the medial half of patellar tendon and the tibia

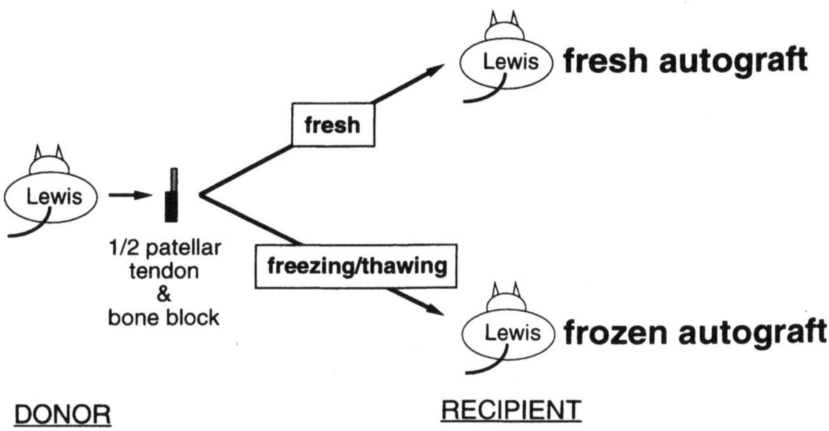

Fig. 1. The scheme of the experimental design

bilaterally. The graft was then placed in the defect with the distal bone block fixed with a 0.7-mm Kirschner wire. The proximal and lateral ends of the graft were sutured with 6-0 nylon thread. The grafts harvested from the right knee were transplanted into the right knee of the recipients, and the grafts from the left were transplanted into the left. After transplantation, the animals were allowed free cage activity.

Mechanical Evaluation

The recipient rats were killed at 4, 8, 12, and 24 weeks after transplantation. The medial half of the patellar tendon, which was transplanted from the donor, was harvested with the patella and the tibia. The specimen was cleared of all other structures except the patella–graft–tibia complex. Sixteen fresh patella–tendon–tibia complexes from eight 16-week-old male Lewis rats were also prepared and were used as normal controls. Initially, the cross-sectional area of the transplanted tendons, of which the cross-sectional area had been already measured before transplantation, were measured to determine the changes. The approximately 1-mm-wide midportions of the grafts or normal tendons were used for tensile-to-failure testing. The residual portions of the grafts were available for histological evaluation. The cross-sectional area of the specimens was measured by the method reported by Butler et al. [7]. The graft was first placed in a 1-mm slot and compressed at 0.12 MPa. The cross-sectional area was defined by multiplying the width of the slot by the thickness of the tendon after 120 s of compression. Measurement was repeated twice at the center of the specimen. The resulting two sets of data were then averaged.

Proximal suture site and distal tendon insertion were reinforced with chemical glue (cyanoacrylate, Aron Alpha, Toagosei Chemical Industry, Tokyo, Japan) to avoid failure at the suture or insertion sites. The dimensions of the midsubstance of the specimen were approximately 5 mm long and 1 mm wide, and the aspect ratio was about 1:5. Two parallel lines, approximately 3 mm apart, were drawn on the central graft perpendicular to its long axis with Voerhoff's stain as gauge length marks for tensile strain determination. The patella and the tibia of the specimen were then clamped with a specially designed apparatus and mounted in an Instron-type material testing machine (Model AO75001, Tokyokoki Seizosho, Tokyo, Japan). Each specimen was preloaded with a force of 0.2 N and then subjected ten times to a loading-unloading cycle between 0 and 0.1 mm deformation at an extension rate of 6 mm/min. The specimens then underwent tensile-to-failure testing at a cross-head speed of 10 mm/min, corresponding to a 3.3%/s strain rate. The movement of the two lines on the tendon surface during tensile testing was recorded by a video system. The deformation between the two lines was analyzed with a video dimension analyzer (Percept Scope C3160, Hamamatsu Photonics, Hamamatsu, Japan) to obtain deformation–time curves. The data from the load cell were transfered to a computer (Macintosh SE/30, Apple Computer, CA, USA) through an A/D converter (Mac ADIOS-8ain, GW Instruments, MA, USA) and stored as load–time curves. Load–time and deformation–time curves were then combined into a load–deformation curve. Only the tendons that failed at their mid substances were used for biomechanical analysis. Stress–strain curves were obtained with the measurement system just described, and three parameters were calculated from these curves: (1) tensile strength, (2) tangent modulus, and (3) strain at failure. The data were statistically analyzed by a two-way analysis of variance (ANOVA) among two different groups and four posttransplantation periods. Subsequent post hoc comparisons were performed for significant differences ($P < 0.05$) resulting from a two-way ANOVA. The methods of comparison were Scheffe's F-test and Student's *t*-test.

Results

All the recipient animals could run without lameness and no signs of infection were observed. The cross-sectional area (CSA) of the grafts after transplantation was about twice as large as that before transplantation at each posttransplantation period for each type of graft (Fig. 2). There was no significant difference in the CSA ratio between the fresh and the frozen grafts at each posttransplantation period. The failure patterns of the grafts and normal controls in the tensile-to-failure test are shown in Table 1. Fifty-nine of 72 specimens that failed at the midsubstance were used for data analysis. The mechanical properties for the experimental groups and controls are shown in Table 2. The changes in the tensile strength of the grafts over time are shown in Fig. 3. At 4 weeks, the fresh grafts showed tensile strengths comparable to that of the frozen grafts. After the twelfth week, however, the tensile strength of the fresh grafts was significantly greater than that of the frozen grafts. At 24

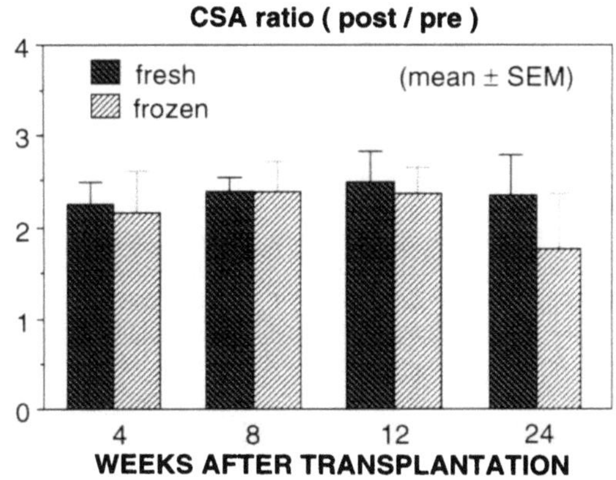

Fig. 2. The cross-sectional area (*CSA*) ratio of the grafts between posttransplantation and pretransplantation. There were no statistically significant differences in the CSA ratio between the two types of the grafts at each period (*SEM*, standard error of mean)

Table 1. Failure patterns of the grafts and normal controls in the tensile-to-failure testing

Weeks	Group	Total	Tear site		
			Midsubstance	Suture site	Bone insertion
0	Normal control	16	14 (88)	0	2
4	Fresh	10	8 (80)[a]	2	0
	Frozen	10	9 (90)	1	0
8	Fresh	6	6 (100)	0	0
	Frozen	6	4 (67)	1	1
12	Fresh	6	5 (83)	1	0
	Frozen	6	5 (83)	0	1
24	Fresh	6	4 (67)	1	1
	Frozen	6	4 (67)	2	0

[a] Numbers in parentheses are the percentage of midsubstance tear.

weeks, the mean tensile strengths of the fresh and the frozen grafts were 44.2 and 30.5 MPa, respectively, and these values were 80% and 55%, respectively, of those of the normal controls.

The changing patterns of the tangent modulus for the three groups were similar to the changes in tensile strength (Fig. 4). The tangent modulus of the fresh grafts tended to be larger than that of the frozen grafts after the eighth week, but significant differences were not found.

The mean strain-at-failure of each group ranged from 13% to 23%. There was no significant difference in the strain-at-failure among the two types of grafts at each posttransplantation period (see Table 2).

Table 2. The material properties of the grafts and controls

Weeks	Group	n	Tensile strength (MPa)[a]	Tangent modulus (MPa)[a]	Strain at failure (%)[a,b]
0	Normal control	14	55.3 ± 4.0	510.0 ± 28.3	19.7 ± 1.7
4	Fresh	8	16.2 ± 2.3	126.3 ± 14.7	18.5 ± 2.8
	Frozen	9	15.5 ± 1.7	117.1 ± 16.3	22.3 ± 3.5
8	Fresh	6	21.5 ± 2.1	274.2 ± 69.9	15.6 ± 2.9
	Frozen	4	14.3 ± 3.0	93.9 ± 24.6	21.6 ± 1.7
12	Fresh	5	33.2 ± 5.0[c]	452.5 ± 143.0[b]	15.4 ± 2.7
	Frozen	5	19.3 ± 2.4[d]	201.0 ± 43.9	19.3 ± 4.1
24	Fresh	4	44.2 ± 2.9[b,e]	458.0 ± 92.3[b]	13.5 ± 3.1
	Frozen	4	30.5 ± 4.7[d,e]	277.7 ± 49.9[e]	17.9 ± 5.0

[a] Values are presented as mean ± SEM.
[b] Not significantly different from the control values ($P < 0.05$).
[c] Significantly different from 4-week value ($P < 0.05$).
[d] Significantly different from the value for the fresh grafts at the same period ($P < 0.05$).
[e] Significantly different from 4- and 8-week values ($P < 0.05$).

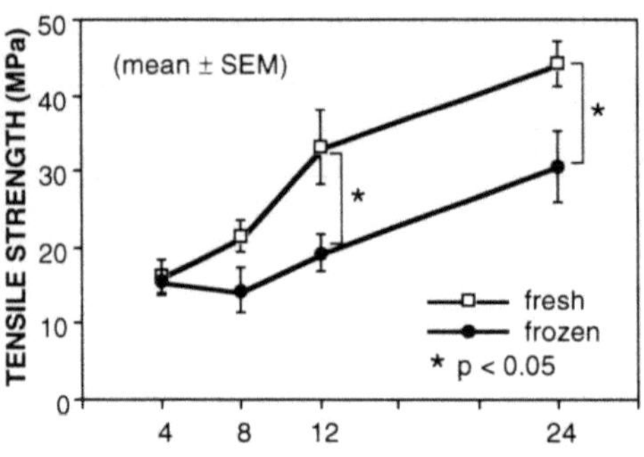

3

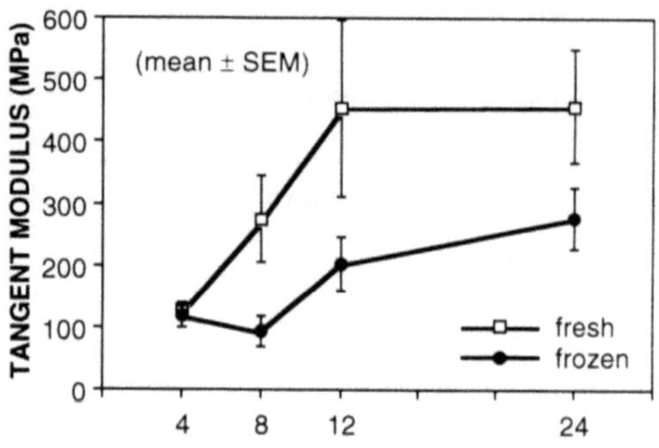

4

Fig. 3. The tensile strength of the patellar tendon grafts after transplantation in a rat model. Note that the fresh autografts showed significantly larger tensile strength than the fresh-frozen autografts at 12 and 24 weeks

Fig. 4. The tangent modulus of the patellar tendon grafts after transplantation in a rat model

Discussion

Advantages of This Study's Experimental Model

Many studies regarding graft remodeling after ligament reconstruction have been reported in the literature [3–6,8–10]. However, animal models used in these previous studies were thought to be failure models, because the animals could not be controlled, and the grafted tissue tended to be overstressed. In this study, however, replacing only one-half of the patellar tendon with the graft was intended to protect the graft against overstress and failure. Actually, the tensile strength of the fresh autografts in 24 weeks regained up to 80% of that of the normal control. In this study, moreover, an isograft was adopted as an autograft model, and therefore there were no other differences in the grafting conditions between the fresh and the frozen groups. For these reasons, this study model was appropriate for the investigations performed in this study.

Fig. 5. Light micrographs of the fresh and the frozen grafts at 12 weeks after transplantation. **A** A fresh graft. **B** A frozen graft (hematoxylin and eosin). Note that the fresh graft had more regularly aligned collagen fibers than did the frozen graft

The Effect of Freezing on Mechanical Properties

At 4 weeks, there were no significant differences in the tensile strengths and the tangent moduli between the fresh and frozen autografts. The tensile strength of the fresh autografts was greater than that of the frozen autografts at 8 and 12 weeks, and the tangent modulus of the fresh autografts was greater than that of the frozen autografts at 8 weeks. Histologically, the fresh autograft showed better alignment of collagen fibers and a smaller number of cells than the frozen autograft at 12 weeks (Fig. 5).

King et al. [11] also compared fresh autografts to frozen and thawed autografts in a whole rabbit MCL transplantation model. The frozen autografts

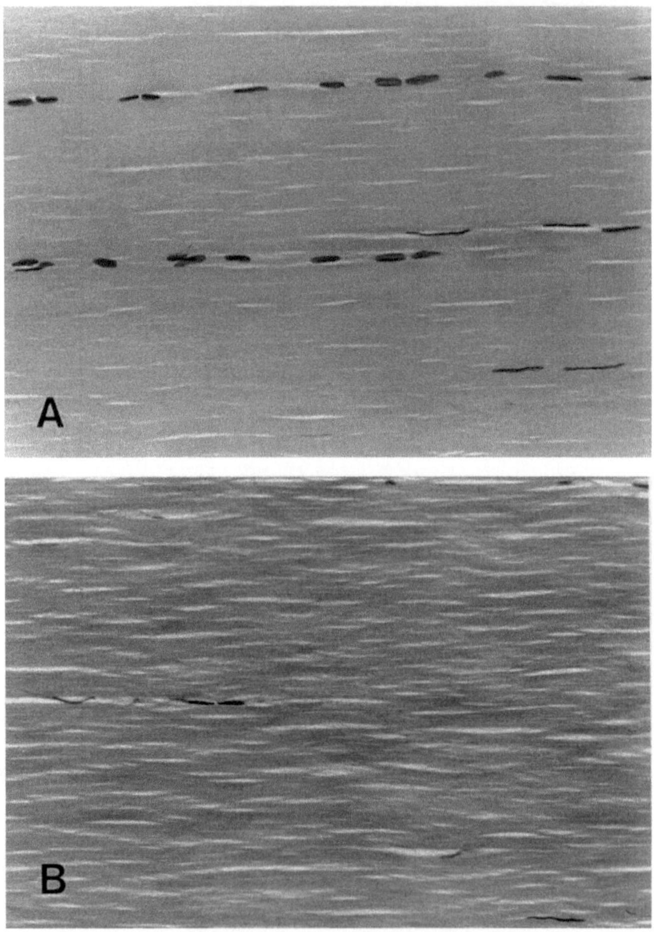

Fig. 6. Light micrographs of the fresh and the frozen grafts before transplantation. **A** A fresh graft. **B** A frozen graft (hematoxylin and eosin). Note that fewer cells and larger interfibriller spaces occurred in the frozen graft than in the fresh graft

were made by repetitive short-duration freeze/thaw cycles with liquid nitrogen. Up to 48 weeks after transplantation, the frozen autografts showed lower tensile strengths than did the fresh grafts. Although there were some differences in transplantation model and freezing technique for the grafts, this study and our current study showed similar results. These studies suggest that freezing has adverse effects on graft remodeling in terms of mechanical properties of the grafts.

It was reported that the formation of ice crystals during the freezing process expanded spaces among collagen bundles in the grafts, while the collagen fibers themselves remained intact [12]. Our histological results also showed that the fresh-frozen tendon decreased in number of cells and increased in interfibrillar space (Fig. 6). In light of these results, the biomechanical and histological differences between the two types of autografts in this study could be caused by (1) differences in cell viability at the time of transplantation or (2) changes of the extracellular matrix from freezing and thawing, for example, widening in the interfibrillar spaces of collagen fibers.

The results in this study cannot be directly adapted to a clinical situation such as an ACL reconstruction, because the ACL is an intraarticular ligament. Regarding extraarticular ligament reconstruction, however, freezing could have an adverse effects on graft remodeling.

Conclusion

Using a patellar tendon transplantation model in rats, we evaluated differences in remodeling processes, represented by mechanical properties, between fresh and frozen autografts. This study showed that deep-freezing for graft preservation had an adverse effect on the biomechanical remodeling process in the extraarticular graft.

References

1. Shino K, Inoue M, Horibe S, Hamada M, Ono K (1990) Reconstruction of the anterior cruciate ligament using allogeneic tendon. Long-term followup. Am J Sports Med 18:457–465
2. Indelicato PA, Bittar ES, Prevot TJ, Woods GA, Branch TP, Huegel M (1990) Clinical comparison of freeze-dried and fresh frozen patellar tendon allografts for anterior cruciate ligament reconstruction of the knee. Am J Sports Med 18:335–342
3. Shino K, Kawasaki T, Hirose H, Gotoh I, Inoue M, Ono K (1984) Replacement of the anterior cruciate ligament by an allogeneic tendon graft. An experimental study in the dog. J Bone Joint Surg 5B:672–681
4. Sabiston P, Frank C, Lam T, Shrive N (1990) Transplantation of the rabbit medial collateral ligament. I. Biomechanical evaluation of fresh autografts. J Orthop Res 8:35–45

5. Sabiston P, Frank C, Lam T, Shrive N (1990) Transplantation of the rabbit medial collateral ligament. II. Biomechanical evaluation of frozen/thawed allografts. J Orthop Res 8:46–56
6. Jackson DW, Grood ES, Goldstein JD, Rosen MA, Kurzweil PR, Cummings JF, Simon TM (1993) A comparison of patellar tendon autograft and allograft used for anterior cruciate ligament reconstruction in the goat model. Am J Sports Med 21:176–185
7. Butler DL, Grood ES, Noyes FR, Zernicke RF, Brackett K (1984) Effects of structure and strain measurement technique on the material properties of young human tendons and fascia. J Biomech 17:579–596
8. Nikolaou PK, Seaber AV, Glisson RR, Ribbeck BM, Bassett FH (1986) Anterior cruciate ligament allograft transplantation. Long-term function, histology, revascularization, and operative technique. Am J Sports Med 14:348–360
9. Vasseur PB, Rodrigo JJ, Stevenson S, Clark G, Sharkey N (1987) Replacement of the anterior cruciate ligament with a bone-ligament-bone anterior cruciate ligament allograft in dogs. Clin Orthop 219:268–277
10. Thorson E, Rodrigo JJ, Vasseur P, Sharkey N, Heitter D (1989) Replacement of the anterior cruciate ligament. Acta Orthop Scand 60:555–560
11. King G, Edwards P, Brant R, Shrive N, Frank C (1992) Freeze-thawing impairs long term healing of a rabbit medial collateral ligament autograft model. Trans ORS 17:660
12. Turner WD, Vasseur P, Gorek JE, Rodrigo JJ, Wedell JR (1988) An in vitro study of the structural properties of deep-frozen versus freeze-dried, ethylene oxide-sterilized canine anterior cruciate ligament bone-ligament-bone preparations. Clin Orthop 230:251–256

Effects of Stress Shielding on Autografts in Augmentation Procedures: Experimental Studies Using the In Situ Frozen Patellar Tendon

KAZUNORI YASUDA[1] and KOZABURO HAYASHI[2]

Summary. Stress shielding reduces the strength of the in situ frozen patellar tendon (PT), depending on the degree of stress shielding, and increases the cross-sectional area, compensating for the decrease of tensile strength and keeping the maximum failure load at a normal level. However, there is some limit in the compenstion effect, depending on the extent of stress shielding. Mechanical response of the PT to stress shielding occurs independent of living cells, and therefore the remodeling is associated with noncellular components that include collagen. With regard to the ultramicrostructure of the frozen PT, complete stress shielding reduces the total number of collagen fibrils and the ratio of the total area of the fibrils to the whole visualized area while it increases the number of thick fibrils. With respect to effects of restressing, if a PT that has undergone the in situ freezing treatment is completely shielded from stress for 1 week or more, subsequent restressing cannot produce complete recovery of tensile strength and the maximum load reduced by stress shielding, although it prevents the marked reduction of those parameters caused by continuous stress shielding. Moreover, restressing has no effect on the tensile strength and the maximum load of the previously frozen PT if stress is completely shielded for 6 weeks before restressing.

Key words: Stress shielding—Restressing—Patellar tendon—In situ freezing—Autograft—Augmentation—Mechanical properties—Ultramicrostructure—Remodeling

[1] Department of Orthopedic Surgery, Hokkaido University School of Medicine, Kita 14 Nishi 5, Kita-ku, Sapporo, Hokkaido, 060 Japan
[2] Department of Mechanical Engineering, Faculty of Engineering Science, Osaka University, Machikaneyama, Toyonaka, Osaka, 560 Japan

Introduction

Patellar tendon (PT) autografts have been commonly used for cruciate ligament reconstruction. Therefore, the mechanical properties of PT grafts have been the focus of a number of recent studies in sports medicine. Histological studies on these autografts have reported that fibroblast necrosis sequentially occurs during the early period after surgery [1–5]. Mechanically, many studies have demonstrated great reduction of the graft strength soon after reconstruction, although it gradually recovers with time as remodeling proceeds [5–8]. The early weakness of the graft is a serious problem because a risk of failure increases if a graft is exposed to high loads during physical activities in the early period after surgery. To prevent failure of the PT grafts, they often have been reinforced with synthetic prostheses [9–19]. It has been believed that these augmentation procedures would protect the grafts from excessive loads. In this mechanical condition, however, the devices shield the stress exerted on the grafts. Only a few experimental studies have been conducted to mechanically clarify these effects of stress shielding on biological grafts. Andrish and Woods [9] reported that autogenous PT grafts augmented with Dacron devices used in anterior cruciate ligament (ACL) reconstruction were absorbed because of stress shielding. Yoshiya et al. [18] suggested that stress shielding had adverse effects on the strength of the graft in augmented ACL reconstruction. McCarthy et al. [14] found that stress shielding by synthetic augmentation prolonged the remodeling of the graft. Conversely, Fowler and Amendola [10] described the normal remodeling process that occurred after augmented ACL reconstruction on the basis of their findings that the collagen content in the graft was not altered by the presence of synthetic devices. Thus, effects of stress shielding on PT grafts remain controversial.

The following criticism has been applied to the previous studies. First, the degree of the stress shielding induced by augmentation was not quantitatively documented in those studies. Second, although the maximum failure load of composite structures consisting of biological grafts and synthetic devices has been determined in most of the previous studies, the mechanical properties of the individual materials have not been studied. Third, few studies have been done on the relationship between the degree of stress shielding and the effect on the mechanical properties of augmented autografts. These shortcomings strongly suggest conducting detailed biomechanical studies using adequate animal autograft models and new experimental techniques with which the degree of stress shielding can be quantitatively controlled. In addition, experimental studies on animals have demonstrated that autografting does not improve the anteroposterior joint laxity in ACL-deficient knees [10,18,19], although favorable clinical results have been reported for ACL reconstruction with PT autografts in human patients. Therefore, animal experiments with the conventional intraarticular ACL reconstruction may be inadequate for detailed study of the graft remodeling process.

To obtain a solution to these problems, we first developed an extraarticular PT autograft model using rabbits as an alternative to intraarticular ACL reconstruction [20–22]. This model was made with a technique of freezing the PT in situ to kill the fibroblasts in the PT, because fibroblast necrosis is the most essential change in autografts transplanted for ACL reconstruction. In this model, fibroblast necrosis is easily and quickly stimulated without disturbance of the tendon attachment sites or alteration of the mechanical environment of the PT. Second, we developed a novel technique to shield the PT from stress. In this technique, tension is released completely [21,23] or partially [24,25] from the PT throughout the experimental period without immobilization of the knee joint. Using these techniques, we have conducted a series of precise biomechanical studies to resolve the described controversy concerning the effect of stress shielding on the PT autograft.

In this chapter, we discuss first the effects of stress shielding on mechanical properties, histology, and ultrastructure of the in situ frozen PT as an autograft model, and second the effects of resumption of stress (restressing) on the mechanical properties of the PT shielded completely from stress for various periods.

Biomechanical Testing

In a series of our experiments [20–30], an identical measuring system was used to determine mechanical properties of the PT. The cross-sectional area of the whole PT was measured using an area micrometer similar to those reported by previous investigators [7]. To determine the tensile strength of PT substance, the central one-third of the tendon was used for the tensile testing [30]. A bone–tendon preparation mounted onto a set of specially designed grips was attached to a tensile tester with the tibia inclined at 45° in the sagittal plane against the patella and the PT. The specimen and the grips were immersed in saline solution of 37°C during testing. Two parallel lines were drawn transversely on the PT surface using a stain. Strain in the PT substance was determined with a video dimension analyzer using these lines as gauge-length markers [31]. After applying a preload of 0.5 N for 10 min, the specimen was subjected to ten cycles of loading and unloading at a deformation between 0 and 0.5 mm (0.5% strain) at a rate of 20 mm/min. Following this preconditioning, the specimen was stretched to failure at the same rate. Stress was calculated by dividing tensile load by the initial cross-sectional area, and strain from dividing displacement by the initial gauge length. Based on this results, the maximum load of the whole PT was estimated by multiplying the tensile strength times the whole cross-sectional area.

Effects of In Situ Freezing on the Patellar Tendon

The in situ freezing technique [20–22] for the PT is summarized as follows. After the anterior part of the rabbit's right knee was exposed through a midline longitudinal skin incision, subcutaneous retinacula were incised longitudinally along the PT edges. The posterior surface of the PT was then separated from the infrapatellar fat pad. A silicone rubber sheet was inserted between the PT and the fat pad, and the edges of the sheet were clamped with four graspers to make a bath with the PT inside it. Liquid nitrogen was poured into the bath; the PT was frozen for 1 min. The frozen PT was then thawed by pouring a physiological saline solution into the bath.

Effects of the in situ freezing treatment (Table 1) were observed for 24 weeks [20]. The cross-sectional area of the PT started to increase by 3 weeks, reaching a plateau at 12 weeks. The modulus and tensile strength of the tendon began decreasing by 3 weeks, although the maximum load decreased to 81% and 62% of the control values at 12 and 24 weeks, respectively. Histologically, cells were absent until 2 weeks, while a large number of new fibroblasts appeared by 3 weeks. Although the previously frozen PT had apparently normal cells at 24 weeks, it did not contain dense collagen bundles such as are normally seen. This study showed that fibroblast necrosis itself did not change the biomechanical properties of the PT immediately, but that the frozen PT were weakened during the subsequent period of tissue remodeling. Our results also indicated that the in situ frozen PT can be used as an autograft model.

Table 1. Effects of in situ freezing on the patellar tendon

Parameters	Periods (weeks)					
	0	2	3	6	12	24
Cross-sectional area (mm^2)						
Frozen	13.6 ± 0.5	14.1 ± 0.4	18.3 ± 0.9*	21.1 ± 1.3*	24.6 ± 1.5*	23.7 ± 2.6*
Control	13.4 ± 0.3	13.7 ± 0.3	14.3 ± 0.4	13.8 ± 0.5	13.4 ± 0.5	13.4 ± 0.3
Elastic modulus (MPa)						
Frozen	1510 ± 98	1602 ± 135	932 ± 135*	636 ± 51*	449 ± 26*	427 ± 77*
Control	1284 ± 76	1490 ± 150	1457 ± 131	1267 ± 90	1465 ± 81	1369 ± 147
Tensile strength (MPa)						
Frozen	63.8 ± 2.6	72.0 ± 6.3	48.4 ± 3.2*	35.3 ± 3.5*	29.8 ± 1.2*	24.6 ± 6.2*
Control	63.2 ± 1.8	58.0 ± 7.9	61.7 ± 4.1	63.6 ± 2.7	67.0 ± 2.4	64.6 ± 2.1
Maximum load (N)						
Frozen	867 ± 48	1016 ± 96	876 ± 37	747 ± 83	728 ± 33*	540 ± 99*
Control	848 ± 37	803 ± 122	882 ± 59	880 ± 56	894 ± 34	866 ± 31

Mean ± SEM. $n = 5$ for each period.
* Significantly different from the control group at the corresponding period ($P < 0.05$).
Frozen, In situ frozen patellar tendons.
Control, Control patellar tendons.

Effects of Stress Shielding on the In Situ Frozen Patellar Tendon

Surgical Techniques in Experimentation

We conducted two studies on effects of complete stress shielding the in situ frozen PTs using two artificial materials [22,25], a flexible stainless steel wire 0.97 mm in diameter and a 10-mm-wide, commercially available Leeds-Keio (LK) artificial ligament, as augmentation devices of different stiffness. Stress shielding techniques [21–23] used in the two studies were almost identical: Two stainless steel pin markers were embedded into the patella and the tibial tubercle. In addition, a stainless steel screw was inserted transversely into the tibial tubercle and a stainless steel pin was placed in the middle part of the patella (Fig. 1). The PT was slackened by drawing the patella toward the tibial tubercle, and this position was maintained using one of the two artificial materials, installed between the screw and the pin. The distance between the two markers was shortened by approximately 6 mm, which was approximately 20% of the original distance. It was confirmed that the PT was relaxed at all knee angles not only immediately after surgery but also immediately before the animal was killed.

Moreover, we also developed a unique experimental method for controlled partial stress shielding [24,25]: a pin, and a screw, were inserted into the bone in the same manner as the complete stress shielding procedure. A buckle transducer [30,32] was attached to the PT while the knee was flexed 90°. The patella was then manually pulled to the proximal direction using a wire. The load applied to the PT was monitored with a load cell attached to the wire. After applying ten cycles of loads from 0 to 80 N, a load (load cell output) displacement (buckle transducer output) curve was recorded. The output obtained from the buckle transducer when applying a 78 N load to the PT was defined as a reference value. An LK ligament was then installed between the

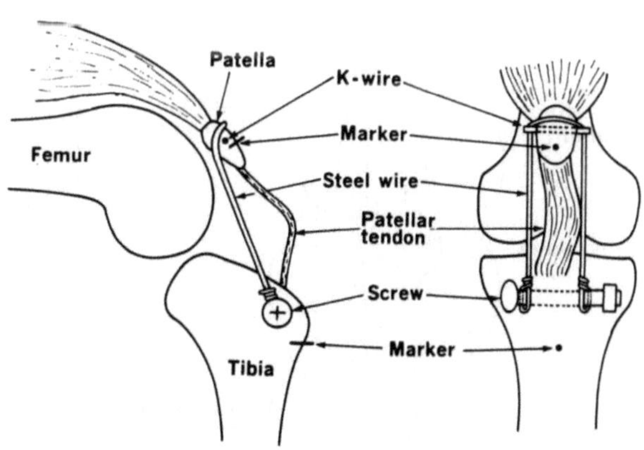

Fig. 1. Complete stress shielding technique

patella and the tibial tubercle while applying a 78 N load to the PT. Monitoring the output from the buckle transducer the LK ligament was stretched until the output from the buckle transducer decreased to 23 N (30% of the reference value), and the LK artificial ligament was then fixed firmly with a clamp (Fig. 2). After ten cycles of loading between 0 and 80 N were applied to the PT–LK ligament complex, a load displacement curve was again recorded. The ratio of the load shared between the PT and the LK ligament was calculated using the two load displacement curves. In our study, 27.3 ± 0.9% (mean ± SE) of the total force was applied to the PT at the operation, while 72.7% of the total load was applied to the LK artificial ligament [24,25].

Mechanical Properties

To clarify the effects of complete and partial stress shielding, we compared the in situ frozen PTs (sham group) with the PTs released completely from tension after the freezing treatment [completely shielded from stress (CSS) group] and the PTs released partially from tension after the treatment [partially shielded from stress (PSS) group]. Figure 3 demonstrates the averaged stress–strain curves for the control, sham (Fr in the figure), and CSS (Fr/CSS in the figure) groups. In the sham group, the slope of the curve started to decrease at 3 weeks. On the other hand, the slope for the CSS group started to decrease at the first week and reached a maximum at 3 and 6 weeks. The slope of the CSS group was significantly different from that of the sham group at each period.

To compare quantitatively the initial slope of the stress–strain curve between the CSS and sham groups, the experimental stress–strain data in strain increments of 0.5% up to 3.0% were analyzed [28] using the constitutive equation derived by Gomez [33].

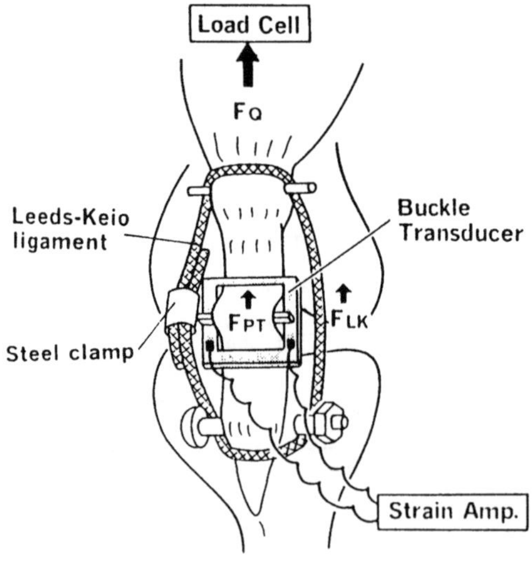

Fig. 2. Experimental method for controlled partial stress shielding. $F_Q = F_{PT} + F_{LK}$, where F_Q is the load in the quadriceps, F_{PT} is the load in the PT, and F_{LK} is the load in the Leeds-Keio (LK) artificial ligament. The ratio of the load shared between the PT and the LK ligament was calculated by measuring the F_Q and the F_{PT} with a load cell and a buckle transducer, respectively

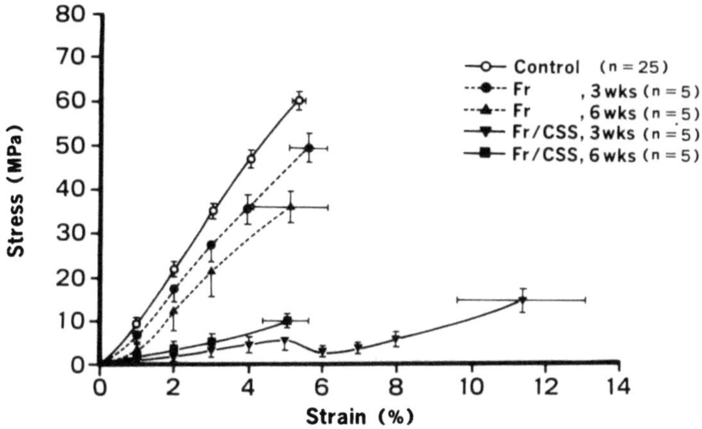

Fig. 3. The stress–strain curves for the control, sham (*Fr*, frozen in situ), and CSS (*Fr/CSS*, completely shielded from stress after the freezing treatment) groups. The slope of the Fr group was significantly steeper than that of the Fr/CSS group at each period

$$\sigma = C\varepsilon \exp(a\varepsilon + b\varepsilon^2)$$

where σ is the stress, ε is the strain, and C, a, and b are material constants to be determined using a nonlinear least squares fitting routine. The material constants showed that the initial slope of the stress–strain curve of the group without stress shielding was significantly steeper than that of the group with stress shielding at 3 weeks, but not at 6 weeks. The increase in the tangent modulus (the gradient of the stress–strain curve) of the group without stress shielding was significantly more than that of the group with complete stress shielding at both 3 and 6 weeks.

Stress shielding significantly affected the mechanical properties of the in situ frozen PTs. Figure 4 demonstrates effects of complete and partial stress shielding on the tensile strength. There were no significant differences on effects of complete stress shielding between the two experiments described earlier [22,25] using different artificial materials for augmentation devices. Partial or complete stress shielding significantly decreased the tensile strength within 1 week after treatment; the mean strength in the PSS and CSS groups at 1 week were 60% and 55% of that of the sham group, respectively. On the other hand, in the sham group to which only the freezing treatment was applied, the tensile strength started decreasing between 2 and 3 weeks. This result indicates that stress shielding enhanced reduction of the strength in the previously frozen PT. With respect to difference between complete and partial stress shielding, the tensile strength and the tangent modulus of the partially stress-shielded PT were generally lower than those of the completely stress-shielded PT; the differences were significant at 3 and at 6 weeks. These results indicate that the strength and modulus of the PT change depending on the degree of stress shielding.

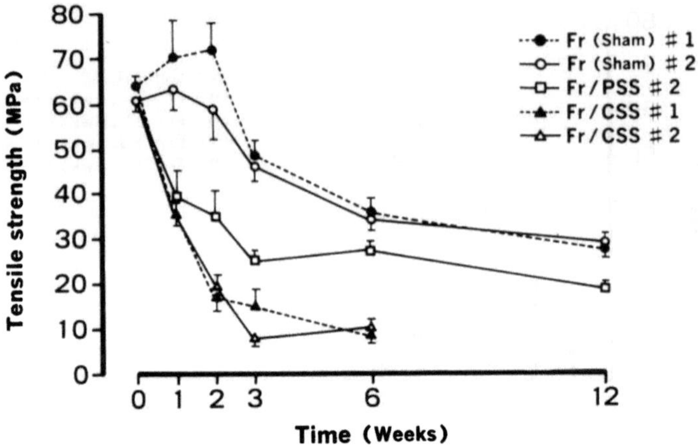

Fig. 4. Reduction of the tensile strength in proportion to the grade of stress shielding. *Fr*, Frozen; *Fr/PSS*, partially stress shielded after the freezing treatment; *Fr/CSS*, completely stress shielded after the freezing treatment. #1, Data reported by Ohno et al. [22]. #2, Data reported by Majima et al. [25]

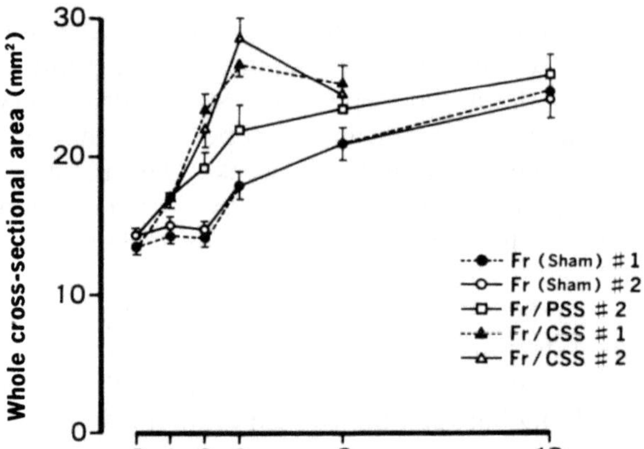

Fig. 5. Increase of cross-sectional area in proportion to the grade of stress shielding. *Fr*, Frozen; *Fr/PSS*, partially stress shielded after the freezing treatment; *Fr/CSS*, completely stress shielded after the freezing treatment. #1, Data reported by Ohno et al. [22]. #2, Data reported by Majima et al. [25]

With regard to the cross-sectional area of the whole PT (Fig. 5), the stress shielding significantly increased it in 2 weeks after treatment; the whole cross-sectional areas in the PSS and CSS groups at 2 weeks were 133% and 156% of that in the sham group, respectively. Figure 6 demonstrates effects of complete and partial stress shielding on the maximum load [25]. The estimated maximum load in the PSS group at 1, 2, 3, 6, and 12 weeks was 68%, 78%, 63%, 85%, and 68% of the corresponding load in the sham group, respectively; however, there were no significant differences between the PSS and sham groups at 1, 2, and 6 weeks. The increase of the area appears to be compensating for the decrease of tensile strength and keeping the maximum failure load at the normal level. The maximum load in the CSS group was much lower in comparison with that in the sham group throughout the experimental period. Between the PSS and CSS groups, the maximum load was significantly larger in the former group than in the latter group at 3 and 6 weeks. These results imply that there is some limit in the compensation effect brought about by the increase of the cross-sectional area, depending on the extent of stress shielding.

Histology

No cells were observed within the first 2 weeks in any group; new cells appeared after 3 weeks. There were essentially no differences in the light microscopic structure between the PSS (Fig. 7b) and sham groups in this period; however, the microstructure of the PT in the CSS group was obviously different from that in the other two groups (Fig. 7c); collagen fibers were fragmented in the CSS group but not in the other groups, and the nuclei of fibroblasts were more round in the CSS group than in the other groups. These observations show that the partial and complete stress shielding have different effects on PT microstructure [22,25]. On the other hand, significant reduction of the tensile strength and the tangent modulus was already observed at 1 week in the PSS and CSS groups, although there were no cells at this period. These results demonstrated that mechanical response of the PT to stress shielding

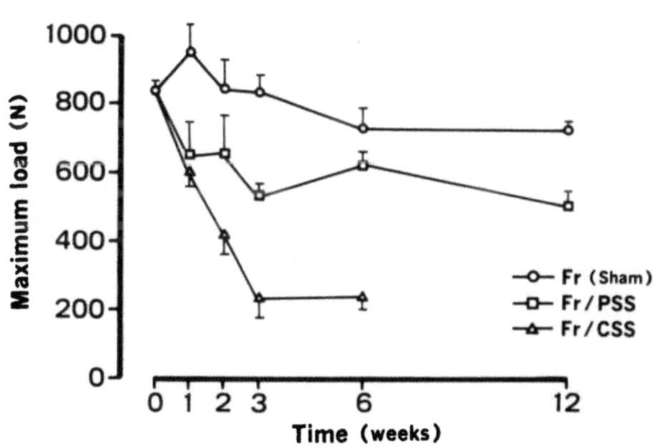

Fig. 6. Reduction of maximum load in proportion to the grade of stress shielding. There were no significant differences between the PSS and sham groups at 1, 2, and 6 weeks. *Fr*, Frozen; *Fr/PSS*, partially stress shielded after the freezing treatment; *Fr/CSS*; completely stress shielded after the freezing treatment

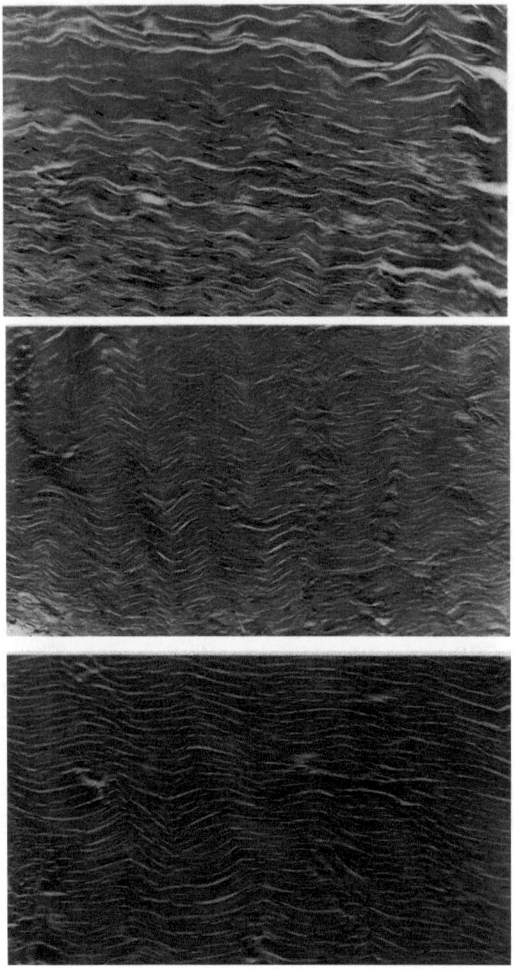

a

b

c

Fig. 7a–c. Histological observations. **a** Control PT. **b** PT shielded partially from stress for 3 weeks. **c** PT shielded completely from stress for 3 weeks. Note that collagen fibers are fragmented

occurs independent of living cells, and therefore that the remodeling occurring during stress shielding is associated with noncellular components that include collagen [22]. Even after cells appeared in the PT, the reduced strength and modulus did not increase, which implies that thost cells were not as effective in tissue remodeling.

Ultramicrostructure

To clarify the effects of complete stress shielding on the ultramicrostructure of the in situ frozen PTs, a study was conducted using the transmission electron microscope [29]. In addition, effects of in situ freezing on the ultramicrostructure of the PTs were analyzed. Ultrathin sections were obtained from the central one-third of each PT at 3 and 6 weeks after the treatment. Diameters of all collagen fibrils in a 9-μm^2 area were measured and histograms of the diameters were determined. In addition, the average of the collagen fibril (CF) numbers in a 1-μm^2 area, and the ratio of the total area of the collagen fibrils to the whole visualized area (%CF), were calculated.

This study clearly demonstrated that freezing and complete stress shielding had a strong influence on the ultrastructure of the PT. In the control specimen (Fig. 8a), fibril diameters ranged from 30 to 360 nm, and the histogram was bimodal. In the sham (in situ frozen) group, the results were almost the same as those in the control at time zero and at 3 weeks. At 6 weeks, however, the number of thin fibrils having a diameter less than 90 mm increased (Fig. 8b),

a 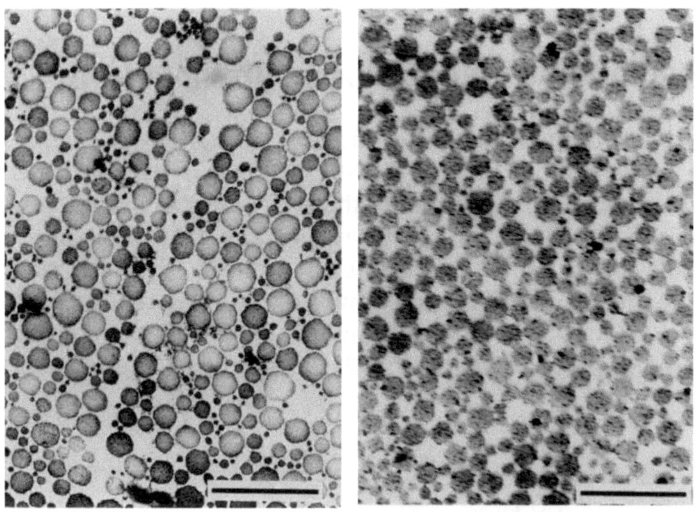 b

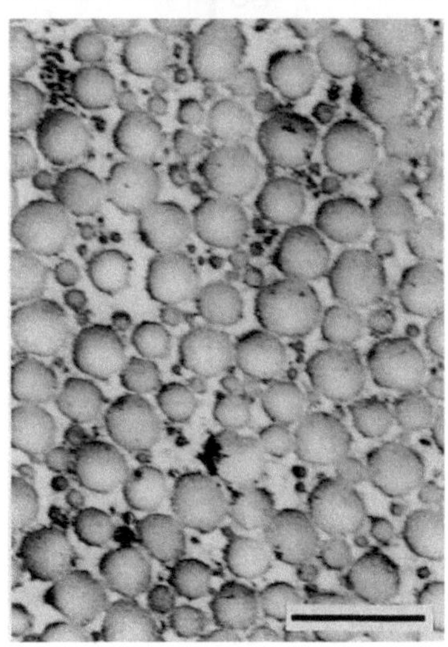 c

Fig. 8a–c. Ultramicrostructure of PTs (cross section of collagen fibrils). **a** Control specimen. **b** Sham-operated (in situ frozen) PT. **c** PT shielded completely from stress for 6 weeks. The bars drawn in the photographs show 1 μm

and the histogram became unimodal. In the CSS group at 3 weeks, the number of thin fibrils decreased, resulting in reduction of the total number of fibrils. It was also noted that thick fibrils having a diameter ranging from 360 to 420 nm were observed, and that the histogram showed a bimodal distribution. At 6 weeks, the features observed at 3 weeks were more evident (Fig. 8c). With respect to %CF, complete stress shielding reduced this parameter in the in situ frozen PT, while freezing did not change it (Fig. 9). These observations provide useful information for better understanding of the cause of reduced tensile strength in the PT caused by freezing and stress shielding. Reduction of the total area of fibrils in the cross section, reduction of the frictional force developed between fibrils, and decreased tensile strength of collagen fibril itself were considered to be the reasons that the tensile strength of the PT was reduced.

Effects of Restressing After Complete Stress Shielding

On the basis of results in the above-described studies, this question is asked: Can early resumption of loading (restressing) to a stress-shielded autograft maintain the mechanical properties of the graft? Or, can this restressing provide recovery of the mechanical properties reduced by stress shielding? The following study was conducted to answer this question [26,27]. After complete stress shielding treatments, rabbits were divided into the following four groups: "CSS-1W," "CSS-2W," "CSS-3W," and "CSS-6W," in which the tension in the PT was completely shielded for 1, 2, 3, and 6 weeks, respectively. The wire

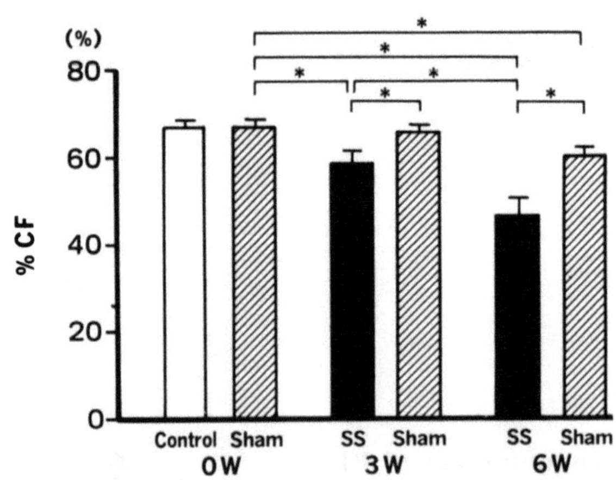

Fig. 9. Ratio of total area of the fibrils to the whole visualized area (percent collagen fibers, %*CF*). Complete stress shielding (SS) reduced it in the in situ frozen PT; freezing (sham) did not change it. *Asterisk**, significant difference between the two groups (ANOVA, $P < 0.01$; $n = 5$)

then was cut through a small skin incision to restress the PT. This procedure allowed the PT to be subjected again to normal stress.

This study demonstrated that if the PT that has undergone the in situ freezing treatment is completely shielded from stress for 1 week or more, restressing after that time cannot completely recover the tensile strength and the maximum load reduced by the stress shielding, although it prevents the marked reduction of those parameters that is caused by continuous stress shielding (Figs. 10 and 11). Moreover, restressing has no effects on the tensile strength and the maximum load of the previously frozen PT if stress is completely shielded for 6 weeks before restressing. These results suggest that long-term complete stress shielding permanently changes the structure of collagen fibers in the frozen PT or the mechanical properties of one collagen fiber in itself.

It was noted that significant differences in tensile strength among the CSS-1W, CSS-2W, and CSS-3W groups disappeared at 3 and 6 weeks after restressing although these differences were observed immediately before restressing (Fig. 10). These complex change patterns in tensile strength may be explained by a hypothesis that the mechanical properties may tend to converge into a constant level regardless of the length of period of the stress shielding when the previously frozen PT is restressed after complete stress shielding for 3 weeks or less. On the other hand, the data obtained from the CSS-6W group were markedly different from those in the other groups (Fig. 10); the tensile strength in the CSS-6W group was significantly smaller than that in the other groups, and did not react to restressing. This suggests that the frozen PT that

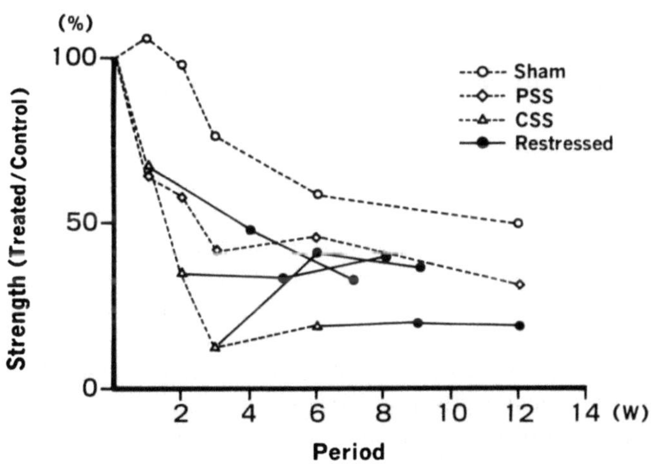

Fig. 10. Changes of tensile strength caused by restressing after complete stress shielding (*CSS*) are represented as the average ratio to the control value. Changes associated with the freezing treatment (sham) and partial stress shielding (*PSS*) [25] are superimposed on the figure. Restressing after CSS cannot provide complete recovery of tensile strength once it has been reduced

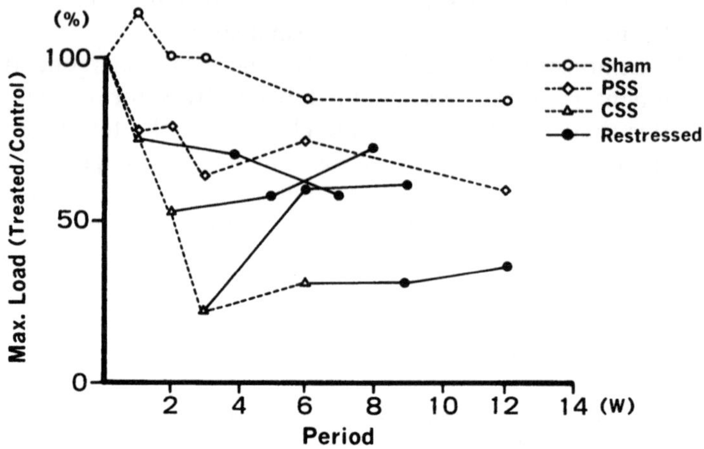

Fig. 11. Changes of maximum load caused by restressing after complete stress shielding. Changes associated with the freezing treatment (sham) and partial stress shielding (PSS) [25] are superimposed on the figure. Patterns of change are similar to changes in the tensile strength

has undergone complete stress shielding for a long period is not subject to this hypothesis. When we observe the effects of restressing for a long time period, they seem to depend on the period of stress shielding before restressing. The structure or the mechanical properties of collagen fibers may be affected, depending on the period of stress shielding.

Conclusion and Future Directions in Research

Our studies clearly demonstrated that stress shielding has a large influence on the mechanical properties, histology, and ultramicrostructure of collagen fibrils in the in situ frozen PT, and that the effects are dependent on the degree of stress shielding. This remodeling occurred independent of cells regardless of the extent of stress shielding. These are extremely interesting phenomena; however, the mechanisms are not sufficiently clarified. Further biomechanical, histological, and biochemical studies should be conducted. As to the clinical relevance, our studies imply that complete stress shielding yields adverse effects on augmented biological grafts at a very early stage after augmentation surgery. Therefore, we suggest that clinicians avoid complete stress shielding when they apply augmentation techniques to ligament reconstruction. Our studies however also demonstrated that reduction of the PT maximum load was minimal in the case of partial stress shielding, suggesting that the concept of augmentation itself should not be discarded.

Early resumption of loading (restressing) to the autograft has been expected to prevent adverse effects caused by stress shielding in augmentation procedures. Our studies demonstrated that this strategy is not completely wrong when trying to reduce the adverse effects of continuous stress shielding. However, the tensile strength and the maximum load of the frozen PT, once reduced by complete stress shielding, cannot be completely recovered by restressing. Moreover, restressing has no effects on these mechanical parameters if the stress is completely shielded for 6 weeks or more. Therefore, we must conclude that the clinical utility of early resumption of loading is extremely low in augmentation procedures. As we reported [25], partial stress shielding may have some possibility in searching for solutions. More suitable stress conditions, that is, magnitude of applied stress, duration of stress shielding, and timing of reloading, for the remodeling of biological grafts must be determined for the development of clinically acceptable augmentation techniques. Thus, further studies should be conducted to establish clinically useful augmentation techniques for ligament reconstruction surgery.

Acknowledgments. This research work was done in the Department of Biomedical Control, Research Institute of Applied Electricity, (presently, Department of Scientific Instrumentation and Control, Research Institute for Electronic Science), Hokkaido University. Animal experiments were carried out in the Institute of Animal Experimentation, Hokkaido University School of Medicine, under the Rules and Regulations of the Animal Care and Use Committee, Hokkaido University School of Medicine. Observations of the ultrastructure of the PT was performed in Central Laboratory for Research and Education, Asahikawa Medical College. The authors appreciate their collaborators, Drs. Kiyoshi Kaneda, Noritaka Yamanoto, Kazunori Ohno, Harukazu Toyama, Motoharu Keira, Tokifumi Majima, Takamasa Tsuchida, Hirohide Ishida, Kurio Tanaka, Mr. Kiyoshi Miyakawa, Takahide Ishizaka, and Ms. Yoshie Tanabe, as well as their former students, Hiroyuki Kuriyama, Fumihiro Hayashi, and Takashi Fujii. This work was supported financially in part by the Grant-in-Aids for Scientific Research of Priority Areas (nos. 04237102, 04237104 and 06671433) from the Ministry of Education, Science and Culture in Japan.

References

1. Alm A, Ekstrom H, Gillquist J, Stromberg B (1974) The anterior cruciate ligament. A clinical and experimental study on tensile strength, morphology and replacement by patellar ligament. Acta Chir Scand 140(suppl 445):4–49
2. Amiel D, Kleiner JB, Akeson WH (1986) The natural history of the anterior cruciate ligament autograft of patellar tendon origin. Am J Sports Med 14:449–462
3. Arnoczky SP, Tarvin GB, Marshall JL (1982) Anterior cruciate ligament replacement using patellar tendon. J Bone Joint Surg 64A:217–224
4. Chiroff RT (1975) Experimental replacement of the anterior cruciate ligament. A histological and microradiographic study. J Bone Joint Surg 57A:1124–1127

5. Clancy WG, Narechania RG, Rosenberg TD, Gmeiner JG, Wisnefske DD, Lange TA (1981) Anterior and posterior cruciate ligament reconstruction in rhesus monkeys. J Bone Joint Surg 63A:1270–1284
6. Ballock RT, Woo SL-Y, Lyon RM, Hokkis JM, Akeson WH (1989) Use of patellar tendon autograft for anterior cruciate ligament reconstruction in the rabbit: a long-term histologic and biomechanical study. J Orthop Res 7:474–485
7. Butler DL, Grood ES, Noyes FR, Olmstead ML, Hohn RB, Arnoczky SP, Siegel MG (1989) Mechanical properties of primate vascularized vs. nonvascularized patellar tendon grafts; changes over time. J Orthop Res 7:68–79
8. Noyes F, Butler DL, Paulos LE, Grood ES (1983) Intra-articular cruciate reconstruction. I: Perspectives on graft strength, vascularization, and immediate motion after replacement. Clin Orthop 172:71–77
9. Andrish JT, Woods LD (1984) Dacron augmentation in anterior cruciate ligament reconstruction in dogs. Clin Orthop 183:298–302
10. Fowler P, Amendola A (1990) Allograft ACL reconstruction in a sheep model: the effect of synthetic augmentation. Trans Orthop Res Soc 15:80
11. Jackson DW, Grood ES, Arnoczky SP, Butler DL, Simon TM (1987) Cruciate reconstruction using freeze dried anterior cruciate ligament allograft and a ligament augmentation device (LAD): an experimental study in a goat model. Am J Sports Med 15:528–538
12. Kennedy JC, Roth JH, Mendenhall HV, Sanford JB (1980) Presidential address: intraarticular replacement in the anterior cruciate ligament-deficient knee. Am J Sports Med 8:1–8
13. Kennedy JC (1983) Application of prosthetics to anterior cruciate ligament reconstruction and repair. Clin Orthop 172:125–128
14. McCarthy JA, Steadman JR, Dunlap J, Shively R, Stonebrook S (1990) A nonparallel, nonisometric synthetic graft augmentation of a patellar tendon anterior cruciate ligament reconstruction: a model for assessment of stress shielding. Am J Sports Med 18:43–49
15. McPherson GK, Mendenhall HV, Gibbons DF, Plenk H, Rottmann W, Sanford JB, Kennedy JC, Roth JH (1985) Experimental mechanical and histologic evaluation of the Kennedy ligament augmentation device. Clin Orthop 196:186–195
16. Park JP, Grana WA, Chitwood JS (1985) A high-strength Dacron augmentation for cruciate ligament reconstruction: a two-year canine study. Clin Orthop 196:175–185
17. Roth JH, Kennedy JC, Lockstadt H, McCallum CL, Cunning LA (1985) Polypropylene braid augmented and nonaugmented intraarticular anterior cruciate ligament reconstruction. Am J Sports Med 13:321–336
18. Yoshiya S, Andrish JT, Manley MT, Kurosaka M (1986) Augmentation of anterior cruciate ligament reconstruction in dogs with prostheses of different stiffness. J Orthop Res 4:475–485
19. Yoshiya S, Andrish JT, Manley MT, Bauer TW (1987) Graft tension in anterior cruciate ligament reconstruction. An in vivo study in dogs. Am J Sports Med 15:464–470
20. Ohno K, Yasuda K, Yamamoto N, Hayashi K (1991) Effects of in situ freezing on the mechanical and histological properties of the patellar tendon. J Jpn Orthop Assoc 65:S1081
21. Ohno K, Yasuda K, Yamamoto N, Hayashi K (1991) Effects of stress shielding on the mechanical properties of normal and in situ frozen patellar tendon. Trans Orthop Res Soc 16:134

22. Ohno K, Yasuda K, Yamamoto N, Hayashi K (1991) Effects of complete stress shielding on the mechanical properties and histology of in situ frozen patellar tendon. J Orthop Res 11:592–602
23. Yamamoto N, Ohno K, Hayashi K, Kuriyama H, Yasuda K, Kaneda K (1993) Effects of stress shielding on the mechanical properties of rabbit patellar tendon. Trans ASME J Biomech Eng 115:23–28
24. Majima T, Yasuda K, Yamamoto N, Hayashi K, Kaneda K (1992) Effects of stress shielding on the mechanical properties of in situ frozen patellar tcndon augmented with polyester artificial ligament. Trans Orthop Res Soc 17:674
25. Majima T, Yasuda K, Yamamoto N, Kaneda K, Hayashi K (1994) Deterioration of mechanical properties of autograft in controlled stress-shielded augmentation procedures. An experimental study. Am J Sports Med (in press)
26. Ishida H, Yasuda K, Majima T, Kaneda K, Yamamoto N, Hayashi K (1993) Effects of restressing on the mechanical properties of the in situ frozen patella tendon in the rabbit. Jpn J Orthop Assoc 67:S1514
27. Ishida H, Yasuda K, Hayashi K, Yamamoto N, Kaneda K (1994) Effects of resumption of loading on stress-shielded autografts in augmentation procedures. An experimental study. Presented at the annual meeting of the AOSSM, Palm Desert, CA, June 1994
28. Toyama H, Ohno K, Yamamoto N, Hayashi K, Yasuda K, Kaneda K (1992) Stress-strain characteristics of in situ frozen and stress-shielded rabbit patellar tendon. Clin Biomech 7:226–230
29. Tsuchida T, Yasuda K, Hayashi K, Majima T, Yamamoto N, Kaneda K, Miyakawa K, Tanaka K (1994) Effects of stress shielding on the ultrastructure of the in situ frozen patellar tendon in the rabbit. Trans Orthop Res Soc 19:638
30. Yamamoto N, Hayashi K, Kuriyama H, Yasuda K, Kaneda K (1992) Mechanical properties of the rabbit patellar tendon. Trans ASME J Biomech Eng 114:332–337
31. Hayashi K, Nakamura T (1985) Material test system for the evaluation of mechanical properties of biomaterials. J Biomed Mater Res 19:133–144
32. Lewis JL, Lew WD, Schmidt J (1982) A note on the application and evaluation of the buckle transducer for knee ligament force measurement. Trans ASME J Biomech Eng 104: 125–128
33. Gomez MA (1988) The effect of tension on normal and healing medial collateral ligaments. PhD dissertation, University of California, San Diego

Part 11. Three Dimensional Motion Analysis

Definition of Three-Dimensional Motion Parameters for the Analysis of Knee Joint Motion

KAZUHIRO TERAJIMA, SHOUJIRO TERASHIMA, TOSHIAKI HARA, YOSHINORI ISHII, and YOSHIO KOGA

Summary. A technique based on the new definition of 6 degree-of-freedom motion parameters has been developed to improve the applicability of biomechanical joint motion analysis to clinical estimation. Two methods of define femoral translation and rotation angle are used to analyze motion data for the Genesis total knee arthroplasty (TKA) knee with the 6 degree-of-freedom electric goniometer and biplanar computed radiography (CR) X-ray system. One method is obtained from the virtual contact point located in the femoral-sagittal plane, and the other corresponds to the screw displacement vector. It was found that the discrepancy between the current and conventional motion analysis is insignificant.

Key words: Biomechanics—Knee joint—Three-dimensional motin analysis—Translation—Rotation

Nomenclature

H, translation vector
w, screw displacement vector
$\mathbf{P}_c$, position vector to the point of contact
R_{pd}, radius of proximal-distal direction axis of the best fitting oval of femoral condyle
P_{ap}, radius of anteroposterior direction axis of the best fitting oval of femoral condyle
α, flexion angle
θ_0, angle between the X_f axis and the line that passes through the femoral origin and contact point

Graduate School of Science and Technology, Niigata University, 8050 Ikarashi 2-nocho, Niigata, 950-21 Japan

(X_f, Y_f, Z_f), fixed coordinate system connected with the femur
(X_t, Y_t, Z_t), moving coordinate system connected with the tibia
(x_c, y_c, z_c), coordinates of position vector **Pc**

Introduction

We developed a three-dimensional motion analysis system (using a 6 degree-of-freedom electrogoniometer, instrumented spatial linkage [ISL], and a biplanar computed radiography [CR] X-ray system) [1] and analyzed the knee motion of osteoarthritic patients [2] and patients after total knee joint replacement [3]. In our motion analysis system, the ISL consisted of connecting six potentiometers fixed by a strap at the lateral thigh and calf. Knee motion during active knee flexion or gait is measured in a noninvasive manner, and biplanar X-rays are taken with the knee in full extension to transfer the ISL coordinates to the femur. This is handled automatically by computer, and the entire calculation requires about only 20 min.

The knee motion can be defined by three angles and three translations. Although these parameters were calculated according to the Grood definition [4] (the coordinate systems were determined by our biplanar X-ray method, shown in Fig. 1), the rollback did not coincide with the results of analysis of patients after Genesis total knee arthroplasty, as shown in Fig. 2a. While rollback was induced with knee flexion in the normal knee, our calculated results showed motion in the opposite direction. That is, if the tibia moved posterior in curvature toward flexion, in our calculation from anatomicaly shaped component geometry the femoral origin moved forward in spite of observing rollback in the X-ray. To explain the reason, we simulated the femoral origin motion as shown in Fig. 3. The ellipses are the shape that best

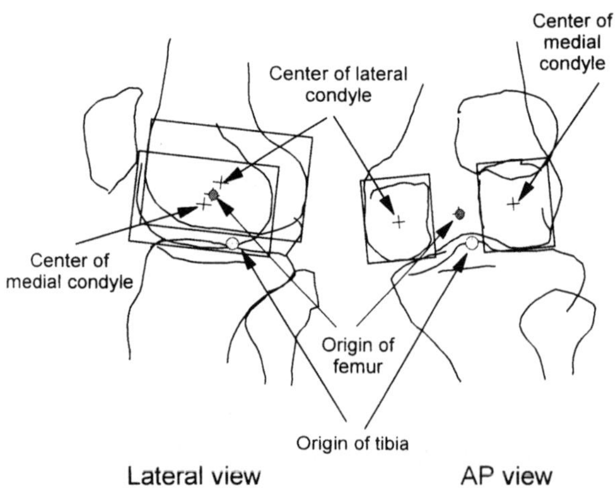

Fig. 1. The origin of the bone coordinate system

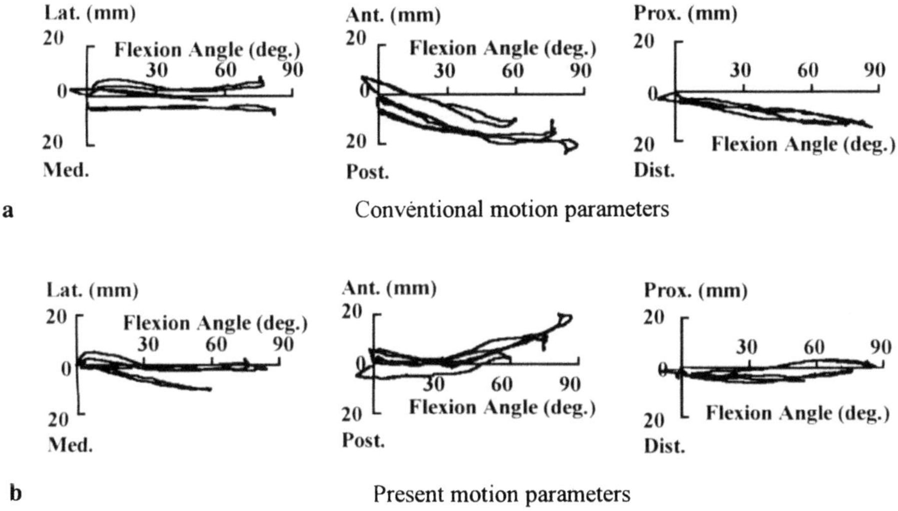

Fig. 2. Comparison of **a** conventional motion parameters and **b** current motion parameters in TKA knee motion

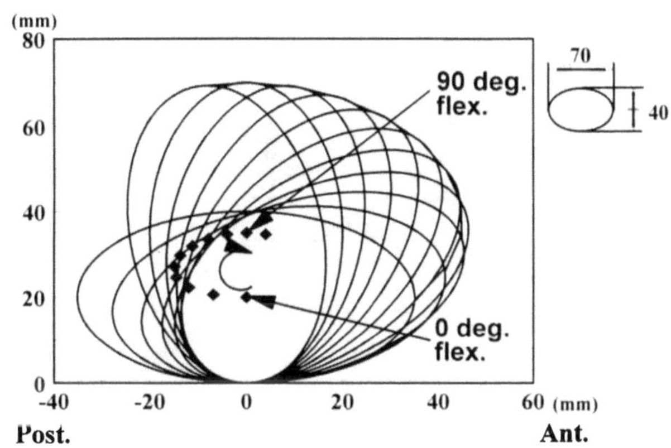

Fig. 3. Simulated locus of the femoral origin relative to tibial coordinate system

fits the oval of the femoral condyle. If the knee joint flexion motion consists only of a sliding movement, contact points relative to the tibial coordinate system are not moved but the femoral origin is moved. To eliminate this effect of the origin motion, we selected the reference point at the joint contact point. Because our study was in vivo, we could not dynamically measure the anteroposterior (AP) contact points. Instead, in this study the femoral condyle in the sagittal plane was simulated as an oval shape, and the origin was transferred to the contact point to compensate for the difference between the rollback phenomena and the results.

Materials and Methods

Description of the Motion Parameter

It is necessary to adopt a set of mathematical definitions that correspond as closely as possible to existing clinical terminology. The clinical terms we usually adopt are six parameters that are based on the kinematics that were described by Grood et al. [4]. The motion of the knee joint is characterized by the relative motion of the coordinate systems fixed on the femur and tibia.

Translation

The mediolateral tibial thrust or shift is motion along the X_f-axis, the anteroposterior tibial pull is a motion along the floating Y'-axis, and the proximal-distal motion is a movement along the Z_t-axis.

Mathematically, the three clinical translations are defined as the projections of the translation vector **H** along each of the axes of the joint coordinate system. The translations as defined by the internal production of the direction vector along each axis and the translation vector **H** are shown in Fig. 4. Usually

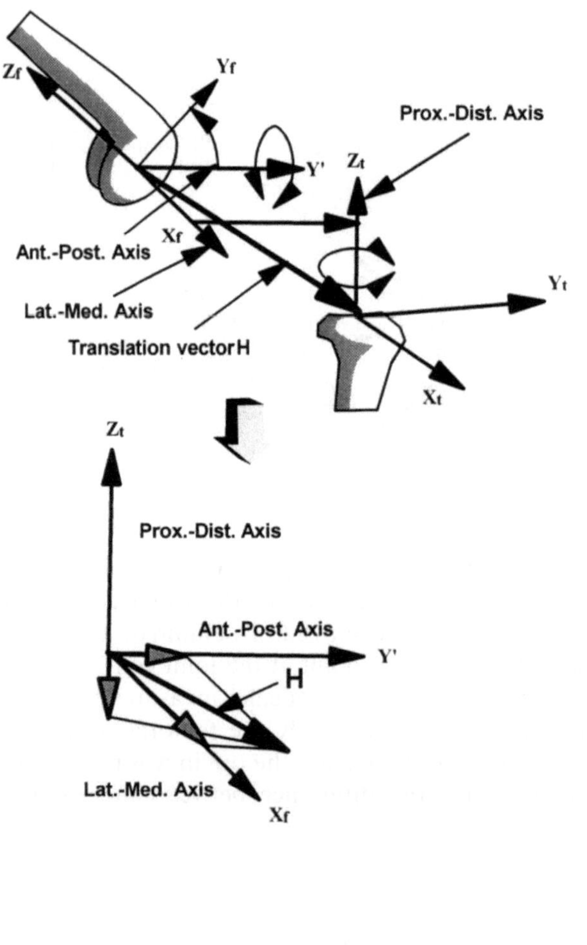

Fig. 4. Clinical translations defined by vector H and the projection of H onto the three axes

the translation vector **H** is directed from the femoral origin to the tibial origin, but we take a new reference point to determine the translation vector **H**. The new reference points are discussed later.

Rotation

For rotation angles, the same method as for translation can be adapted. According to clinical convention, we define the rotation parameters as follows (Fig. 5): flexion-extension of the knee is a rotation around the X_t-axis; adduction-abduction is a rotation around the floating Y_t-axis; and internal-external rotation is a rotation around the Z_t-axis.

Mathematically, the three clinical rotation angles are defined as the projections of screw displacement vector **w** of the joint coordinate system [5].

Femoral and Tibial Reference Points

Anteroposterior translation is measured to estimate the instability in an anterior cruciate ligament (ACL)-deficient knee (anteroposterior tibial adductor). The AP adductor is usually measured by stress X-rays in which reference points are selected from the contact point of the femoral-tibial joint in lateral projection. To compare these translations with our data, we have to select the reference points on the contact point in the three-dimensional space. We are not able to precisely determine the contact point, however, because the contact occurs over a finite area. An approximation method was determined as follows.

The origin of femoral coordinate system, shown in Fig. 6, is determined by the center of a rectangle containing the medial and lateral condyle, taken from the X-ray projection of lateral and AP views. We calculate points on the oval that fits the rectangle of each condyle in the sagittal plane (which is through the origin of the femoral coordinate system). We assumed the contact point (Fig. 7) is the closest point to the tibial joint plane, which is perpendicular to the tibial bone axis. The point depends on knee joint flexion angle and the reference point of the femur. The angle θ_0 (Fig. 8) is given as a function of flexion angle α:

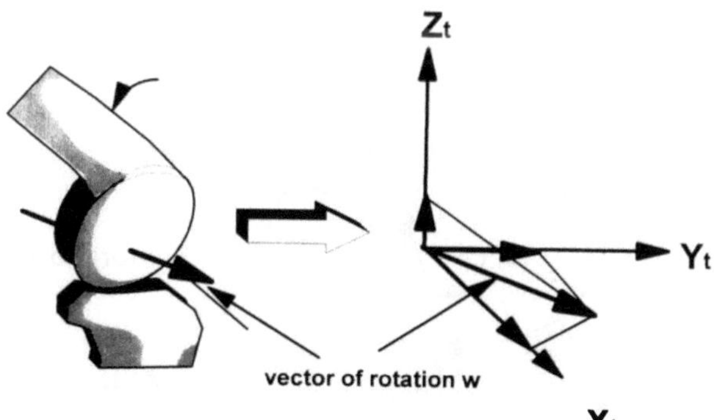

Fig. 5. Definition of clinical rotations and rotation vector **w** projected to the three axes in the tibial coordinate system

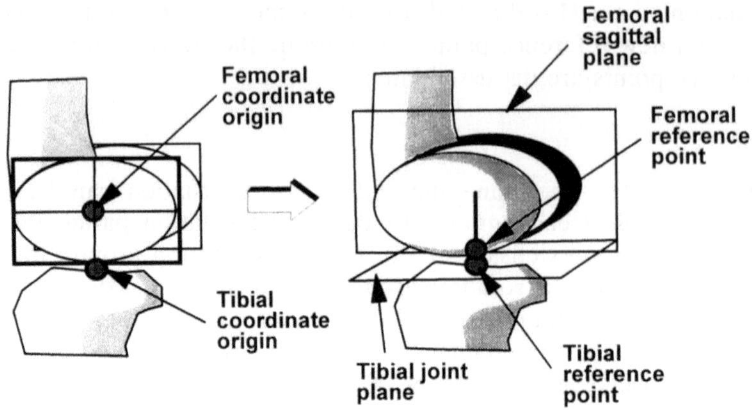

Fig. 6. Reference points for translation

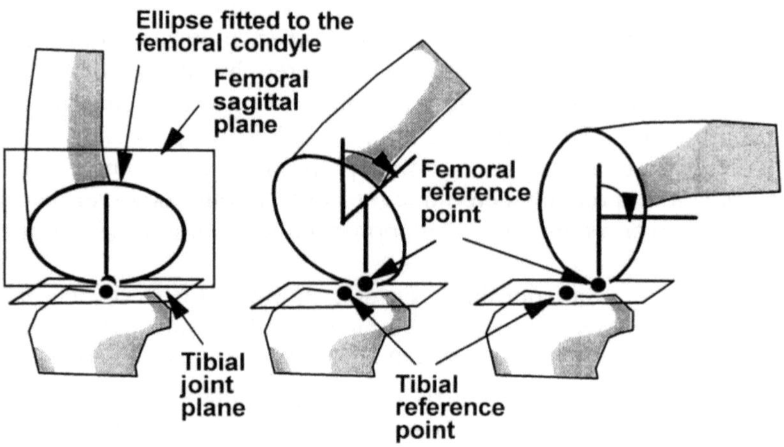

Fig. 7. Schematic depiction of knee joint motion

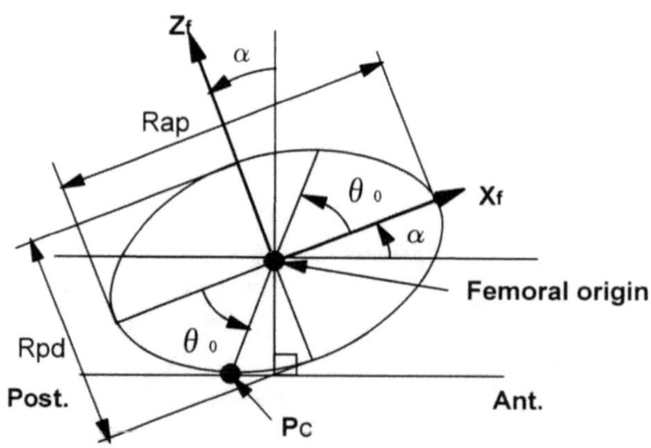

Fig. 8. The location of the contact point *Pc* expressed by flexion angle α and θ_0

$$\theta_0 = \tan^{-1}((R_{pd} \cdot \cos\alpha)/(R_{ap} \cdot \sin\alpha)) + \pi \tag{1}$$

Using the angle θ_0 and the flexion angle α, the position vector $P_c = (x_c, y_c, z_c)$ to the contact point is denoted.

$$\begin{aligned} xc &= 0 \\ yc &= R_{ap} \cdot \cos\alpha \cdot \cos\theta_0 - R_{pd} \cdot \sin\alpha \cdot \sin\theta_0 \\ zc &= R_{ap} \cdot \sin\alpha \cdot \cos\theta_0 + R_{pd} \cdot \cos\alpha \cdot \sin\theta_0. \end{aligned} \tag{2}$$

where xc is a coordinate of the X_f axis; yc is a coordinate of the Y_f axis, and zc is a coordinate of the Z_f axis.

The tibial reference point is located at the contact point in full flexion, and moves with the tibial coordinate system.

Results

The translations during active motion of three patients with Genesis TKA are shown in Fig. 2. Conventional motion parameters are shown in Fig. 2a and our current motion parameters in Fig. 2b. In Fig. 2b, the translation of the proximal-distal direction is decreased and the rollback phenomenon is recognized.

The rotation parameters of active motion of patients with TKA are shown in Fig. 9: conventional motion parameters in Fig. 9a, and the current motion parameters in Fig. 9b. In Fig. 9b the dispersion of the pattern is decreased.

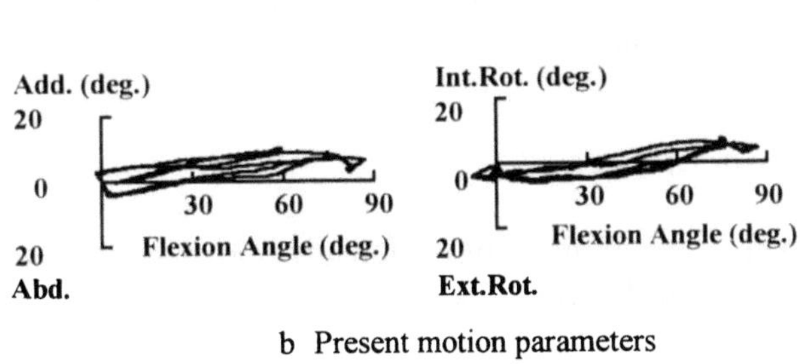

Fig. 9. Comparison of rotation parameters on the basis of the different definition

Discussion

The results of three-dimensional knee kinematic motion analysis depend on how the coordinate system or reference points were defined. The most important matter in setting up the coordinate system or reference points is the choice of the primary motion parameter.

Although motion described in different coordinate systems cannot be compared, such comparisons are often reported in the literature. Therefore, it is important to standardize the definitions of motion description and coordinate systems to allow comparisons among data of different researchers. By designing the standardized motion parameter close to the result of the observation, discussion between bioengineers and physicians is facilitated in clinical applications.

In our method, the contact point between the femur and tibia is assumed as a reference point although accurately speaking the joint contact place has a finite area. The parameter is decided by the reference point and is not be affected by the deformity of joint surface on the data. This approximation has much less influence on the data than the influence of individual differences or deformity of joint surfaces, and thus we can readily consider the roll-back phenomenon.

Grood designed the "screw axis" for a more clinical analysis of knee kinematics. First, the screw axis was set between 0° and 90° during passive motion. The reference point in the femur was defined as the intersect between the screw axis and the sagittal plane of the femur, and the tibial reference point was defined as the midpoint of femoral reference points between 0° and 90°. It was difficult to reproduce the Grood parameter in vivo when the knee had ligamentous instability or deformity of geometry. The current method requires only the geometry of the joint. In reference to rotational angle, the helical angle system was used because the Euler angle system is well known to include some bias.

Because methodology for the analysis of knee motion has not been established, the results depend on how the coordinate system or reference point is defined, which is the reason that the results of three-dimensional motion analysis are not in common clinical use. We wished to make it clear how the coordinate system or reference point was defined and how the data should be analyzed if there are some differences between the results of analysis and observed phenomena. In addition, the design of the parameter that facilitates the standing of rollback is important.

Conclusions

The new definition of a clinically relevant, three-dimensional motion parameter has been designed, developed, and discussed.

References

1. Terajima K, Hara T, Koga Y, Nagata T (1991) Development of three dimensional knee motion analysis using CR system. Jpn Soc Orthop Biomech 13:213–217
2. Ishii T, Koga Y, Takahashi Y, Terajiam K, Hara T (1991) Gait analysis of the osteoarthritic knee using new motion analyzing system. Jpn Soc Orthop Biomech 13:143–146
3. Terajima K, Tsuchiya Y, Shimizu M, Hara T, Koga Y, Ishii Y (1992) 3-dimensional motion analysis after total knee arthroplasty. Jpn Soc Orthop Biomech 14:371–374
4. Grood ES, Suntay WJ (1983) A joint coordinate system for the clinical description of three-dimensional motions: application to the knee. J Biomech Eng 105:136–144
5. McCarthy JM (1990) Introduction to theoretical kinematics. Massachusetts Institute of Technology Press, Cambridge

Gait Analysis of Patients with Varus Osteoarthritic Knees

Takashi Komura[1], Nobuhiro Tsumura[1], Seishi Sawamura[1], Masahiro Kurosaka[2], and Kosaku Mizuno[2]

Summary. We evaluated the lateral thrust of the knees in patients with varus osteoarthritis (OA). In 46 knees of 28 patients with OA and 20 knees of 10 healthy volunteers without knee symptoms, we measured the distance and the velocity of lateral knee movement in the stance phase of level walking by using a gait analysis system. The distance of lateral knee movement was 19.6 ± 7.4 mm in patients with OA and 13.6 ± 5.5 mm in healthy volunteers, and this difference was statistically significant ($P < 0.01$). When a laterally wedged insole was used, the distance of lateral knee movement decreased to 16.1 ± 7.6 mm in OA patients ($P < 0.05$). The value of lateral knee movement velocity showed no significant difference between OA patients and healthy volunteers. Measurement of the distance of lateral knee movement seemed to be a useful technique for quantitative analysis of lateral thrust of the knee. The decrease of lateral thrust may provide a rationale for using a laterally wedged insole in the treatment of varus osteoarthritic knees.

Key words: Gait analysis—Osteoarthritic knee—Wedged insole—Lateral thrust

Introduction

Varus knee is the most common type of deformity of osteoarthritic (OA) knees, and it has been reported that patients with varus OA knees show abnormal lateral movement of the knees in the stance phase of level walking. Although this phenomenon is well known as lateral thrust, it has never been

[1] Department of Orthopedic Surgery, Hyogo Rehabilitation Center, 1070 Akebono, Nishi-ku, Kobe, Hyogo, 651-21 Japan
[2] Department of Orthopedic Surgery, Kobe University, School of Medicine, 7-5-1, Kusunoki, Chuo-ku, Kobe, Hyogo, 650 Japan

analyzed quantitatively. Using the gait analysis system [1], we measured the lateral thrust of the knee in OA patients.

Materials and Methods

Forty-six knees of 28 patients with varus OA were evaluated in this study. Four men and 24 women, ranging in age from 57 to 80 years old (average, 67.3 years), were included in our study. The control group consisted of 20 knees of 10 healthy male volunteers. The average age of this group was 27.4 years, ranging from 22 to 33 years old.

An ELITE SYSTEM (Bioengineering Technology and Systems, Italy [2]), two force platforms (Kistler, Switzerland), and a personal computer (IBM, USA) were used for gait analysis. The ELITE SYSTEM consisted of four television cameras with an infrared stroboscope, a television monitor, and MOTION ELITE, which calculates three-dimensional coordinates of the markers attached to the body of the patient. Two cameras were installed on the right-side wall of the walkway and two other cameras on the left-side wall (Fig. 1). The patients were asked to walk straight on the two force platforms in the center of the walkway with ten small hemispherical passive markers attached on the surface skin of both their ears, acromions, greater trochanters of the femurs, lateral joint space of the knees, and lateral malleoli (Fig. 2). While the patients walked along the walkway, infrared rays were emitted from the

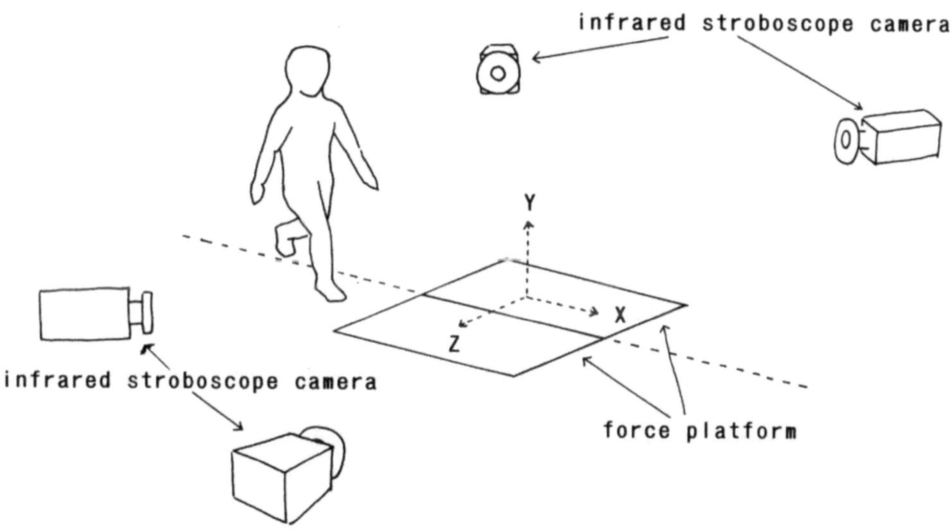

Fig. 1. Two cameras were installed on the right-side wall of the walkway and two cameras on the left-side wall. Gait analysis was conducted by recording the patients walking on two force platforms in the center of the walkway

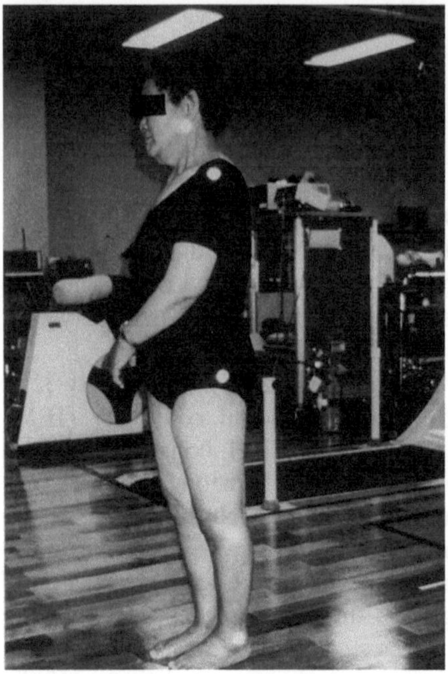

Fig. 2. Ten small hemispherical passive markers were attached on both ears, acromions, greater trochanters of the femurs, lateral joint space of the knees, and lateral malleoli

cameras and the reflections of the rays from the markers were detected. The original point of the three-dimensional coordinates was determined on the center of the force platforms, and the three axes (X, Y, Z) were drawn as illustrated in Fig. 1. MOTION ELITE calculated the position of each marker in three-dimensional coordinates (x, y, z) from the signals of the cameras. Kinematic data from the ELITE SYSTEM were matched with kinetic data obtained from the measurements from the force platforms. Four parameters were investigated: gait speed, cadence, distance of lateral knee movement (Zd), and peak velocity of lateral knee movement (Vp).

The line graphed in Fig. 3 shows the relationship between time and the X coordinates of the marker attached to the right ear. Gait speed was calculated from the inclination of this line.

The curve in Fig. 4 shows the relationship between time and the Z coordinates of the marker attached to the lateral joint space of the right knee. Z_0 is the Z coordinate of the marker at heel strike, and Z_1 is the maximum Z coordinate observed in the stance phase. Zd was defined as the distance of lateral knee movement in the stance phase and was calculated from the formula $Zd = Z_1 - Z_0$.

The curve in Fig. 5 represents the relationship between time and the transverse (Z) components of velocity of the marker attached to the lateral joint space of the right knee. Vp is the peak velocity of the lateral knee movement in the stance phase.

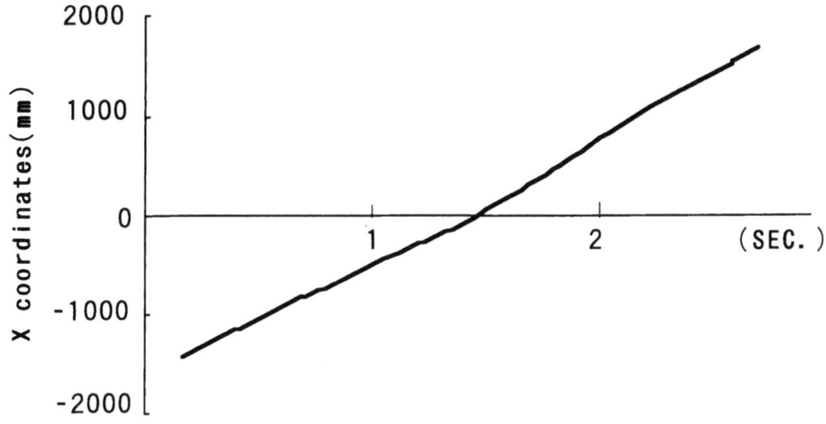

Fig. 3. The line shows the relationship between time and X coordinates of the marker attached to the right ear. Inclination of this line indicates gait speed

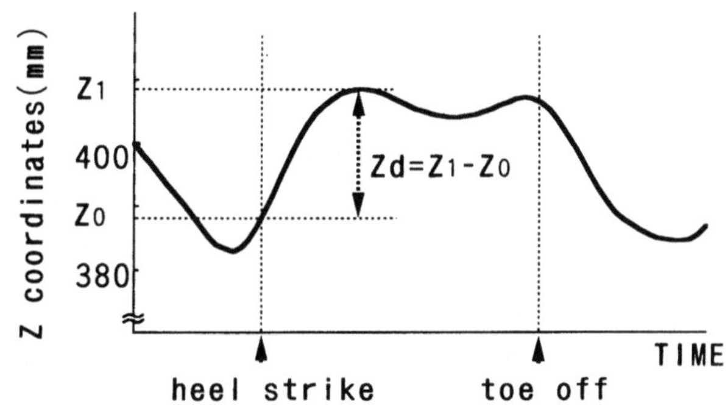

Fig. 4. Relationship between time and Z coordinates of the marker attached to the lateral joint space of the knee. Zd was calculated from the formula $Zd = Z_1 - Z_0$

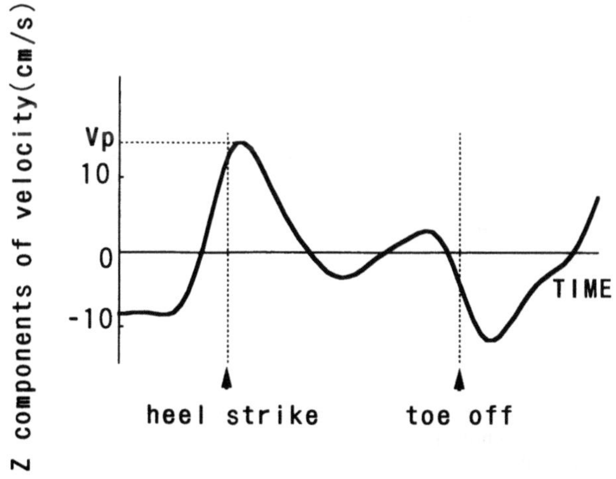

Fig. 5. Relationship between time and transverse (Z) components of velocity of the marker attached to lateral joint space of the knee. Vp is the peak velocity of the lateral knee movement in stance phase

Both OA patients and the healthy volunteers were asked to walk in the walkway on the force platforms. Three repetitions of free gait level walking were recorded, and each of the four parameters described was analyzed. The same gait analysis was repeated with a 10-mm-thick, laterally wedged insole, and the same parameters were again investigated.

Results

Gait Speed

The average gait speed (Table 1) of OA patients in free gait was 88.6 ± 19.4 cm/s, ranging from 41.1 to 114.9 cm/s. Little difference was found in gait with the laterally wedged insole, which ranged from 47.5 to 121.6 cm/s (average, 88.7 ± 18.6 cm/s). The average gait speed of the healthy volunteers in free gait was 139.8 ± 17.9 cm/s (range, 117.2–166.2 cm/s), and the laterally wedged insole also had little influence on their gait speed, the average speed being 142.7 ± 18.1 cm/s (range, 123.5–164.7 cm/s). The average gait speed was significantly greater in healthy volunteers compared to OA patients ($P < 0.01$).

Cadence

The cadence (Table 2) of free gait of OA patients ranged from 89.6 to 127.7 steps per minute (average, 108.2 ± 10.8 steps/min). The cadence of the gait did not change with the laterally wedged insole, with the average 107.8 ± 10.5 steps per minute ranging from 90.3 to 130.5 steps per minute. The cadence of the free gait of the healthy volunteers ranged from 115.4 to 126.1 steps per minute (average, 121.2 ± 4.2 steps/min). Again, the insole was found to have little influence on the cadence of the gait, the average being 120.5 ± 4.8 steps per minute (range, 111.9–125.1 steps/min). The OA patients walked at a slower pace than did healthy volunteers ($P < 0.01$), and the laterally wedged insole essentially had no influence on the cadence.

Table 1. Gait speed of osteoarthritis (OA) patients and healthy volunteers

	OA patients in free gait	OA patients with insole	Healthy volunteers in free gait	Healthy volunteers with insole
Gait speed (cm/s)	88.6 ± 19.4	88.7 ± 18.6	139.8 ± 17.9	142.7 ± 18.1

OA patients in free gait vs. OA patients with insole: N.S.
Healthy volunteers in free gait vs. Healthy volunteers with insole: N.S.
OA patients in free gait vs. Healthy volunteers in free gait: **

N.S., not significant.
**, $P < 0.01$.

Table 2. Cadence of OA patients and healthy volunteers

	OA patients in free gait	OA patients with insole	Healthy volunteers in free gait	Healthy volunteers with insole
Cadence (steps/min)	108.2 ± 10.8	107.8 ± 10.5	121.2 ± 4.2	120.5 ± 4.8

OA patients in free gait vs. OA patients with insole: N.S.; Healthy volunteers in free gait vs. Healthy volunteers with insole: N.S.; OA patients in free gait vs. Healthy volunteers in free gait: **

N.S., not significant.
**, $P < 0.01$.

Table 3. Distance of lateral knee movement (Zd) of OA patients and healthy volunteers

	OA patients in free gait	OA patients with insole	Healthy volunteers in free gait	Healthy volunteers with insole
Zd (mm)	19.7 ± 7.3	16.1 ± 7.6	13.6 ± 5.5	11.9 ± 6.1

OA patients in free gait vs. OA patients with insole: *; Healthy volunteers in free gait vs. Healthy volunteers with insole: N.S.; OA patients in free gait vs. Healthy volunteers in free gait: **

N.S., not significant.
*, $P < 0.05$; **, $P < 0.01$.

Distance of Lateral Knee Movement (Zd)

The distance of lateral knee movement (Table 3) of OA patients in free gait was 19.7 ± 7.3 mm, ranging from 5.6 to 33.6 mm. This lateral movement was significantly less (16.1 ± 7.6 mm) with the use of laterally wedged insole ($P < 0.05$). Healthy volunteers showed a significantly smaller distance of lateral knee movement compared to OA patients in free gait (13.6 ± 5.5 mm; $P < 0.01$). This lateral knee movement tended to be less with the laterally wedged insole, although the difference was not statistically significant (11.9 ± 6.1 mm with wedged insole).

Peak Velocity of Lateral Knee Movement (Vp)

The peak velocity of lateral knee movement (Table 4) tended to be higher in healthy volunteers (17.3 ± 14.8 cm/s) than in the OA patients (12.1 ± 6.5 cm/s). However, this difference was not statistically significant, nor was the comparison statistically significant between the peak velocity of the lateral knee movement of free gait and the gait with a laterally wedged insole, both in OA patients (10.1 ± 6.8 cm/s) and in healthy volunteers (18.3 ± 16.3 cm/s).

Table 4. Peak velocity of lateral knee movement (Vp) of OA patients and healthy volunteers

	OA patients in free gait	OA patients with insole	Healthy volunteers in free gait	Healthy volunteers with insole
Vp (cm/s)	12.1 ± 6.5	10.1 ± 6.8	17.3 ± 14.8	18.3 ± 16.3

OA patients in free gait vs. OA patients with insole: N.S.
Healthy volunteers in free gait vs. healthy volunteers with insole: N.S.
OA patients in free gait vs. healthy volunteers in free gait: N.S.

N.S., not significant.

Discussion

Kinematic analysis and kinetic analysis are generally conducted to investigate the gait. Photographs, cinematographics, videorecordings, and electric goniometers are usually utilized for kinematic analysis, and electromyography; force platforms, footprints, and accelerometers for kinetic study [3]. Although there is some difficulty in combining kinematic and kinetic studies, and a relatively small number of such studies have been reported, ideally both analyses are done at the same time. The use of an infrared stroboscope for kinematic analysis and of force platforms for kinetic analysis made this combination possible in our study. Also, reflection of infrared rays from the hemispherical passive markers that were used to detect the position of the reference points of the body enabled us to eliminate cables attached to the patients and therefore disturbance of the natural gait was avoided.

Although several studies of kinetic analysis of OA patients using force platforms have been reported [4,5], there are few kinematic analyses, especially of lateral thrust. Lateral thrust of the knee joint could be assessed by analyzing the distance, velocity, and acceleration of lateral knee movement. Studying the change of the femorotibial angle (FTA) during the stance phase of the gait is another way to evaluate the lateral thrust. Ishii et al. [6], using an electric goniometer and computed radiography, measured not only the lateral thrust in varus movement but also the external rotational translation of the tibia in relation to the femur. Using accelerometers, Ogata et al. [7] measured the acceleration of lateral knee movement and evaluated the effect of a laterally wedged insole. Although these studies showed the importance of the phenomenon of lateral thrust, the analyses of these studies were qualitative and quantitative analysis remained unclear. An electric goniometer used in Ishii's study had an undesirable amount of weight and several cables, which may have disturbed the natural gait. An accelerometer such as the one used in Ogata's study is so sensitive that even a slight change in gait speed greatly influences the results. Therefore, this kind of study seemed useful for qualitative analysis, or preoperative and postoperative evaluation, but had less apparent value for quantitative analysis.

When an optical instrument such as the infrared stroboscope is used for kinematic analysis, the change in FTA, the distance of lateral knee movement, and the peak velocity of lateral knee movement could be easily assessed. However, in this type of analysis the change of bony alignment such as the FTA could not be precisely analyzed because the surface markers attached to the skin were detected as the reference points.

Our study indicated that the patients with varus OA walked slowly with a smaller cadence when compared to healthy volunteers. Both OA patients and healthy volunteers showed lateral knee movement, and the peak velocity of lateral knee movement tended to be higher in healthy volunteers. Measurement of the distance of lateral knee movement was significantly larger in the OA patients, suggesting that this measurement may correlate with the so-called lateral thrust. Because the use of the laterally wedged insole yielded significantly less lateral movement, this might mechanically explain the function of the wedged insole in the level walking gait of OA patients.

In summary, lateral knee movement was slow in OA patients compared to healthy volunteers, but a significant amount of lateral movement was observed. Although use of the wedged insole did not influence the speed of lateral knee movement, the total amount of lateral knee movement was reduced. Further study, such as analysis of the lateral movement of ankle and hip joints and acceleration of lateral knee movement, should be required to elucidate the real mechanism of lateral thrust of the knee.

Conclusion

Measurement of the distance of lateral knee movement seemed to be useful for quantitative analysis of the lateral thrust. The distance of lateral knee movement of the OA patients was significantly larger than that of the healthy volunteers in free gait. With the use of a laterally wedged insole, gait speed, cadence, and peak velocity of lateral knee movement did not change but lateral knee movement decreased.

References

1. Komura T, Tsumura N, Nakagawa A, Sawamura S (1992) A gait analysis system and clinical application. J Hyogo Rehab Center 1:119–122
2. Ferringo G, Pedotti A (1985) ELITE: a digital dedicated hardware system for moment analysis via real-time TV signal processing. IEEE Trans Biomed Eng BME-32(11):943–950
3. Norimatsu T (1993) Gait analysis. J Joint Surg 12:1129–1141
4. Berman AT, Zarro VJ, Bossacco SJ (1987) Quantitative gait analysis after unilateral or bilateral total knee replacement. J Bone Joint Surg 69A:1340–1345

5. Steiner ME, Simon SR, Pisciotta JC (1989) Early changes in gait and maximum knee torque following knee arthroplasty. Clin Orthop 238:174–182
6. Ishii S, Koga Y, Takanashi Y, Terashima K, Hara T (1991) Gait analysis of the osteoarthritic knee using new motion analyzing system. Orthop Biomech 13: 143–146
7. Ogata K, Miura H, Kawamura H, Sugioka Y (1992) The effect of insole for osteoarthritic knee—an evaluation of the lateral thrust by accelerography. J Jpn Orthop Assoc 66:S706

Kinematic Gait Analysis of Patients After Total Knee Arthroplasty

Yasuhiko Hatanaka[1], Shinro Takai[2], Hitoshi Hase[1], Shuichi Kubo[1], and Yasasuke Hirasawa[2]

Summary. We examined the gait pattern of the lower extremities, using a three-dimensional motion analysis system, before and after total knee arthroplasty (TKA) in ten female patients with osteoarthritis (OA) of the knee. The gait analysis was also performed in five healthy female subjects. Two video sequences (frontal and lateral) were converted to digitized body location data, and then transformed into three-dimensional images. After smoothing, the joint angle displacement data were graphed and used to calculate the joint angle of the hip, knee, and ankle. Double knee action was observed in the control subjects, but it was not obvious in the OA group. The movement of the knee was related to the joint angles of the hip and ankle. Compared to that before surgery, the extensional angle during the early stance phase 4 weeks after the operation was significantly increased, and the flexion range at mid-swing 8 and 12 weeks after the operation was also increased. Thus, it was considered that the improvement of gait function observed in the patients after TKA resulted from pain relief and improvement of the alignment of the lower extremities obtained after arthroplasty, combined with recovery of muscle strength and increase of range of motion and proprioceptive and sensorimotor activities affected by postoperation theraputic exercises.

Key words: Kinematics—Gait analysis—Osteoarthritis—Total knee arthroplasty—Rehabilitation

Introduction

The gait abnormalities in patients with osteoarthritis (OA) of the knee are a biomechanical response to degeneration of the joint cartilage, pain, and

[1] Division of Rehabilitation, Kyoto Prefectural University of Medicine, Kawaramachi-hirokoji, Kamigyo-ku, Kyoto, 602 Japan
[2] Department of Orthopedic Surgery, Kyoto Prefectural University of Medicine, Kawaramachi-hirokoji, Kamigyo-ku, Kyoto, 602 Japan

secondary impairments. The main purpose of total knee arthroplasty (TKA) is the relief of pain and improvement of ambulatory function.

Previous gait analysis studies of patients with OA have been concerned with the ground reaction force or such time–distance factor measurements as cadence or step length. However, these kinetic data do not reflect the cause of the limp but rather the work done or the result of compensatory abnormal gait patterns. For a clinical understanding of the gait abnormality in a patient with OA, information regarding the movement or displacement of each individual joint is required. Three-dimensional motion analysis is now possible because of the recent development of image processing techniques.

This study was designed to examine the preoperative and postoperative three-dimensional motion of the lower extremities during walking, using a three-dimensional motion analysis system.

Materials and Methods

Ten female patients with unilateral osteoarthritis of the knee and a control group of five healthy female subjects were studied at Kyoto Prefectural University of Medicine. The patients ranged in age from 66 to 74 years, with a mean of 70.0 years; the control subjects ranged in age from 52 to 55 years, with an average age of 53.0 years. None of the women had any other skeletal or neuromuscular disease that affects gait.

Prosthetic Design and Postoperative Physical Therapy Programs

The Miller–Galante prosthesis, a design that permits retention of the posterior cruciate ligament, was used. On the day following TKA, passive range of motion (ROM) exercise with no restriction was begun, and a continuous passive motion unit was also employed. Isometric quadriceps sets and straight leg-raise exercises were performed to prevent muscle atrophy and to decrease swelling. Two weeks after TKA, weight bearing at 30% of maximum was permitted. Four weeks after the operation, full weight-bearing ambulation and stair climbing were performed, and all the patients were able to leave the hospital with no assistance at the time of discharge.

Gait Evaluations

Gait evaluations were conducted just before surgery and at 4, 8, and 12 weeks postoperatively. Each subject was instructed to walk at a self-selected speed on a 6-m walkway. Two video cameras were used to record the gait pattern; one camera was placed in front of the walkway and the other to the side of the walkway. The cameras were placed exactly 4 m apart for precise analysis.

The Ariel Performance Analysis System was used to process the data (Fig. 1). This system consists of an AST Premium 286 computer, graphic monitors, video cameras, a video playback unit, and a mouse. The two video sequences from the VCR were captured, and the video images were then converted to digitized body joint location data using the mouse. The digitized data were saved for the transformation module, which converted them into true three—dimensional images. The transformation process consisted of the two distinct phases of time synchronization of the two video sequences used to record the motion, followed by computation of the true image coordinates for the body joints in each frame from multiple sets of digitized coordinates. Because the digitized data generally contain random error or noise from imprecision in locating the positions of the body joints, we applied a smoothing module to eliminate such errors and noise from the computed image coordinates by a cubic spline method. The smoothed data were graphed as images of joint angle displacement, and the waveforms and amplitudes of the graphed data for the patients were compared with those for the control subjects.

Differences between the groups in mean data and within the patient group at various times postoperatively were analyzed with the Student's *t* test, and a *P* value of less than 0.05 was accepted as significant.

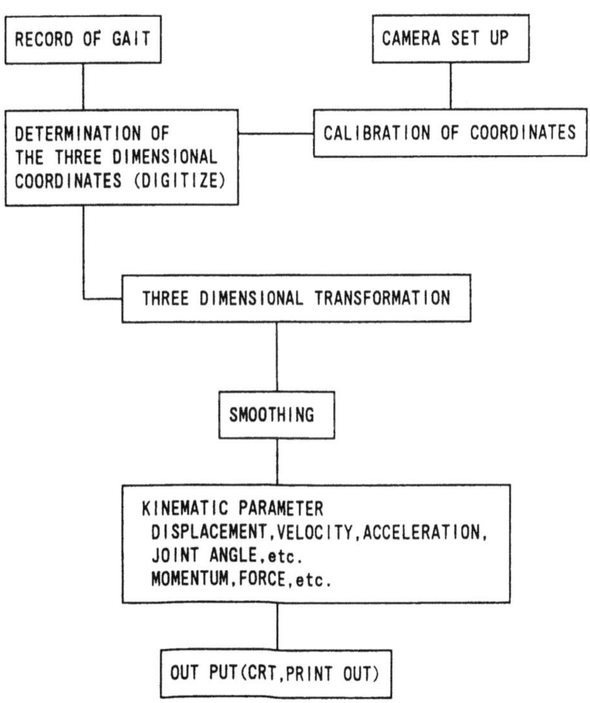

Fig. 1. Procedure of three-dimensional gait analysis

Results

Preoperatively, the gait patterns of the patients showed many variations and each subject showed a different waveform. Nevertheless, several common gait patterns, as shown in Table 1, could be observed (Figs. 2 and 3).

Heel Contact

In the control group, the mean flexion angle of the knee in the sagittal plane at heel contact was 2.5° ± 1.6° and that of the hip was 23.1° ± 4.3°. In the patient

Table 1. Joint angles of the hip and knee in osteoarthritis (OA) group and control subjects

	Control (n = 5), degrees	OA patient (n = 10), degrees
Heel contact		
Knee	2.5 ± 1.6	30.1 ± 13.1**
Hip	23.2 ± 4.3	23.8 ± 17.3
Midstance		
Knee	2.9 ± 2.0	22.7 ± 12.8**
Hip	·2.5 ± 2.1	18.5 ± 17.4**
Push-off		
Hip	15.8 ± 3.2	−14.6 ± 16.9**
Midswing		
Knee	52.1 ± 4.3	42.6 ± 18.9*

**, $P < 0.01$; *, $P < 0.05$.

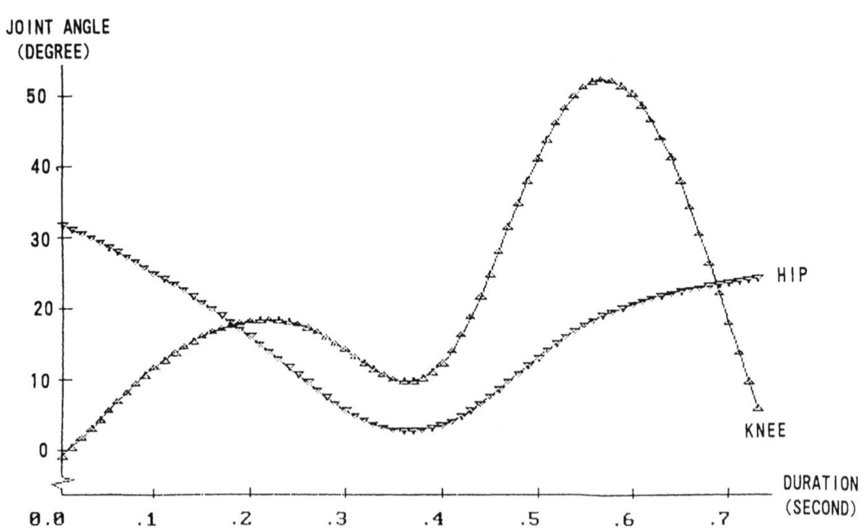

Fig. 2. Joint angles of hip and knee of typical case in control subjects (55-year-old woman) during gait

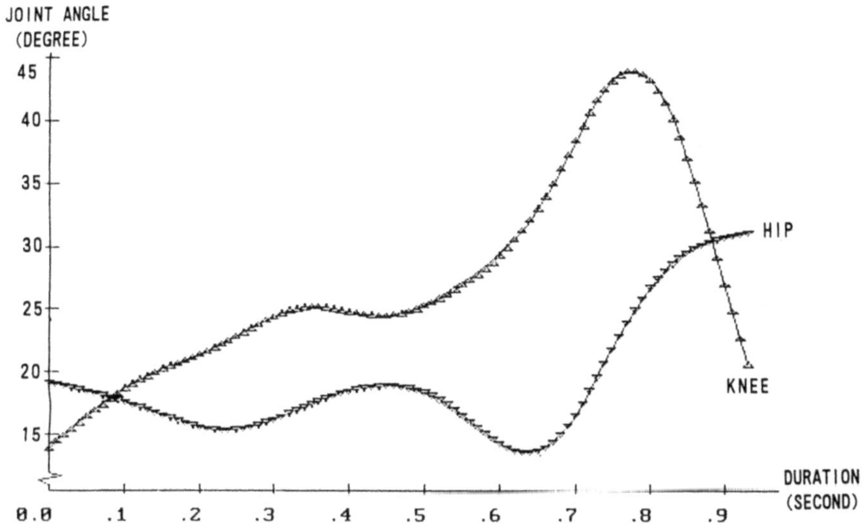

Fig. 3. Joint angles of hip and knee of typical case of patient with osteoarthritis (OA) (70-year-old woman) during gait

group, these angles were 30.1° ± 13.1° and 23.7° + 17.3°, respectively. There was no difference between the groups in the angle of the hip. However, the angle of the knee in the patient group was significantly greater than that in the control subjects ($P < 0.01$).

Midstance

In the control group, the mean flexion angle of the knee in the sagittal plane at midstance was 2.9° ± 2.0° and that of the hip was 2.5° ± 2.1°. In the patient group, these angles were 18.5° ± 17.4° and 22.7° ± 12.8°, respectively. The angles of both the knee and the hip in the patient group were greater than those of the control subjects (for both, $P < 0.01$).

Acceleration

During acceleration in swing phase, the mean maximum flexion angle of the hip in the sagittal plane was 14.7° ± 16.9° in the control group and 42.7° ± 18.9° in the patient group ($P < 0.01$).

Midswing

At midswing, the mean flexion angle of the knee in the sagittal plane was 52.1° ± 4.3° in the control group and 42.7° ± 18.9° in the patient group ($P < 0.05$).

Joint Angle of the Ankle

In each group, the change of the joint angle of the ankle in the sagittal plane was synchronized with that of the knee, and two repetitions of dorsiflexion and plantarflexion were seen during a single gait cycle. However, in the patient group, this change was less marked than that in the control group. In addition, during the entire gait cycle, the joint angle of the ankle in the OA patients was rather dorsiflexed compared with that in the control subjects (Fig. 4).

Postoperative Characteristics

The most characteristic feature of the postoperative change in the joint angle of the knee in the patient group was a decrease in the values of the extension angles at heel contact and midstance and of the flexion angle during the initial swing phase. As shown in Table 2, the joint angles during walking were markedly increased compared to those preoperatively (Fig. 5).

Heel Contact

At 4 weeks postoperative, the flexion angle of the knee at heel contact in the sagittal plane was decreased compared to that seen preoperatively ($P < 0.05$), and was further decreased at 8 weeks ($P < 0.05$) and 12 weeks postoperatively ($P < 0.05$). Twelve weeks postoperative, the flexion angle of the knee had decreased to a value similar to that in the control subjects.

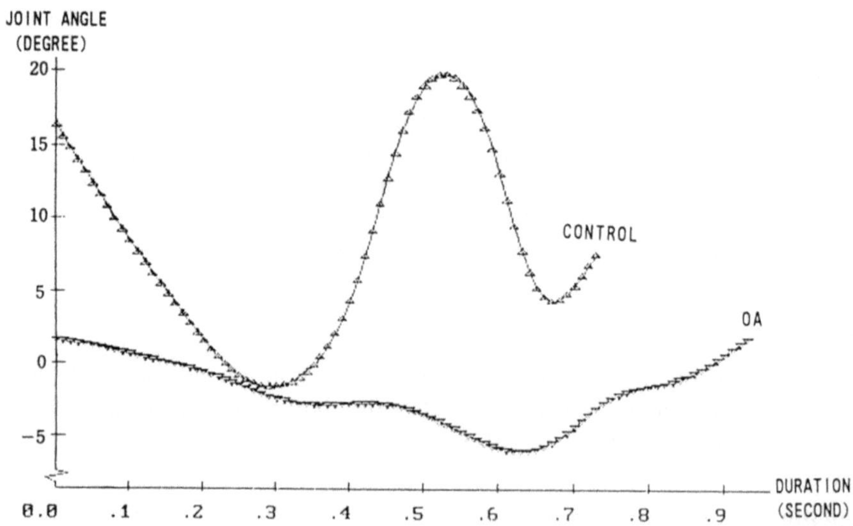

Fig. 4. Comparison of the joint angle of the ankle between typical OA patient (70-year-old woman) and control subject (55-year-old woman)

Table 2. Preoperative and postoperative changes of joint angles of the knee in OA group ($n = 10$)

	Heel contact	Midstance	Midswing
Before operation	30.1 ± 13.1	22.7 ± 12.8	42.6 ± 18.9
Postoperative			
4 Weeks	25.1 ± 12.3*	16.4 ± 2.9*	43.9 ± 9.8
8 Weeks	21.7 ± 10.3*	12.2 ± 3.4*	46.2 ± 10.8*
12 Weeks	17.9 ± 9.3*	9.7 ± 5.8*	50.9 ± 5.6*

*, $P < 0.05$.

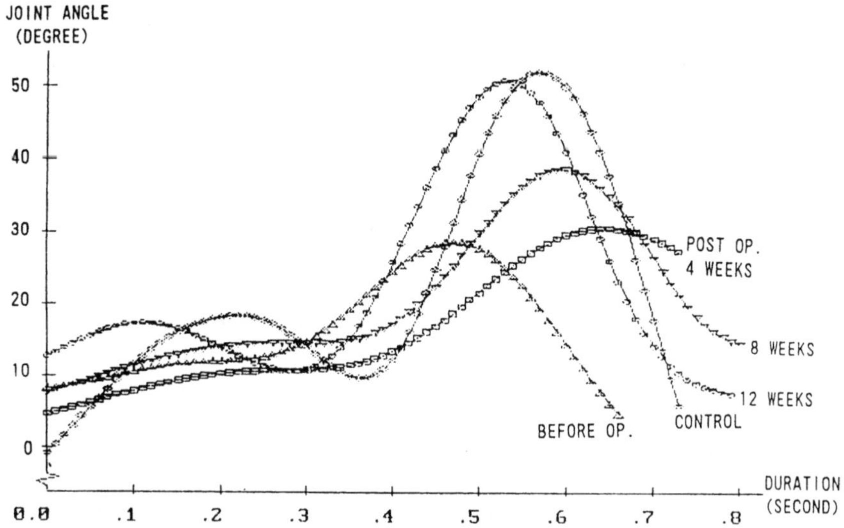

Fig. 5. Comparison of the joint angle of the knee between typical OA patient (70-year-old woman) and control subject (55-year-old woman)

Midstance

The flexion angle of the knee at midstance 4 weeks postoperatively, like that at heel contact, decreased compared to that before the operation ($P < 0.05$) and further decreased at 8 weeks ($P < 0.05$) and at 12 weeks ($P < 0.05$). The flexion angle of the knee 12 weeks after TKA was similar to that in the control subjects.

Swing Phase

There was no difference between the flexion angle of the knee during swing phase at 4 weeks postoperatively and that before the operation. That same measurement taken at 8 weeks, however, was increased compared to that before surgery ($P < 0.05$).

Discussion

Limitation of the range of motion, especially that of extension, has an influence on the gait of patients with osteoarthritis. In normal gait, the knee shows two flexion and two extension peaks during each gait cycle. It is fully extended before heel contact, flexes early in the stance phase, extends again at about midstance, then flexes again, reaching a peak early in the swing phase. It extends again before the next heel contact. These actions contribute to the weight support and smooth anterior shift of the gravity center of the body in the early stance phase and to the generation of the propulsive force during pushing backward against the ground in the late stance phase. In patients with osteoarthritis, these functions are reduced, mainly because of the limitation from the pain of weight support during the stance phase and the change of the alignment of the lower extremities as the result of joint degeneration, secondary joint contracture, and weakness of the quadriceps muscle. The compensative position of hip flexion or anterior flexion of the trunk results in the reduction of the time–distance factors of the gait pattern (i.e., stride length and step length), and a secondary pathological gait pattern such as circumduction of the hip or elevation of the pelvis can also be observed.

Four weeks postoperative, improvement not only of these extensional limitations caused by to pain, joint degeneration, and contracture but also of the gait abnormalities during the stance phase was observed. However, muscular activity and proprioceptive sensation had been reduced because of preoperative and postoperative inactivity. Also, patients showed a short step length or could not swing the leg forward sufficiently. The patients who had undergone bilateral TKA frequently showed a waddling gait. Some patients showed reduced walking speed and prolonged cadence, which might have been caused by insufficient knee flexion during the early swing phase after the terminal rocker.

Eight and 12 weeks postoperatively, these functions during swing phase were improved. The reasons for improvement were considered to be increases in activity level, walking speed, and the strength of quadriceps muscle.

Walking speed and cadence depend on the duration of the stance phase of the involved leg or the function of extensional weight support. In the normal subjects, the single stance time was 40% of the gait cycle, but that in the patient group was decreased; the duration of the swing phase of the contralateral leg was also decreased, and the patients could not advance the uninvolved leg adequately. Furthermore, in the normal subjects the anterior inertia moment of the gravity center of the body was preserved and transferred efficiently. However, in the OA patients preoperatively and 4 weeks postoperatively these inertia moments themselves were decreased and the range of movement of the upper extremities was also decreased. As regards smoothness, the normal subjects showed so-called jointed dual pendulum-like movement and could walk with only trace activity of the muscles around the knee. In the patient group, however, this dual pendulum-like movement, which may be helpful in acceleration in the early swing phase, was not observed, and the gait pattern was quite inefficient and antalgic. The improvement during the swing phase observed at 8 and 12 weeks might be related to the recovery of function during this phase and the improvement of the efficiency of the gait pattern.

Andriacchi et al. [1] compared the gait patterns of patients with prostheses of five different designs. They investigated the characteristics of abnormal gait in patients with knee replacement by evaluating the pattern of external flexion-extension moment about the fixed center of the knee joint. The pattern characterized as normal was that found to occur in 80% of the control subjects. In the two abnormal moment patterns, extrinsic moment was maintained predominantly by either flexion of the joint (flexional moment pattern) or extension of the joint (extensional moment pattern) throughout the stance phase. The abnormality in the flexional and extensional moment patterns was observed during the middle portion of the stance phase. One explanation for these abnormal characteristics of gait could be that, postoperatively, the patients continue to walk using the pattern that they learned before treatment. Another possible cause for these abnormalities in walking is abnormal muscle function. Abnormal muscle function could be caused by a partial loss of proprioceptive control or by a reduction or imbalance in muscle capacity after joint replacement. Both proprioceptive control and muscle capacity could be influenced by interaction of the joint kinematics with the surrounding soft tissues and muscle. In fact, proprioceptive exercise and closed kinetic training have been adopted in muscle-strengthening exercises, but their quantitative evaluation has rarely been undertaken.

Berman et al. [5] evaluated the time–distance factors of stance time and step length preoperatively and postoperatively in 35 patients. They divided these patients into three groups consisting of those who underwent a unilateral TKA and showed no signs or symptoms of OA in the contralateral knee (group I), those who underwent a unilateral TKA and showed asymptomatic OA in the contralateral knee (group II), and those who had undergone bilateral TKA (Group III). Postoperatively, the degree of improvement in group II was not as marked as that in groups I and III. They concluded that even asymptomatic OA can impair gait, that bilateral TKA can yield excellent results, and that the use of the lower limbs in OA patients should be as symmetrical as possible. They also concluded that 4 months are required for the recovery of the gait pattern in the majority of patients who undergo TKA. The patients in our study, who are comparable to group II patients in the study by Berman et al., showed a good recovery 3 months after TKA.

Many investigators have studied the gait function after TKA using various apparatuses and methods, including force platform measurement [6–9], radiographical static alignment [11], and measurement of abduction-adduction moment [10]. There is a consensus that improvement can be obtained in almost all patients [12–15]. The inferred reasons for improvement are the complicated interaction of pain relief and improvement of the alignment of the lower extremities after TKA, together with recovery of muscle strength and increase of range of motion as well as proprioceptive or sensorimotor activities as a result of the theraputic exercises.

In this study, we performed quantitative analysis of these pathological characteristics and found that this method can be useful clinically and in the assessment of the effectiveness of treatment.

Conclusions

1. Gait analysis was performed in ten patients with osteoarthritis (OA) of the knee and five healthy subjects using a three-dimensional motion analysis system.
2. Double knee action was observed in the control subjects, but it was not obvious in the OA group before TKA. A relationship between the movement of the knee and the joint angles of the hip and ankle was found.
3. Four weeks after TKA, the extensional limitation caused by pain, joint degeneration, and contracture in the OA group was improved as were the gait abnormalities during the stance phase.
4. Eight and 12 weeks after TKA, the flexional range during the swing phase was improved. The improvement was considered to result from increases of activity, walking speed, and strength of the quadriceps muscle.
5. The method employed in this study is believed to be useful in clinical kinematic gait analysis of patients with OA of the knee.

References

1. Andriacchi TP (1979) Three dimensinal coordinates data processing in human motion analysis. J Biomech Eng 101:279–283
2. Whittle M (1990) Normal gait. In: Gait analysis: an introduction. Butterworth Heinemann, London, pp 48–90
3. Andriacchi TP (1982) The influence of total knee-replacement design on walking and stair-climbing. J Bone Joint Surg 64A:1328–1335
4. Knott M (1968) Proprioceptive neuromuscular facilitation. Hoeber Medical, New York
5. Berman A (1987) Quantitative gait analysis after unilateral or bilateral total knee replacement. J Bone Joint Surg 69A:1340–1345
6. Khodadadeh S (1987) Quantitative approach to osteoarthritic gait assessment. Eng Med 16(1):9–14
7. Andriacchi TP (1988) Biomechanics and gait analysis in total knee replacement. Orthop Rev 17(5):470–473
8. Wang JW (1990) The influence of walking mechanics and time on the result of proximal tibial osteotomy. J Bone Joint Surg 72A:905–909
9. Chao EYS (1983) Nomative data of knee joint motion and ground reaction force in adult level walking. J Biomech 16:219–233

10. Herzog W (1988) The relation between the resultant moments at a joint and the moments measured by an isokinetic dynamometer. J Biomech 21:5–12
11. Laughman RK (1984) Functional evaluation of total knee replacement. J Orthop Res 2:307–313
12. Kim W (1993) Biomechanics of the knee. In: Total knee replacement. Raven, New York, pp 9–57
13. Winter DA (1987) The biomechanics and motor control of human gait. University of Waterloo Press, Waterloo, Ontario
14. Winter DA (1985) Concerning the specific basis for the diagnosis of pathological gait and for rehabilitation protocols. Physiother Can 37:245–252
15. Wittle MW (1982) Calibration and performance of a three-dimensional television system for kinematic analysis. J Biomech 15:185–196

Three-Dimensional Motion Analysis After TKA: The Effect of PCL Resection

Y. Ikezawa[1], G. Omori[1], H. Shimagaki[2], Y. Koga[3], S. Terashima[4], K. Terajima[4], and T. Hara[4]

Summary. Utilizing an original goniometer system and biplanar X-ray methods, the three-dimensional (3-D) knee motion after total knee arthroplasty (TKA) was analyzed. Seven patients who underwent Yoshino–Shoji (Y/S) total knee arthroplasty and who had no complaints in their daily activity and could walk without any support were studied. The mean follow-up period of this current study was 32 months after surgery. Clinical and radiological assessments were also evaluated. The following measurements were performed for this study: active knee motion (nonweight and with 1 kg weight), level walking, and going up and down stairs. In active knee motion (nonweight), two types of tibial translation, anterior or posterior, were seen with knee flexion less than 70°. At flexion greater than 70°, all knees showed anterior tibial translation. The result of gait analysis showed no double knee action and diminished flexion angle in swing phase. Going up or down stairs showed a basically similar motion pattern, but knee flexion and extension differed when going down stairs. At more than 70° knee flexion, the tibia moved forward by quadriceps muscle tension. This result demonstrates that quadriceps muscle power is important for compensation of posterior cruciate ligament (PCL) resection. Lack of normal rollback motion, however, will lead to stress concentration and is the cause of wear and loosening of the tibial component.

Key words :3-D motion analysis—Y/S TKA system—Posterior cruciate ligament—Sagging—Femoral rollback

[1] Department of Orthopaedic Surgery, Niigata University School of Medicine, Asahimachidori, Niigata, 951 Japan
[2] Department of Orthopedic Surgery, Tsubame Rousai Hospital, Japan
[3] Department of Orthopedic Surgery, Niigata Kobari Hospital, Niigata, Japan
[4] Faculty of Engineering, Niigata University, Niigata, Japan

Introduction

Prosthetic design and the function of the cruciate ligaments are the most significant factors in knee kinetics after total knee arthroplasty (TKA). The posterior cruciate ligament (PCL) is sometimes divided in TKA because it is thought to be one of the causes of flexion contracture. The PCL, however, is an essential structure for rollback of the femoral condyles in the normal knee, and therefore when the PCL is inevitably sacrificed during TKA the use of a cruciate-substituting design such as the posterior-stabilized type is generally recommended. There is concern, however, whether saving the PCL will improve the function and durability of a TKA.

Shoji et al. [1] reported that normal femoral rollback occured even after the transection of both cruciate ligaments by sequential lateral radiographs. They also indicated that posterior rollback was not caused by the cruciate ligaments. Andriacchi et al. [2] reported that patients with less constrained, cruciate retaining designs of TKA have a more normal gait during stair climbing than patients with a more constrained, cruciate-sacrificing design.

In this investigation, we performed motion analysis after the Yoshino–Shoji (Y/S) II TKA system because it has a less constrained, cruciate-sacrificing design, and estimated the effect of PCL resection in TKA using the three-dimensional (3-D) motion analysis system developed in our laboratory.

Materials and Methods

Materials

Four women (seven knees) who underwent Y/S TKA (Fig. 1) were studied. The mean follow-up period to the current study was 32 months after the initial operation. All patients have had clinical and radiographic evaluation on their clinic visit.

All patients could walk without any support, and their range of motion of the knee joint was 0/124 on average. Although sagging of the joint was seen in all patients in clinical evaluation, none of the patients complained about it. In radiographic evaluation, the mean angle of each component according to the Knee Society [3] was measured: α, 94.4; β, 87.9; γ, 5.7; δ, 88.4.

Three-dimensional (3-D) Motion Analysis

The electrogoniometer with 6-degree of freedom that was developed at Niigata University was attached to the patient's thigh and leg with a strap (Fig. 2). The linear accuracy of this instrument is $\pm 500\,\mu m$ and the angular accuracy is $\pm 0.5°$. The conductive rubber sensor, which was also developed at Niigata University, was attached to the sole of the big toe and heel to record the stance phase and the swing phase.

The biplanar radiograph was taken to define the coordinate system. Two coordinate origins, origin of femur (OF) and origin of tibia (OT), were also

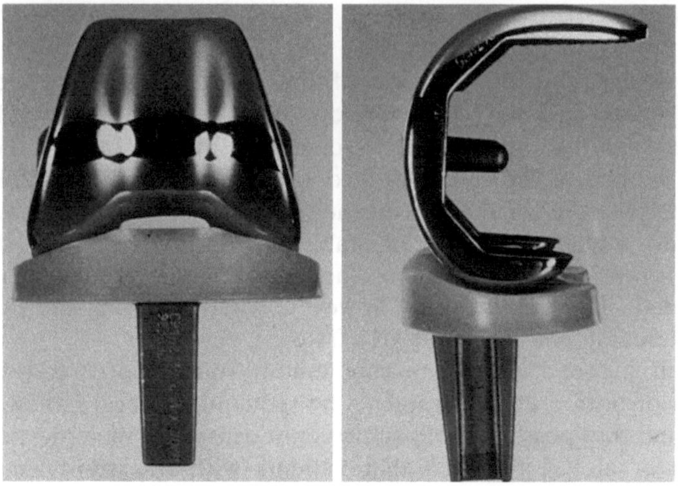

Fig. 1. Y/S knee components. Symmetrical femoral component. The intercondylar notch of the tibial component is cut out at the posterior third and the articular surface is flattened posteriorly

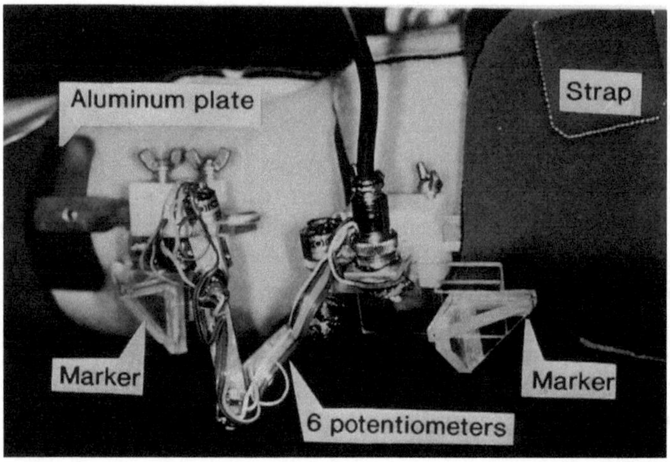

Fig. 2. The 6-degree of freedom electrogoniometer was attached to the patient's thigh and leg with strap

defined and constructed on the femoral and tibial components, respectively (Fig. 3).

The following measurements were made: (1) active knee motion in chair sitting (nonweight and with 1 kg weight) ($n = 7$); and (2) level walking ($n = 7$) and going up and down stairs ($n = 2$). The height of the stair was 15 cm, as

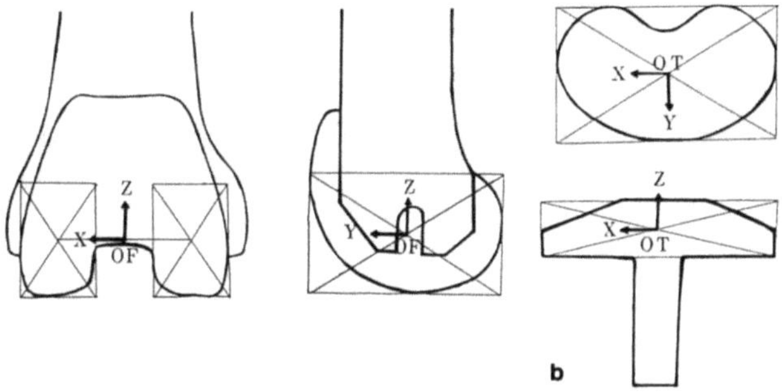

Fig. 3a,b. The coordinate system. **a** *Front view and side view*; *OF*, origin of the femoral component. X–Z OF was defined from front view; Y–Z OF was defined from side view; 3-D OF was constructed from these 2-D OF. **b** *Top view and front view*; *OT*, origin of the tibial component. 3-D OT was defined in the same manner

used for physiotherapy. All measurements were recorded and analyzed with a computer according to our three-dimensional motion analysis system [4].

The 3-D knee motion was displayed in two different ways: one used 6-degree of motion parameters (extension–flexion, abduction–adduction, external–internal rotation, anterposterior translation, mediolateral translation, and proximal-distal translation), and the other used helical axes [5]. The helical axis display gives us better visualization of the image of rotation axis of the femur relative to the tibia at each knee flexion.

This time we used the helical axes to compare the 3-D motion of Y/S with that of Mark II, which has geometry similar to the Y/S technique but retains the PCL [6].

Sequential Lateral Radiograph

A sequential lateral X-ray was also taken after Y/S TKA to confirm the anteroposterior translation of the tibial component during active knee flexion.

Results

Active Knee Motion

Sagging was seen in all knees objectively during clinical evaluation; two knees under nonweight and three knees under 1 kg of weight in active motion showed sagging resembling acute anterior translation in initial knee flexion (Fig. 4).

In active motion under nonweight, three knees showed anterior tibial translation (anterior translation group) and the other four knees showed posterior tibial translation (posterior translation group) with knee flexion less than 70°. At more than 70°, all knees showed anterior tibial translation.

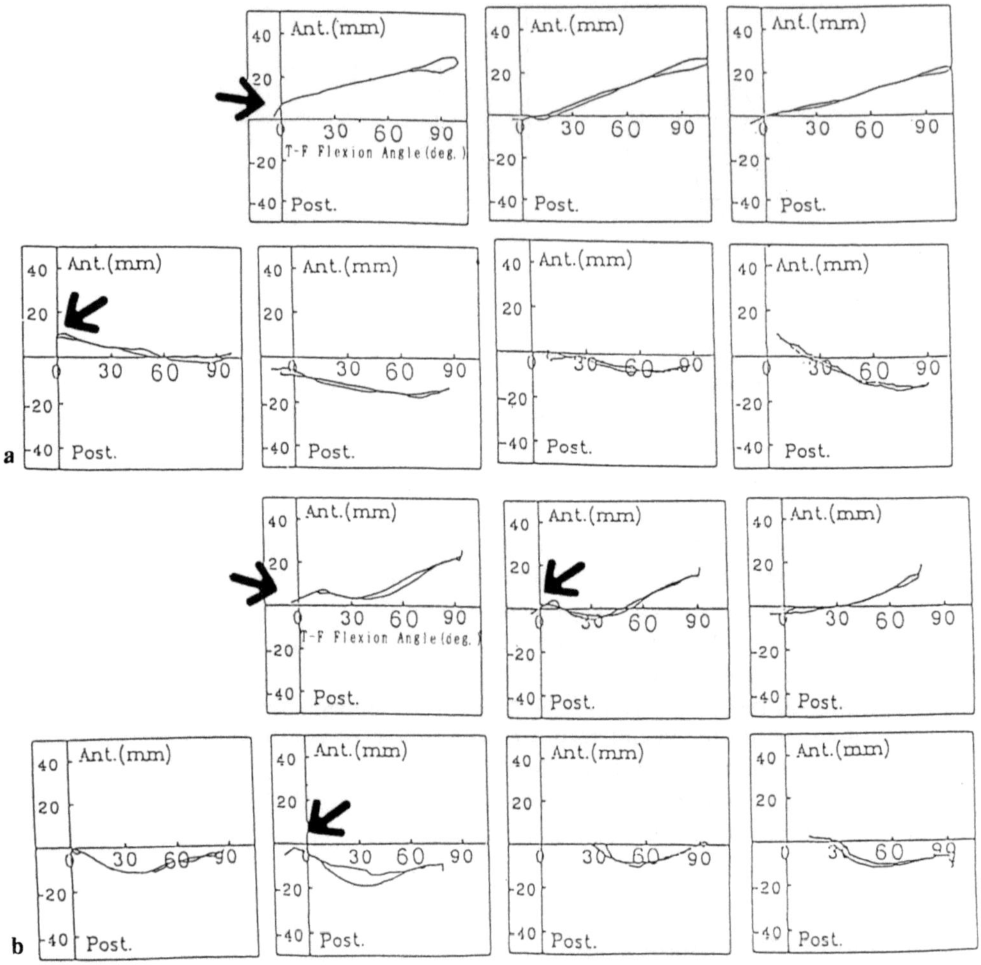

Fig. 4a,b. Anterior (*ant.*)/posterior (*post.*) shift of knee joint in active motion after TKA. **a** Nonweight: two knees showed a sagging motion resembling acute anterior translation (*arrows*). In early flexion, two types of translation were seen: anterior or posterior. **b** With 1 kg weight: three knees showed sagging motion resembling acute anterior translation (*arrows*). All knees showed posterior tibial translation in early flexion

In active motion with 1 kg of weight, all knees showed posterior tibial translation in initial knee flexion followed by anterior tibial translation. This was similar to posterior translation group in active motion without weight.

No screw-home movement was seen, but reverse screw-home movement was found in four of seven knees in active motion without weight (Fig. 5).

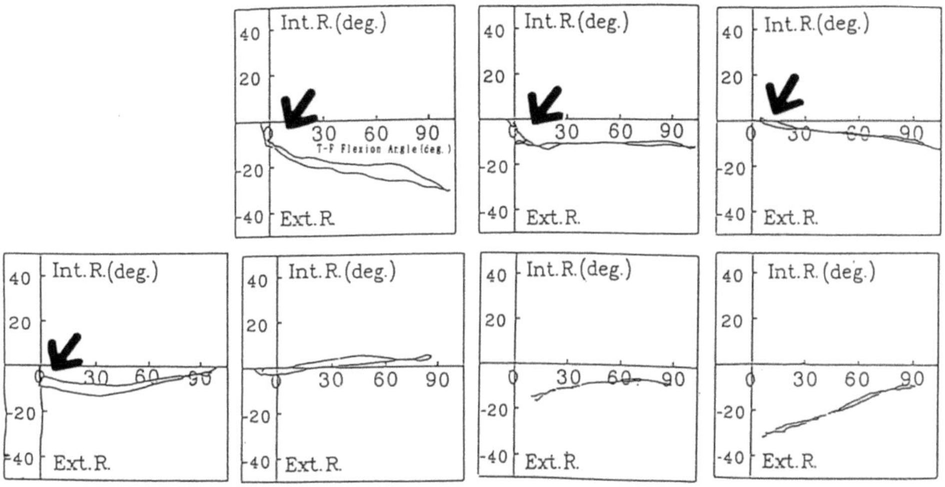

Fig. 5. Rotation of the knee joint in active motion after TKA (nonweight). The screw-home movement disappeared, and four knees showed reverse screw-home movement (*arrows*)

Gait Analysis of Level Walking and Stair Climbing

The result of gait analysis showed no double knee action and diminished flexion angle in swing phase by comparison with a normal knee. The mean maximum knee flexion angle in swing phase was 50°; the other five parameters showed a motion pattern similar to that of normal knee motion (Fig. 6a,b).

In a motion cycle of knee joint during going up or down stairs, both motion patterns were basically similar. When a patient went down stairs, however each motion parameter showed a different course between knee flexion and extension (Fig. 7a,b).

3-D Motion Using Helical Axes

Display of the helical axes of active knee motion in Y/S TKA showed two patterns. In the anterior translation group, the helical axes showed large anteroposterior translation and internal rotation with knee flexion. In the posterior translation group, the helical axes mainly rotated externally and showed a slight anteroposterior translation. Both groups showed a larger variance in comparison with Mark II (Fig. 8).

The Sequential Lateral Radiograph

Sequential X-ray taken after the Y/S TKA system was used showed that the tibia placed posteriorly in extension but went relatively anteriorly over 70°

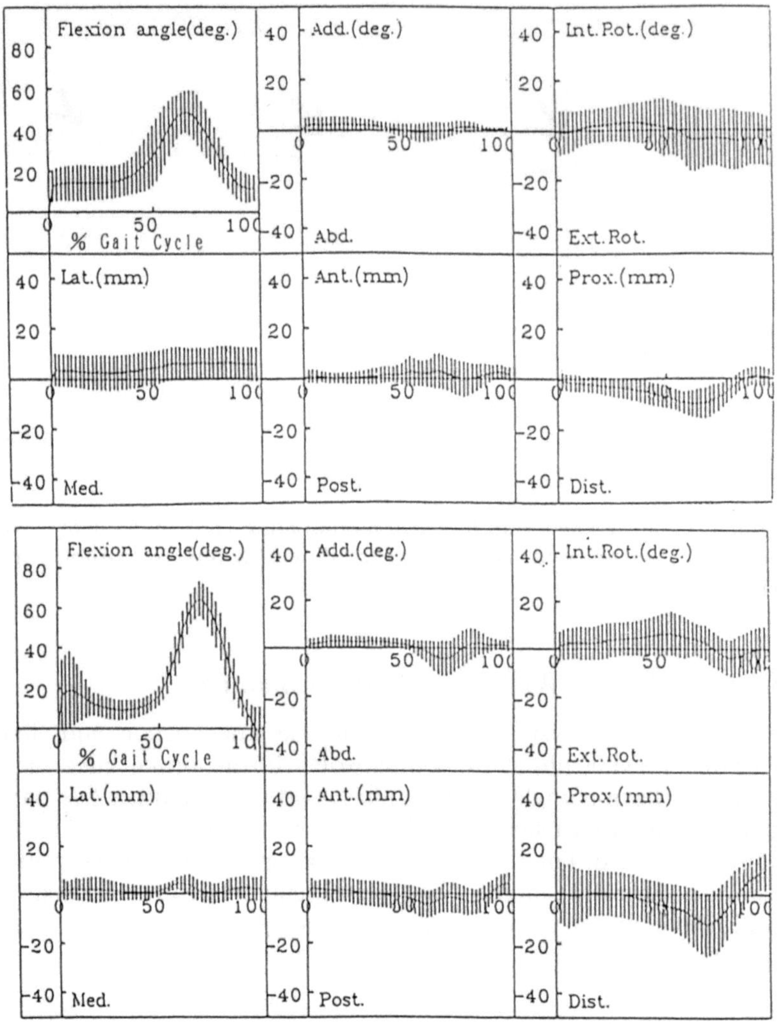

a

b

Fig. 6a,b. The result of gait analysis (average). **a** After TKA (Y/S). The flexion angle decreased to less than 50° and the double knee action vanished. **b** Control (normal study)

knee flexion. It became neutral at 90° of knee flexion and then underwent an anterior translation (Fig. 9).

Discussion

Kinetic studies of function in patients after TKA have been evaluated, and there is a good deal of controversy regarding which type of prosthesis provides

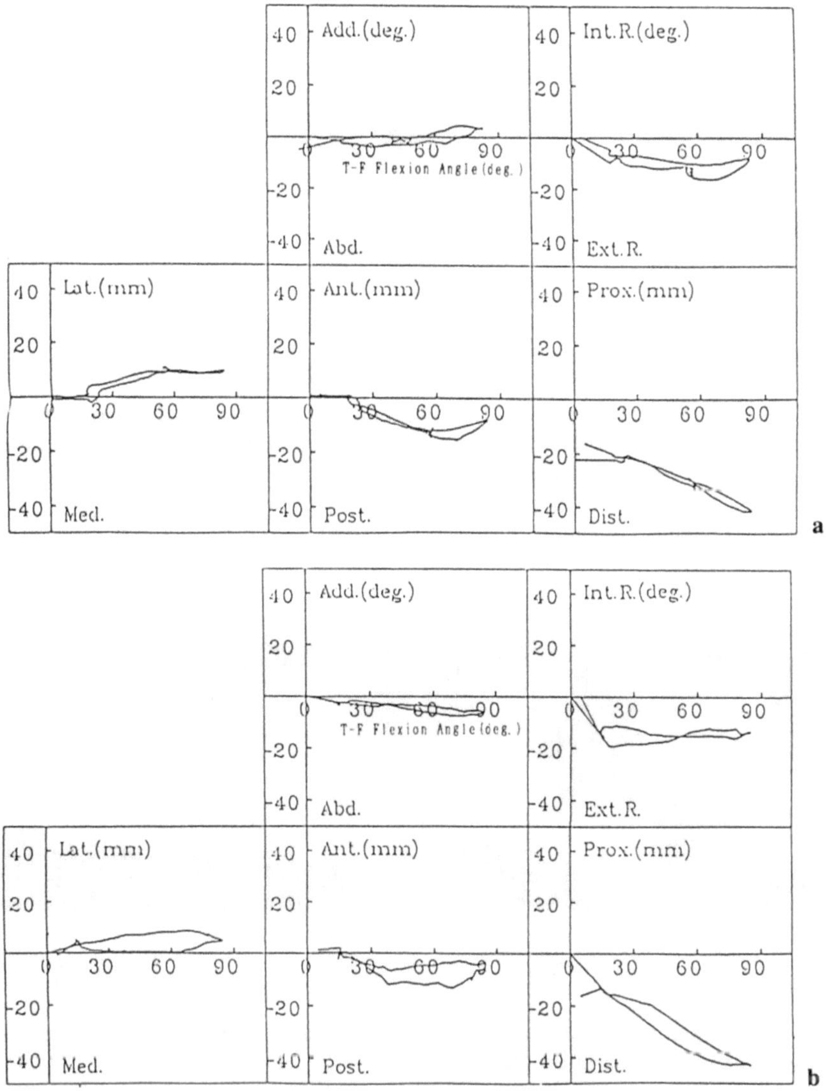

Fig. 7a,b. The result of gait analysis (stair climbing) after TKA (Y/S). **a** Stair up. **b** Stair down. Knee flexion and knee extension have a different motion when going down stairs

a better gait pattern after TKA. The finding that patients who appear to be clinically asymptomatic after TKA have abnormal gait pattern has been reported previously. The same finding was seen in this study. Although there were no complaints, all the patients showed sagging in clinical observations and a low range of motion of the knee while level walking and stair climbing.

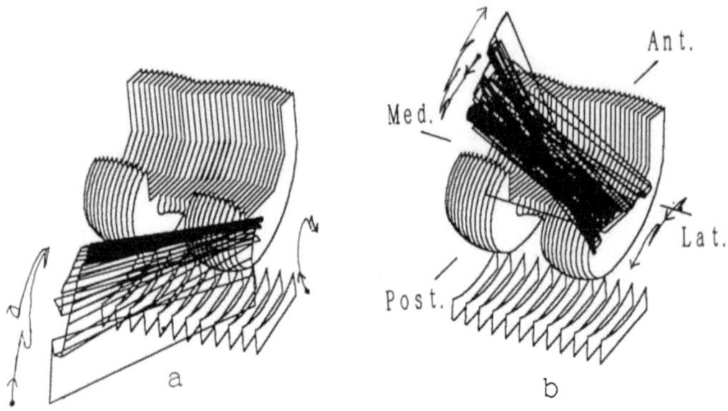

Fig. 8. Display of the helical axes of active knee motion after TKA (nonweight). **a–c** Posterolateral view. *Arrows* show tracking of the medial and lateral ends of the helical axis. **a** Anterior translation group of Y/S TKA system; large anterior/posterior shift of femur during knee flexion. **b** Posterior translation group of Y/S TKA system shows large external rotation of femur during knee flexion. **c** Mark II TKA system. Variation of the helical axis is small

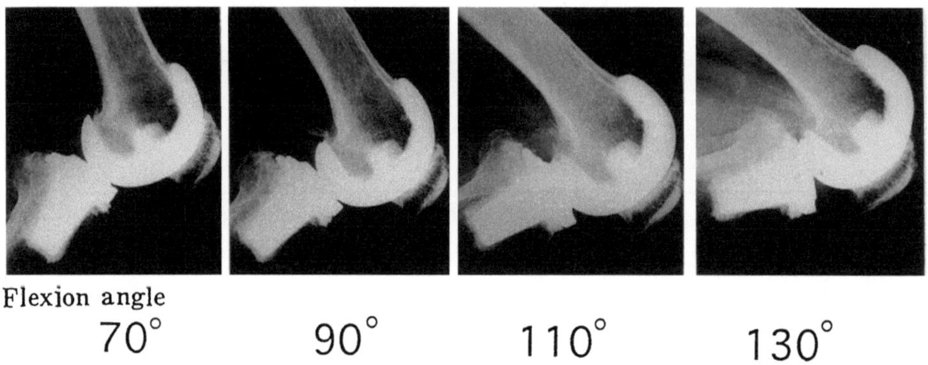

Fig. 9. X-ray study; the quadriceps muscle contracts and draws the tibia forward according to knee flexion after TKA (Y/S)

The prosthetic features of the Y/S TKA system are as follows: (1) The articular surface of the tibial component has posterior slope and is flattened to the posterior end, allowing smooth backward sliding and a rolling motion of the femoral component; and (2) the articular surface of the bilateral femoral condyle are the same size and symmetrical in a mediolateral dimension. The feature of the procedure is that the anterior and posterior ligaments are routinely divided but this system does not have the posterior-stabilized structure as does the transverse income.

The normal screw-home movement was not seen, but the reverse movement as, demonstrating the rotational freedom of the Y/S TKA system. The analysis of the Mark II system showed the limitation of rotation, and the screw-home movement disappeared.

The four-bar linkage model reported by Goodfellow [7] is famous for the femoral rollback. In the Y/S TKA system, of course, the normal femoral rollback according to the cruciate ligaments does not occur because there are no cruciate ligaments. Our analysis showed the anterior translation of the tibia; in other words, a motion resembling femoral rollback was seen at more than 70° of knee flexion.

Vector analysis of patellar ligament pull was reported by Daniel et al. [8,9]. When flexion of the knee begins, the patellar ligament passes anteriorly from the tibial tubercle to the patella, relative to a line drawn perpendicular to the surface of the tibial plateau. When the PCL is divided, the patella ligament is still directed anteriorly in 90° of flexion (Fig. 10), and thus the motion resembling femoral rollback occurs.

According to our definition of coordinate origin of femur in this study, the femoral component goes forward to 90° of knee flexion. At more than 90°, the femoral component goes dramatically backward. The lateral aspect of the femoral component of Y/S TKA is not perfectly spherical; the distal radius is larger than posterior radius. This difference of the radius results in femoral backward translation at more than 90° flexion (Fig. 11).

In active knee motion without weight, two patterns were seen; three tibias went forward and four went backword in early motion. This may be the reason for the instability of the knee because of low axial force resulting from the weak quadriceps muscle power.

The influence of the posterior tilt of the tibial component was considered as one reason for femoral posterior translation. But both δ-angles were the same, 88°. There was no relationship between the direction of translation and the δ-angle of the tibial component.

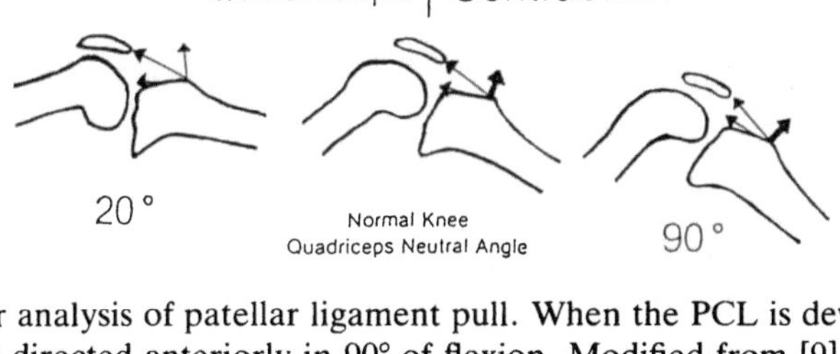

Fig. 10. Vector analysis of patellar ligament pull. When the PCL is devided, the patella ligament is still directed anteriorly in 90° of flexion. Modified from [9]

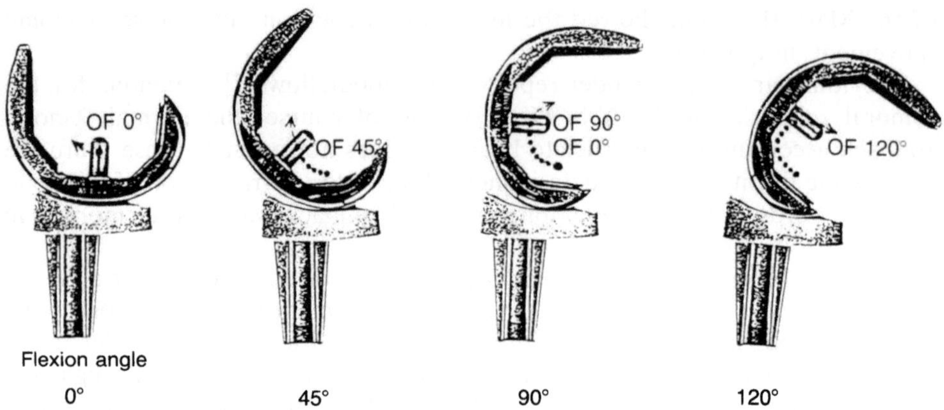

Fig. 11. The femoral component goes forward in our definition of coordinate origin of femur up to 90°. The difference between distal and posterior radius results in femoral backward translation over 90°

Generally the patient that receives TKA complains of instability in going down stairs. Although the number of tests are few, this study also showed this instability of the knee. The course between knee flexion and extension was different in going down stairs. This apparently results from relaxation of the muscular tension in going down stairs.

Dorr et al. [10] reported that the dynamic electromyelograph (EMG) of quadriceps and biceps femoris on level walking after the PCL-sacrificed, (posterior-stabilized) TKA showed more activity than that of a PCL-retained TKA. In addition, the soleus showed more EMG on stair climbing. High EMG means more stress on the muscle in PCL-sacrificed TKA. These results indicate more stress on the muscle will produce the greater varus and flexion moments that may affect durability of the prosthesis.

Delineation of the helical axes showed well the great variance of the rotation axis of the Y/S femoral component in comparison with Mark II. The main difference between Y/S and Mark II is sacrificed or retained PCL. The cause of this variance in Y/S may be the lack of PCL.

The helical axes of the active knee motion will lead to increase shear force and stress concentration on tibial insert [11]. This unpredictable stress concentration is the cause of wear and loosening of tibial component and thereby may affect longevity of the prosthesis.

Conclusion

In conclusion, the PCL has much influence on the stability of the knee joint in TKA, and thus the PCL should be preserved. When PCL is inevitably sacrificed, the posterior-stabilized type TKA should be selected. The current instrument of the posterior-stabilized system needs a lot of improvement in the near future.

References

1. Shoji H, Yoshino S, Komagamine M (1987) Improved range of motion with the Y/S total knee arthroplasty system. Clin Orthop 218:150–163
2. Andriacchi TP, Galante JO, Fermier BS (1982) The influence of total knee-replacement design on walking and stair-climbing. J Bone Joint Surg 64A:1328–1335
3. Ewald FC (1989) The Knee Society total knee arthroplasty roentgenographic evaluation and scoring system. Clin Orthop 248:9–11
4. Terajima K, Hara T, Nagata T (1991) Development of three-dimensional knee motion analysis using C.R. system. Proceedings of 1991 annual meeting of Japan Society of Biomechanics 13:213–217
5. Kinzel GL, Hall AS Jr, Hillberry BM (1972) Measurement of the total motion between two body segments: 1. Analytical development. J Biomech 5:93–105
6. Terajima K, Tsuchiya Y, Shimizu M, Hara T, Koga Y, Ishii Y (1992) Three-dimensional motion analysis after TKA. Proceedings of 1992 annual meeting of Japan Society of Biomechanics 14:371–374
7. Goodfellow J, O'Connor J (1978) The mechanics of the knee and prosthesis design. J Bone Joint Surg 60B(3):358–369
8. Castle TH, Noyes FR, Grood ES (1992) Posterior tibial subluxation of the posterior cruciate-deficient knee. Clin Orthop, 284:193–202
9. Daniel D, Stone ML, Barnet P, Sachs R (1988) Use of the quadriceps active test to diagnose posterior cruciate ligament disruption and measure posterior laxity of the knee. J Bone Joint Surg 70A:386–391
10. Dorr LD, Ochsner JL, Gronley JR, Perry JP (1988) Functional comparison of posterior cruciate retained versus cruciate sacrificed total knee arthroplasty. Clin Orthop 236:36–43
11. Kurosawa H, Soundry M, Walker PS (1984) Forces at the implant–bone interface for condylar knee prosthesis. J Jpn Orthop Assoc 58:11–21

instrument of the posterior stabilized system needs a lot of improvement in the near future.

References

1. Shoji H, Yoshino S, Komagamine M (1987) [illegible]

[illegible]

Subject Index

FSC
www.fsc.org
MIX
Papier aus verantwortungsvollen Quellen
Paper from responsible sources
FSC® C105338

If you have any concerns about our products,
you can contact us on
ProductSafety@springernature.com

In case Publisher is established outside the EU,
the EU authorized representative is:
Springer Nature Customer Service Center GmbH
Europaplatz 3, 69115 Heidelberg, Germany

Printed by Libri Plureos GmbH
in Hamburg, Germany